Antigen Processing and Presentation

Antigen Processing and Presentation

Edited by

Robert E. Humphreys

Department of Pharmacology
University of Massachusetts Medical School
Worcester, Massachusetts

Susan K. Pierce

Department of Biochemistry and Molecular Biology
Northwestern University
Evanston, Illinois

ACADEMIC PRESS

San Diego   New York   Boston   London   Sydney   Tokyo   Toronto

This book is printed on acid-free paper. ∞

Academic Press, Inc.
A Division of Harcourt Brace & Company
525 B Street, Suite 1900, San Diego, California 92101-4495

United Kingdom Edition published by
Academic Press Limited
24-28 Oval Road, London NW1 7DX

Library of Congress Cataloging-in-Publication Data

Antigen processing and presentation / edited by Robert E. Humphreys,
    Susan K. Pierce.
        p.   cm.
    Includes bibliographical references and index.
    ISBN 0-12-361555-0
    1. Antigen presenting cells.   2. Major histocompatibility complex.
I. Humphreys, Robert Edward.   II. Pierce, Susan K.
QR185.8.A59A587   1994
574.2'92--dc20   94-17101
                                                                 CIP

PRINTED IN THE UNITED STATES OF AMERICA
94   95   96   97   98   99   BB   9   8   7   6   5   4   3   2   1

CONTENTS

8. Regulation of MHC Class II Intracellular Transport and Peptide Loading

Paola Romagnoli, Flora Castellino, and Ronald N. Germain

9. Intracellular Events Regulating MHC Class II-Restricted Antigen Presentation

Andrea J. Sant, Alexander Chervonsky, Marta Dykhuizen, John F. Katz, George E. Loss, Christopher Stebbins, and L. J. Tan

10. Regulation of Antigen Processing and Presentation in B Lymphocytes

Susan K. Pierce, Anne E. Faassen, Yin Qiu, Xiao Xing Xu, Peter Schafer, and David Dalke

11. Regulation of Antigen/Class II MHC Complex Formation

*Peter E. Jensen, Melanie A. Sherman, Herbert A. Runnels,
and S. Mark Tompkins*

12. Peptide, Invariant Chain, or Molecular Aggregation Preserves Class II from Functional Inactivation

Scheherazade Sadegh-Nasseri

13. Interaction of MHC Class II Molecules with Endoplasmic Reticulum Resident Stress Proteins

W. Timothy Schaiff and Benjamin D. Schwartz

14. Invariant Chain: Variations in Form and Function

Jim Miller, Mark Anderson, Lynne Arneson, Beatrice Fineschi,
Michelle Morin, Marisa Naujokas, Mary Peterson, Jon Schnorr,
Kevin Swier, and Linda Zuckerman

15. MHC Class II Processing Pathway and a Role of Surface Invariant Chain

Norbert Koch, Gerhard Moldenhauer, and Peter Möller

16. Charging of Peptides to MHC Class II Molecules during Proteolysis of I_i

Minzhen Xu, Masanori Daibata, Sharlene Adams, Robert E. Humphreys,
and Victor E. Reyes

17. Recognition of Class II MHC/Peptide Complexes by T Cell Receptors and Antibodies that Mimic Them
Philip A. Reay and Mark M. Davis

18. Molecular Analysis of MHC–Peptide–TCR Interactions
*Alessandro Sette, Jeff Alexander, Jörg Ruppert, Ken Snoke,
Alessandra Franco, Glenn Y. Ishioka, and Howard M. Grey*

19. Selective Immunosuppression by MHC Class II Blockade
Luciano Adorini and Jean-Charles Guéry

Contributors

Chapter numbers are in parentheses following the names of contributors.

Sharlene Adams (16), Department of Pharmacology, University of Massachusetts Medical School, 55 Lake Avenue North, Worcester, MA 01655

Luciano Adorini (19), Roche Milano Ricerche, Via Olgettina, 58, I-20132 Milano, Italy

Jeff Alexander (18), Cytel Corporation, 3525 John Hopkins Court, San Diego, CA 92121

Mark Anderson (14), Committee on Immunology, Department of Molecular Genetics and Cell Biology, University of Chicago, 920 East 58th Street, Chicago, IL 60637

Matthew Androlewicz (3), Section of Immunobiology, Howard Hughes Medical Institute, Yale University School of Medicine, 310 Cedar Street, New Haven, CT 06510

Lynne Arneson (14), Department of Molecular Genetics and Cell Biology, University of Chicago, 920 East 58th Street, Chicago, IL 60637

Igor Bacik (2), Laboratory of Viral Diseases, National Institute of Allergy and Infectious Disease, National Institutes of Health, Bethesda, MD 20892

Jack R. Bennink (2), Laboratory of Viral Diseases, National Institute of Allergy and Infectious Disease, National Institutes of Health, Bethesda, MD 20892

Flora Castellino (8), Lymphocyte Biology Section, Laboratory of Immunology, National Institute of Allergy and Infectious Disease, National Institutes of Health, Bethesda, MD 20892

Stephanie Ceman (4), Laboratory of Genetics, University of Wisconsin, 445 Henry Mall, Madison, WI 53706

Hilde Cheroutre (20), Department of Microbiology and Immunology, Center for Health Sciences, University of California at Los Angeles, 10833 Le Conte Avenue, Los Angeles, CA 90024-1747

Alexander Chervonsky (9), Section of Immunology, Howard Hughes Medical Institute, Yale University School of Medicine, 310 Cedar Street, New Haven, CT 06510

Josephine H. Cox (2), Laboratory of Viral Diseases, National Institute of Allergy and Infectious Disease, National Institutes of Health, Bethesda, MD 20892

Peter Cresswell (3), Section of Immunobiology, Howard Hughes Medical Institute, Yale University School of Medicine, 310 Cedar Street, New Haven, CT 06510

Masanori Daibata (16), Department of Pharmacology, University of Massachusetts Medical School, 55 Lake Avenue North, Worcester, MA 01655

David Dalke (10), Department of Biochemistry, Molecular Biology, and Cell Biology, Northwestern University, 2153 North Campus Drive, Evanston, IL 60208-3500

Mark M. Davis (17), Department of Microbiology and Immunology, Howard Hughes Medical Institute, Stanford University School of Medicine, Stanford, CA 94305

Robert DeMars (4), Departments of Medical Genetics and Genetics, University of Wisconsin, 445 Henry Mall, Madison, WI 53706

Lisa Denzin (3), Section of Immunobiology, Howard Hughes Medical Institute, Yale University School of Medicine, 310 Cedar Street, New Haven, CT 06510

Marta Dykhuizen (9), Department of Pathology, University of Chicago, 5841 S. Maryland Avenue, Chicago, IL 60637

Anne E. Faassen (10), Department of Biochemistry, Molecular Biology, and Cell Biology, Northwestern University, 2153 North Campus Drive, Evanston, IL 60208-3500

Beatrice Fineschi (14), Departments of Molecular Genetics and Cell Biology, and Department of Pharmacology and Cell Physiology, University of Chicago, 920 East 58th Street, Chicago, IL 60637

Steven Fling (7), Department of Pediatrics, University of Washington, Seattle, WA 98195-0001

Alessandra Franco (18), Cytel Corporation, 3525 John Hopkins Court, San Diego, CA 92121

Ronald N. Germain (8), Lymphocyte Biology Section, Laboratory of Immunology, National Institute of Allergy and Infectious Disease, National Institutes of Health, Bethesda, MD 20892

Roberta Greenwood (4), Laboratory of Genetics, University of Wisconsin, 445 Henry Mall, Madison, WI 53706

Iqbal S. Grewal (22), Department of Microbiology and Molecular Genetics, College of Letter and Science, University of California, Los Angeles, CA 90024-1489

Howard M. Grey (18), Cytel Corporation, 3525 John Hopkins Court, San Diego, CA 92121

Jean-Charles Guéry (19), Roche Milano Ricerche, Via Olgettina, 58, I-20132 Milano, Italy

Clifford V. Harding (6), Institute of Pathology, Case Western Reserve University School of Medicine, University Hospitals of Cleveland, 2085 Adelbert Road, Cleveland, OH 44106

Elizabeth Hiltbold (21), Department of Microbiology and Immunology, Emory University School of Medicine, Rollins Research Center, Atlanta, GA 30322

Robert E. Humphreys (16), Department of Pharmacology, University of Massachusetts Medical School, 55 Lake Avenue North, Worcester, MA 01655

Glenn Y. Ishioka (18), Cytel Corporation, 3525 John Hopkins Court, San Diego, CA 92121

Michael R. Jackson (1), Department of Immunology, The Scripps Research Institute, 10666 North Torrey Pines Road, La Jolla, CA 92037

Peter E. Jensen (11), Department of Pathology and Laboratory Medicine, Emory University School of Medicine, Woodruff Memorial Building, Room 709, Atlanta, GA 30322

John F. Katz (9), Department of Pathology, University of Chicago, 5841 S. Maryland Avenue, Chicago, IL 60637

Norbert Koch (15), Abt. Immunbiologie, Zoologisches Institut, der Universität Bonn, Römerstrasse 164, D-53117 Bonn 1, Germany

Michell Kronenbert (20), Department of Microbiology and Immunology, Center for Health Sciences, University of California at Los Angeles, 10833 Le Conte Avenue, Los Angeles, CA 90024-1747

Timothy LaVaute (5), Laboratory of Immunogenetics, National Institute of Allergy and Infectious Disease, National Institutes of Health, 12441 Parklawn Drive, Rockville, MD 20852

Heidi Link (2), Laboratory of Viral Diseases, National Institute of Allergy and Infectious Disease, National Institutes of Health, Bethesda, MD 20892

Eric O. Long (5), Molecular and Cellular Immunology, National Institute of Allergy and Infectious Disease, National Institutes of Health, 12441 Parklawn Drive, Rockville, MD 20852

George E. Loss (9), Department of Surgery, University of Chicago, 5841 S. Maryland Avenue, Chicago, IL 60637

Mauro S. Malnati (5), Laboratory of Immunogenetics, National Institute of Allergy and Infectious Disease, National Institutes of Health, 12441 Parklawn Drive, Rockville, MD 20852

Jim Miller (14), Committee on Immunology, Department of Molecular Genetics and Cell Biology, University of Chicago, 920 East 58th Street, Chicago, IL 60637

Gerhard Moldenhauer (15), Deutsches Krebsforschungszentrum, Tumor Immunology Program, Im Neuenheimer Feld 280, D69120 Heidelberg, Germany

Peter Möller (15), Pathologisches Institut, Universität Heidelberg, Im Neuenheimer Feld 220, D69120 Heidelberg, Germany

Tatsue Monji (7), Department of Microbiology, University of Washington, Seattle, WA 98195-0001

Michelle Morin (14), Department of Molecular Genetics and Cell Biology, University of Chicago, 920 East 58th Street, Chicago, IL 60637

Kamal D. Moudgil (22), Department of Microbiology and Molecular Genetics, College of Letter and Science, University of California, Los Angeles, CA 90024-1489

Marisa Naujokas (14), Department of Molecular Genetics and Cell Biology, University of Chicago, 920 East 58th Street, Chicago, IL 60637

Bodo Ortmann (3), Section of Immunobiology, Howard Hughes Medical Institute, Yale University School of Medicine, 310 Cedar Street, New Haven, CT 06510

Jean Petersen (4), Laboratory of Genetics, University of Wisconsin, 445 Henry Mall, Madison, WI 53706

Mary Peterson (14), Department of Molecular Genetics and Cell Biology, University of Chicago, 920 East 58th Street, Chicago, IL 60637

Per A. Peterson (1), Department of Immunology, The Scripps Research Institute, 10666 North Torrey Pines Road, La Jolla, CA 92037

Susan K. Pierce (10), Department of Biochemistry, Molecular Biology and Cell Biology, Northwestern University, 2153 North Campus Drive, Evanston, IL 60208-3500

Valérie Pinet (5), Laboratory of Immunogenetics, National Institute of Allergy and Infectious Disease, National Institutes of Health, 12441 Parklawn Drive, Rockville, MD 20852

Donald Pious (7), Departments of Pediatrics Immunology and Genetics, University of Washington, Seattle, WA 98195-0001

Yin Qiu (10), Department of Biochemistry, Molecular Biology, and Cell Biology, Northwestern University, 2153 North Campus Drive, Evanston, IL 60208-3500

Philip A. Reay (17), Molecular Immunology Group, Nuffield Department of Medicine, Institute for Molecular Medicine, John Radcliffe Hospital, Oxford University, Headington, Oxford OX3 9DU, England

Victor E. Reyes (16), Child Health Center, University of Texas Medical Branch, Galveston, TX 77550

Paul A. Roche (5), Laboratory of Immunogenetics, National Institute of Allergy and Infectious Disease, National Institutes of Health, 12441 Parklawn Drive, Rockville, MD 20852

Paola Romagnoli (8), Lymphocyte Biology Section, Laboratory of Immunology, National Institute of Allergy and Infectious Disease, National Institutes of Health, Bethesda, MD 20892

Richard Rudersdorf (4), Laboratory of Genetics, University of Wisconsin, 445 Henry Mall, Madison, WI 53706

Herbert A. Runnels (11), Department of Pathology and Laboratory Medicine, Emory University School of Medicine, Woodruff Memorial Building, Room 709, Atlanta, GA 30322

Jörg Ruppert (18), Cytel Corporation, 3525 John Hopkins Court, San Diego, CA 92121

Bhanu Sadasivan (3), Section of Immunobiology, Howard Hughes Medical Institute, Yale University School of Medicine, 310 Cedar Street, New Haven, CT 06510

Scheherazade Sadegh-Nasseri (12), Laboratory of Immunology, National Institute of Allergy and Infectious Diseases, National Institutes of Health, Bethesda, Maryland 20892, and Ora Vax, Inc., Walter Reed Army Institute for Research, Washington, D.C. 20307

Susan A. Safley (21), Department of Pathology, Emory University, Winship Cancer, 1327 Clifton Road, N.E., Atlanta, GA 30322

Andrea J. Sant (9), Department of Pathology, University of Chicago, 5841 South Maryland Avenue, Chicago, IL 60637

Peter Schafer (10), Department of Biochemistry, Molecular Biology, and Cell Biology, Northwestern University, 2153 North Campus Drive, Evanston, IL 60208-3500

W. Timothy Schaiff (13), Division of Immunology, Department of Medicine, Washington University School of Medicine, Box 8045, St. Louis, MO 63110

Jon Schnorr (14), Department of Molecular Genetics and Cell Biology, University of Chicago, 920 East 58th Street, Chicago, IL 60637

Benjamin D. Schwartz (13), Monsanto Corporation Research, Mail Zone AA4A, St. Louis, MO 63198-0001

Eli E. Sercarz (22), Department of Microbiology and Molecular Genetics, College of Letter and Science, University of California, Los Angeles, CA 90024-1489

Alessandro Sette (18), Cytel Corporation, 3525 John Hopkins Court, San Diego, CA 92121

Melanie A. Sherman (11), Department of Pathology and Laboratory Medicine, Emory University School of Medicine, Woodruff Memorial Building, Room 709, Atlanta, GA 30322

Yoji Shimizu (4), Department of Microbiology and Immunology, University of Michigan Medical School, Ann Arbor, Michigan 48109

Ken Snoke (18), Cytel Corporation, 3525 John Hopkins Court, San Diego, CA 92121

Chistopher Stebbins (9), Department of Pathology, University of Chicago, 5841 S. Maryland Avenue, Chicago, IL 60637

Kevin Swier (14), Department of Molecular Genetics and Cell Biology, University of Chicago, 920 East 58th Street, Chicago, IL 60637

L. J. Tan (9), Department of Pathology, University of Chicago, 5841 S. Maryland Avenue, Chicago, IL 60637

S. Mark Tompkins (11), Department of Pathology and Laboratory Medicine, Emory University School of Medicine, Woodruff Memorial Building, Room 709, Atlanta, GA 30322

Minzhen Xu (16), Department of Pharmacology, University of Massachusetts Medical School, 55 Lake Avenue North, Worcester, MA 01655

Xiao Xing Xu (10), Department of Biochemistry, Molecular Biology, and Cell Biology, Northwestern University, 2153 North Campus Drive, Evanston, IL 60208-3500

Jonathan W. Yewdell (2), Laboratory of Viral Diseases, National Institute of Allergy and Infectious Disease, National Institutes of Health, Bethesda, MD 20892

H. Kirk Ziegler (21), Department of Microbiology and Immunology, Emory University School of Medicine, Rollins Research Center, Atlanta, GA 30322

Linda Zuckerman (14), Committee on Immunology, Department of Molecular Genetics and Cell Biology, University of Chicago, 920 East 58th Street, Chicago, IL 60637

PREFACE

Processing and presentation of antigen are central to the initiation of the immune response. New therapeutics to control immune responses will come from a more complete understanding of the mechanisms involved in antigen processing and presentation at cellular and molecular levels. One goal is to block antigen presentation in an antigen-specific fashion or in a major histocompatibility complex allele-related way to control autoimmune diseases. New vaccine strategies for infectious diseases will also be developed. Enhancing the presentation of endogenous tumor-associated antigens could lead to immune rejection of cancers. Achieving these clinical goals will become possible, in part, through the fundamental work of our colleagues that is presented in this book.

In *Antigen Processing and Presentation*, the principal investigators in this field present the development of one of their central research questions toward understanding mechanisms in the regulation of antigen processing. They outline the pathways of deductive and inductive logic leading to their current findings and pose the key questions they will address in the near future. The chapters are not reviews of the literature, but are focused reports by individual groups on current targets of their research. Collectively, these contributions give a fine overview to students and investigators of antigen processing and presentation, as well as to others interested in the application of this knowledge to related fields. The chapters, especially in the summary sections on future directions, provide ideas for experimental designs in the coming years.

We have enjoyed coordinating this project and thank our colleagues for their efforts in preparing sparkling reports in this format, which has allowed accelerated publication.

Robert E. Humphreys
Susan K. Pierce

1

APPROACHING AN UNDERSTANDING OF MHC CLASS I ANTIGEN PRESENTATION

Michael R. Jackson and Per A. Peterson

OVERVIEW OF MHC CLASS I ANTIGEN PRESENTATION

Cytotoxic T lymphocytes (CTL) play an important role in the elimination of intracellular pathogens. A typical CTL response requires that receptors on the T-cell specifically recognize on the target cell antigen derived from the pathogen bound to cell surface expressed MHC class I molecules. The finding that synthetic peptides fragments of viral proteins could sensitize a target cell to CTL lysis provided the first clue as to the nature of the bound antigen. However it was not until the determination of the crystal structure of the human MHC class I molecule HLA-A2.1 (1) that immunologist had a molecular picture of how class I molecules function. The structure showed that the two N-terminal domains (α1 and α2) of the class I heavy chain folded together to form a pocket. This pocket was filled with extra electron density, consistent with the presence of co-crystallized antigenic peptide. Furthermore, almost all the hyper variable amino acids within the family of class I molecules were clustered around this binding pocket presumably altering its shape and thereby the spectrum of peptides bound. This structure thus provided an explanation for both the phenomenon of MHC restriction and explained the extensive polymorphism of class I molecules.

It is now widely accepted that the role of class I molecules is that of a reporter, providing an up to date record at the cell surface of the type and quantities of proteins being synthesized within a cell. In uninfected cells they report that all is normal, since the specific set of peptides displayed are derived from endogenous proteins (self peptides) are thus ignored by T-cells. If, however, a cell is infected by a virus, peptides derived from the newly synthesized viral proteins are rapidly presented by class I molecules at the cell surface leaving the cell at the mercy of the circulating killer T cells.

The basics of the class I presentation pathway are also firmly established (see Figure 1). Antigen is first degraded in the cytoplasm by the proteasome into short peptide fragments. These fragments are transported across the membrane of the endoplasmic reticulum (ER) by the TAP proteins (Transporter associated with Antigen Presentation). Once in the ER, peptide is bound by pre-assembled class I heterodimers (composed of class I heavy chain and ß2 microglobulin (ß2m)). The array of peptides bound by a specific class I molecule is determined by the shape of its binding pocket. It is this tri-molecular complex of peptide-class I heavy chain and ß2m that is subsequently transported via the secretory pathway to the cell surface to be inspected by circulating T-cells.

This article is focused on two main issues. First, we discuss progress in our understanding of how the biochemical and physical properties of a class I molecule dictate the antigen it presents. Second, we outline progress that has been made in understanding each of the various steps in the class I antigen processing and presentation pathway, with particular emphasis on which step is quantitatively most likely to be important in determining the repertoire of peptides presented at the cell surface.

BIOCHEMICAL PROPERTIES OF CLASS I MOLECULES

The Basics And The Questions

Analysis of the naturally processed peptides extracted from different cell surface expressed MHC molecules showed that they were typically eight or nine amino acids long (2). Each class I allele showed distinct preferences for specific amino acids, so called anchor residues, at precise locations within the peptide sequence (see Rammensee et al, 1993 for details). For example, HLA A2.1 molecules prefer to bind nona-peptides with leucine in position 2 and a valine in position 9 whereas H-2K^b molecules preferentially bind octa-peptides with a tyrosine in position 3, tyrosine or phenylalanine in position 5, and leucine in position 8.

These findings showed from the outset that the primary factor governing the peptides displayed on a cells surface is the haplotype of the class I molecules expressed. One of our major interests has been to try to understand at the molecular level the structural and biochemical features of class I molecules that determine these peptide binding characteristics. In order to address these issues we have examined the crystal structures of class I molecules containing single peptide species with the belief that such structures will provide the explanations as to how a specific class I molecule can bind such a variety of different peptides with high affinity, why these peptides are so similar in length, and how the polymorphic residues in the binding groove dictate the allele specific anchor residues. In order to put the information gained from these structural studies in to context we also have carried out extensive biochemical analyses of the properties of class I molecules in vitro. These studies have allowed us to determine the characteristics of peptides that result in high affinity binding and to analyze the importance of peptide in assembling and stabilizing class I molecules.

Our experimental approach to this work has relied heavily on our ability to produce and purify soluble MHC class I molecules devoid of high affinity peptides, so called empty class I. This was achieved by co-transfecting Drosophila cells with cDNAs encoding truncated class I heavy chain and ß2 microglobulin. The class I produced by these Drosophila cells are presumably empty because insect cells do not possess the necessary accessory molecules for generating and/or loading peptide (3).

Structure Of Class I Molecules Containing A Single Peptide Species

Crystal structure analyses of two peptides, an octamer derived from vesicular stomatiatis virus (VSV-8) and a nonamer derived from Sendai virus (SEV 9) complexed with murine MHC class I H-2K^b showed that the antigen binding site is indeed the previously identified groove (4). The binding site is however more of a pocket than a groove, it best accommodates peptides of 8 amino acids. Longer peptides e.g. a nonamer, may also be accommodated by maintaining similar interactions at the amino and carboxy-termini as the octa-peptide and by allowing the central residues of the nona-peptide to bulge out from the pocket (4,5). A comparison of the structures of H-2K^b, HLA-B27 and HLA Aw68 showed that the NH$_2$ and COOH termini of the bound peptide form extensive hydrogen bonds with the residues lining highly conserved pockets at the end of the peptide binding groove, whereas peptide anchor residues are accommodated in deep polymorphic pockets which exhibit structural complementarity to the corresponding anchor side chains. These structural characteristics provide an adequate molecular explanation for the existence of allele specific anchor residues and the size restriction identified in peptides extracted from class I.

Characteristics Of Peptide Binding To Class I Molecules

The availability of purified, empty class I molecules has allowed us to carry out affinity measurements of peptides for K^b molecules. Using a competitive binding assay and a radio-iodinated peptide we found (6) that peptides corresponding to naturally processed epitopes displayed the highest affinities and their K_D values were remarkably similar (2.7-4.1nM). Peptides that were slightly shorter or longer by just 1 or 2 amino acids lowered the affinities by a factor of 2-100. Interestingly a two residue extensions in the NH_2-terminus of VSV-8 or OVA-8 peptides reduced their affinities by a factor of ~100. By contrast, a two residue extensions at the COOH terminus only reduced the affinity by a factor of ~5. These difference most likely reflect the observation from the crystal structure that the amino-terminus is deeply buried within the K^b molecule relative to the carboxy-terminus (4). Interestingly analysis of the binding of a library of random octamer peptides to K^b showed that most peptides did bind (7). However their affinities represented a complete continuum from $1x10^{-10}$-$1x10^{-3}$ M suggesting that although haplotype specific anchor residues may be important for very high affinity binding, their absence from a peptide does not result in a sudden drop in affinity and does not necessarily preclude a peptide from interacting with a class I molecule.

In order to analyze the importance of the anchor residues in high affinity binding to class I we synthesized a series of alanine substituted variants of VSV-8 and OVA-8 (7). Affinity measurements showed that the anchor residues Tyr^3 and Tyr^5 in VSV8 and Phe^5 in Ova 8 were indeed the most important residues for tight binding to K^b. This result was confirmed by the finding that an octa-peptide composed of serine residues in all positions except for two anchor tyrosines in positions 3 and 5, had an affinity for K^b that was only slightly lower than VSV-8 (7). Similarly, the affinity of the OVA-8 peptide could be reconstructed in this polyserine background by introduction of phenylalanine at position 5 and isoleucines at positions 2 and 3. What was surprising from these results was that high affinity binding to K^b did not require the carboxy-terminal anchor residue leucine, (Leu^8) present in both VSV-8 and OVA-8 peptides. Results that contrast with the clear preference for Leu^8 in the pool of peptides naturally processed and presented by K^b at the cell surface (2). Interestingly determination of the K^b restricted peptide binding motif in vitro, by sequencing the pool of peptides selected by empty K^b from a randomly synthesized mixture of octamers showed selection of the expected anchor determinants at positions 3 and 5, but no selection of an anchor leucine at position 8.

Selection by K^b of peptides with a leucine at position 8 could reflects the pool of peptides available in the ER resulting from specificity of the proteasome, TAP proteins etc. However, cell surface expressed empty K^b molecules also preferentially bind peptides with Leu[8] from a mixture of octamers provided in the culture medium. The explanation for these differences is most likely that the affinity constants and the in vitro peptide binding experiments were determined under equilibrium conditions, whereas the pool of peptides bound by a surface expressed class I molecule are selected under non-equilibrium binding conditions, see below.

Class I Assembly And The Importance Of Peptide For Thermostability

It has been previously shown that empty class I molecules are thermolabile, i.e. they disassemble at 37°C, but that they can be stabilized upon binding of high affinity peptides (8). Analysis of the kinetics of peptide binding to K^b and assembly of the class I heavy chain-ß2m-peptide complex (6,7) showed that the interaction between the heavy chain and ß2m and between the heterodimer and the peptide are reversible processes that obey the laws of mass action. In the absence of peptide K^b molecules readily dissociate at elevated temperatures e.g. 44°C. However, upon cooling e.g. to 23°C, they reassemble in a reaction that depends upon the concentration of the heavy chain and ß2m (6). Experimentally the ability of a peptide to thermostabilize class I molecules in vitro only becomes apparent when the concentration of heavy chain and ß2m are low or if detergent is added to the incubation. Presumably detergent interacts in some way with free heavy chains thereby preventing their reassociation with ß2m.

In order to assess the importance of specific residues in a peptide in thermostabilizing K^b we analyzed the ability of the alanine scan series of VSV-8 and OVA-8 peptides to stabilize empty K^b at 44°C in the presence of 1% Triton X-100. These studies (7) provided the surprising result that Leu[8] was as important as the anchor residues at position 3 and 5 in the VSV8 peptide in stabilizing K^b. The inability of the VSV-alanine 8 peptide to thermostabilize K^b must reflect an increase in the dissociation rate of this heavy chain-ß2m-peptide complex. Furthermore, as substituting Ala for Leu at position 8 in the VSV peptide did not change the equilibrium binding constant, this substitution must also have increased the peptides association rate. The structural reason as to why leucine at position 8 imparts thermostability to the complex is not exactly clear. However the thermostability of a heavy chain-ß2m-peptide complex is dictated by both the affinity of the peptide and the ß2m so one possibility is that the side chain of Leu^8 which points down toward the contact surface with ß2m may be altering the affinity for this sub-unit.

These observations have significant biological implications. Peptide affinity is proably most important for peptides binding to class I molecules in the ER, which presumably occurs under equilibrium conditions. Whereas thermostability is presumably especially important in determining the half life of the class I-peptide complex during transport to and most especially at the cell surface where the concentrations of free peptide and ß2m are likely to be significantly lower than in the exocytic pathway.

ANTIGEN SELECTION BY THE CLASS I PROCESSING AND PRESENTATION PATHWAY

Peptide Generation And The Proteasome

The discovery that two subunits (LMP2 and LMP7 in mouse) of the '20S' proteasome (a multicatalytic protein protease) are encoded by genes located in the MHC class II region led to the immediate speculation that the proteasome generates the peptides for class I molecules (9,10). The proteasome is indeed a suitable candidate, it is located both in the cytoplasm and nucleus and has multiple proteolytic activities that preferentially cleave small peptides on the carboxyl side of hydrophobic, basic or acidic residues. Furthermore, the generation of peptides for class I is ubiquitin dependent (11), a known first step in targeting proteins for hydrolysis by 26S proteasome.

Involvement of the proteasome in generating antigen was challenged by the finding that cell lines in which the LMP subunits were absent are fully capable of presenting peptides to CTL (10). However, a comparison of the proteolytic activities of purified proteasomes containing or lacking these subunits showed that LMP-containing proteasome had a higher activity for substrates having hydrophobic or basic residues immediately preceding the cleavage site (12,13). As human and murine class I molecules show a strong preference for COOH-terminal hydrophobic or basic residues (2), the current hypothesis is that the role of these LMP subunits is to modify the proteolytic activity of the proteasome, directing its efforts into the production of peptides most likely to bind to class I molecules.

Peptide Transporters - The TAP Proteins

The identification of two genes TAP1 and TAP2 located in the MHC class II region which share substantial homologies with a family of ATP-dependent transporter proteins led to the speculation that their gene products transport peptides from the cytoplasm to the lumen of the ER (see 14). The role of the TAP proteins in class I antigen presentation was confirmed by the finding that the mutant cell lines, RMA-S and T2 which are defective in the synthesis of TAP subunits, fail to present antigen, and that this defect is restored upon transfection with the corresponding wild type cDNAs (14,15). The data, missing until recently was the direct proof that TAP1 and 2 transport peptides across the ER membrane. The development of assays using radioactive tracer peptides in translocation assays has allowed the activity of the TAP proteins to be measured. Three different assay systems have shown that translocation of peptide into the ER lumen is ATP dependent and requires the presence of both TAP 1 and 2 (16,17,18). Furthermore, the efficiency of transport was found to depend on the length and carboxyterminal amino acid of the peptide. The most extensive analysis has been carried out with the mouse TAP (19). This translocator was found to preferentially transport peptides with a minimal size of 9 residues and where the COOH-terminal amino-acid is hydrophobic. As COOH-terminal hydrophobic amino-acids are known to be peptide anchor determinant for most mouse class I molecules (2), these findings suggested that the mouse TAP translocator and class I molecules may have co-evolved to improve the efficacy of antigen presentation. Assuming this to be the case then the human TAP translocator must differ from the mouse, as HLA B27 and HLA Aw63 have a strong preference for peptides that have basic residues at their carboxy-terminus (2). Indeed, limited analysis of rat TAP showed it had a preference for peptides with COOH-terminal histidine or glutamic acid (17) confirming that TAP from different species prefer to transport different sets of peptides (20).

The above assays also showed that the TAP translocator can efficiently transported peptides longer than 9 amino acids, an observation consistent with the report that the naturally processed pool of antigenic peptides displayed by HLA Aw68 varied in length between 9 and 11 amino acids (2). What is unclear at present is whether peptides can be trimmed after transport into the lumen of the ER in order to convert them to high affinity class I binders. Such a mechanism could arguably increase the chances of transporting fragments containing a peptide that can bind to a class I molecule. However, to date there is no evidence that trimming of peptides bound to class I naturally occurs in the ER.

Regulation Of Class I Intracellular Transport By Peptide

In the absence of ß2m it is well recognized that class I heavy chains are transported poorly if at all, presumably reflecting the fact that such molecules are poorly folded and retained by the quality control machinery of the ER. It is also well recognized that different class I molecules sharing up to 80% sequence identity are exported out of the ER at very different rates. One explanation for these findings appeared to be a rapidly and quantitative association of nascent class I heavy chains with an 88kd ER resident protein (21). This class I-p88 peptide complex existed only transiently and the length of the association was shown to correlate inversely with the known rate of ER to Golgi transport of specific heavy chains. Based on immunological criteria, p88 has been identified as calnexin or p90; a molecular chaperone known to bind transiently to a variety of nascent proteins in the lumen of the ER. The observation that dissociation of calnexin from the heavy chain was not triggered by binding of ß2 microglobulin suggests that calnexin retains class I molecules in the ER until formation of the tri-molecular complex of heavy chain, ß2m and peptide (22). A suggestion which is further supported by the observation that in RMA-S and T2 cells (TAP deficient) there is a prolonged association between calnexin and the empty class I molecules (22). This observation presumably explains why so few class I molecules are expressed at surface of such cells. Similarly the majority of class I/ß2m complexes are not transported in the mutant human cell line T2. Interestingly, if the temperature in which these mutant cells are cultured is reduced then empty murine but not human class I molecules are transported (8,14,23).

In order to try to understand the role of peptide and other accessory molecules in the assembly and transport of class I molecules we have developed a model system based on Drosophila cells that stably express transfected class I molecules (3,24). In such cells both human and murine class I molecules readily assemble and the heavy chain-ßm heterodimers are rapidly transported to the cell surface. Surface expressed heterodimers had all the properties of empty class I molecules in that they were thermolabile but could be stabilized by addition of appropriate peptides in the medium. Furthermore, class I heavy chains expressed in Drosophila cells in the absence of ß2m were also transported to the cell surface where they are able to reassociate with peptide and ß2m. These findings indicated that MHC dedicated accessory proteins are required for the generation and/or loading of peptides onto class I but not for the assembly of the heavy chain-ß2m heterodimer.

Interestingly the relative transport rates of specific class I molecules in these insect cells were found to be quite different from those previously determined in normal mammalian cells or in the peptide loading deficient RMA-S (14,24). Thus for example the $t_{1/2}$ in insect cells for D^b, K^b and L^d transport were 25, 60 and 80 minutes respectively. Whereas in normal mammalian cells (21), they are 55, 20 and 55 minutes, respectively. What appeared to primarily account for the differences in transport rate in insect cells was differences in the thermostability of the complexes (3) as a direct relationship between thermostability of the empty complexes and transport was observed. Furthermore, co-expression of murine class I with human ß2m (murine heavy chains have a higher affinity for human ß2m) resulted in complexes that were more thermostable and more rapidly transported.

In order to directly test whether calnexin can retain incompletely assembled class I molecules in the ER we transfected Drosophila cells with cDNAs encoding murine heavy chain, ß2m and calnexin. Introduction of calnexin was found to drastically changes the transport rates of the class I molecules such that they became similar to those observed in the peptide deficient cell line RMA-S (24). In addition, transfection of calnexin essentially abrogated transport of free heavy chains and substantially protected the free heavy chains from degradation. As heavy chains and ß2m assembled in the absence of calnexin in insect cells, the role of calnexin is implicated in retaining the empty molecules until a peptide is bound rather than maintaining the conformation of the heavy chain. However, preliminary data suggest that calnexin may also catalyze the formation of properly folded class I heavy chain-ß2m complexes especially if the concentration of the subunits is low.

Our present view of the class I antigen processing, assembly and presentation is illustrated in figure 1. A brief description of the six stages of the pathway indicated in the figure is given below.

Stage 1: Antigen is degraded into peptides by the proteasome in the cytoplasm (most probably the 26S proteasome complex). Skewing of the proteasomes output to match the peptides preferred by MHC class I molecules may be accomplished at this stage by incorporation into the proteasome of the γINF inducible LMP subunits. The quantity of a specific peptide generated is presumably dictated by the abundance of the protein from which it is derived. Stage 2 : Peptides are translocated across the ER membrane by TAP. Skewing of the pool of peptides translocated to match the peptides that bind/thermostabilize class I may result from selectivity in the peptides translocated by TAP. Stage 3 : Class I heavy chain and ß2m are targeted to the ER via signal sequences. After removal of the signal peptide, the subunits fold and associate with one another. In the absence of peptide the weakly associated heterodimer is in equilibrium with partially folded free heavy chain and a pool of free correctly folded ß2m. Both partially folded free heavy chains and the majority of the heterodimer are maintained in the ER by association with calnexin. Stage 4 : Release of the heavy chain from calnexin requires it to attain a specific transport competent conformation, one which is highly favored by the presence of peptide and ß2m but not absolutely dependent upon them. The calnexin-binding and the transport competent conformations of the heavy chain are in equilibrium, this equilibrium is most likely different for each allele. Under normal conditions, release of the heavy chain from calnexin is typically achieved by peptide addition, however empty class I molecules, which transiently attain the transport competent conformation may be able to escape calnexin binding. As the supply of high affinity peptides is limiting in the ER , competition between different class I alleles expressed in the same cell for the available peptide will likely occur at this step. Stage 5 : Class I molecules are transported along the secretory pathway to the cell surface. Stage 6 : Residency at the cell surface and possibly also transport to it, depends on the thermostability of the MHC-peptide complex. Once peptide is lost the heavy chain rapidly denatures into a non functional form which is subsequently degraded.

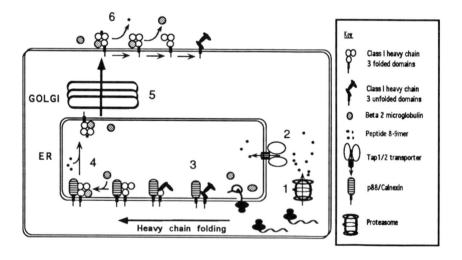

Figure 1
Schematic representation of MHC class I antigen processing and presentation.

Conclusions

The above characterization of the class I antigen processing and presentation pathway has described many factors that appear to play a contributing role in affecting what peptides a given class I molecule presents at the cell surface. Ranking these in order of importance, the structure of the peptide binding pocket is probably the most influential. However there is little doubt that selectivity in both peptide generation and transport affects the pool of peptides available for nascent class I molecules to bind. Such selectivity most likely results in quantitative rather than qualitative differences in the peptides available. One of the processes we know least about is how class I is released from calnexin. Clearly peptide plays a role here, but will any peptide that bind to the class I allow release, or are there special requirements e.g. the peptide has to be of a sufficiently good fit that a specific conformation of the class I molecule is attained. In this respect calnexin might significantly affect the repertoire of peptides displayed. Once a specific class I-peptide complex is displayed at the cell surface the most important factor governing its abundance is the thermostability of the complex. The more stable the complex the longer it survives.

The biological importance of high affinity peptide-class I complexes may be over-stated. There are many experiments to suggest that the supply of high affinity peptides in the ER is limiting and that a significant number of class I molecules reach the cell surface with peptides with low affinity (25,26). As it is the steady state level of a specific class I-peptide complex displayed at the cell surface that is biologically relevant, the short half life of a class I complex may be compensated for by the abundance of the peptide, i.e. the frequency at which such complexes are made. The role of such low affinity or more appropriately low stability peptides-class I complexes has undoubtedly been underestimated and recent data (27) have shown that such complexes are at least biologically relevant in allo-specific CTL reactions.

Future Experiments

Although the class I antigen processing and presentation pathway as we know it to date appears to fairly complete, we are left with the nagging doubt as to whether we have identified all of the MHC dedicated accessory proteins associated with this pathway. Attempts to reconstitute the full class I presentation pathway in Drosophila cells by transfecting them with cDNAs encoding murine TAP1 and TAP2 along with the MHC class I subunits and calnexin have been unsuccessful suggesting that additional components, perhaps not yet identified are required. Furthermore, many step of the pathway are lacking in detail.

One of the more sketchy areas of the pathway is the generation of peptide. We know that peptides derived from nuclear proteins is presented by class I molecules. Are these nuclear proteins degraded by proteasomes located in the nucleus ? If so, how are such peptides transported out of the nucleus to the cytoplasm? Are they protected by a chaperonin-like molecule which docks specifically with the TAP proteins? Little is known about selectivity of substrate for the proteasome. In particular whether all proteins regardless of age are equivalent sources of substrate for class I antigen or are newly synthesized proteins selectivity chosen? Selective use of nascent proteins might ensure that peptides are generated equally well from proteins regardless of their final sub-cellular destination and result in a more rapid and efficient reporting of infection. If nascent proteins are selectively targeted as class I antigen how is this achieved?

Questions also remain about the TAP translocators. Are they physically associated with the proteasomes and/or MHC class I molecules ? How do the TAP translocators work and what confers peptide selectivity? Details of TAP function is important in a wider context as it may help to explain the mechanics of other important members of this family of transporter proteins e.g. multi drug resistance proteins. Other questions concerning the class I presentation pathway involve the role of peptide in class I assembly in the ER. In particular, whether peptide transported into the ER lumen is free or whether it is retained/protect by auxiliary proteins which might also catalyze the peptide loading process.

Now that we are approaching an understanding of the class I antigen processing and presentation pathway our attention is moving away from the events leading up to the presentation of antigen at the cell surface and towards the question of how the cell surface expressed peptides-MHC class I complexes are recognized by the T-cell receptor.

ACKNOWLEDGMENTS

We are especially grateful to our colleagues : A. Brunmark, M. De Bruijn, Y. Yang, E. Song, M. Matsumura, Y. Saito, K. Früh and L. Karlsson for helpful comments. Work in the authors laboratory was supported by the NIH.

REFERENCES

1. Björkman, P. J., M .A. Saper, B. Samraoui, W. S. Bennett, J. L. Strominger, and D. C. Wiley. 1987. Structure of the human histocompatibility antigen, HLA-A2. *Nature 329*: 506.

2. Rammensee, H-G., K. Falk, and O. Rotzschke. 1993. Peptides naturally presented by MHC class I molecules. *Ann. Rev. Immunol. 11*: 213.

3. Jackson, M. R., E. S. Song, Y. Yang, and P. A. Peterson. 1992. Empty and peptide containing conformers of class I MHC molecules expressed in Drosophila melanogaster cells. *Proc. Natl. Acad.Sci.USA. 89*: 12117.

4. Fremont, D. H., M. Matsumura, E. A. Stura, P. A . Peterson, and I. A. Wilson. 1992. Crystal structures of two viral peptides in complex with murine class I H-2K^b MHC. *Science 257*: 919.

5. Matsumura, M., D. H. Fremont, P. A. Peterson, and I. A. Wilson. 1992. Emerging principles for the recognition of peptide antigens by MHC class I molecules. *Science 257*: 927.

6. Matsumura, M., Y. Saito, M. R. Jackson, E. S. Song, and P. A. Peterson. 1992. In vitro peptide binding to soluble empty class I MHC molecules isolated from transfected Drosophila melanogaster cells. *J. Biol. Chem. 267*: 23589.

7. Saito, Y., P. A. Peterson, and M. Matsumura. 1993. Quantitation of peptide anchor residues contributions to class I major histocompatibility complex molecule binding. *J. Biol. Chem. 268*: 21309.

8. Ljunggren, H.-G., N. J. Stan, C. Öhlén, J. J. Neefjes, P. Hoglund, M.-T. Heemels, J. Bastin, T. N. M. Schumacher, A. Townsend, K. Kärre, and H. L. Ploegh. 1990. Empty MHC class I molecules come out in the cold. *Nature 346*: 476.

9. **Monaco, J. J., S. Cho, and M. Attaya.** 1990. Transport protein genes in the murine MHC : possible implications for antigen processing. *Science 250*: 17239.

10. **Goldberg, A. L., and K. L. Rock.** 1992. Proteolysis, proteasomes and antigen presentation. *Nature 357*: 375.

11. **Michalek, M. T., E. P. Grant, C. Gramm, A. L. Goldberg, and K. L. Rock.** 1993. A role for the ubiquitin-dependent proteolytic pathway in MHC class I-restricted antigen presentation. *Nature 363*: 552.

12. **Driscoll, J., M. G. Brown, D. Finley, and J. J. Monaco.** 1993. MHC-linked LMP gene products specifically alter peptidase activities of the proteasome. *Nature 365*: 262.

13. **Gaczynska, M., K. L. Rock, and A. L. Goldberg.** 1993. γ-Interferon and expression of MHC genes regulates peptide hydrolysis by proteasomes. *Nature 365*: 264.

14. **Townsend, A., and J. Trowsdale.** 1993. The transporters associated with antigen presentation. *Semin. Cell Biol. 4*: 53.

15. **Yang, Y., K. Früh, J. Chambers, J. B. Waters, L. Wu, T. Spies, and P. A. Peterson.** 1992. Major Histocompatibility complex (MHC)-encoded HAM2 is necessary for antigenic peptide loading onto class I MHC molecules. *J. Biol. Chem. 267*: 11669.

16. **Shepherd, J. C., T. N. M. Schumacher, P. G. Ashton-Rickardt, S. Imaeda, H. L. Ploegh, C. A. Janeway, and S. Tonegawa.** 1993. TAP-1 dependent peptide translocation in vitro is ATP dependent and peptide selective. *Cell 74*: 577.

17. **Neefjes, J. J., F. Momburg, and G. J. Hämmerling.** 1993. Selective and ATP-dependent translocation of peptides by the MHC encoded transporter. *Science 261*:769.

18. **Androlewicz, M. J., K. S. Anderson, and P. Cresswell.** 1993. Evidence that transporter associated with antigen processing translocate an MHC class I-binding peptide into the endoplasmic reticulum in an ATP-dependent manner. *Proc. Natl. Acad. Sci. USA. 90*: 9130.

19. **Schumacher, T. N. M., D. V. Kantesaria, M.-T. Heemels, P. G. Ashton-Rickardt, J. C. Shepherd , K. Früh, P. A. Peterson, C. A. Janeway, S. Tonegawa, and H. L. Ploegh.** 1993. Peptide length and sequence specificity of the mouse TAP1/TAP2 translocator. *J. Exp Med*, in press.

20. **Howard, J. C.** 1993. Restriction on the use of antigenic peptides by the immune system. *Proc. Natl. Acad. Sci. USA. 90*: 3777.

21. **Degen, E., and D. B. Williams.** 1991. Participation of a novel 88-kD protein in the biogenesis of murine class I histocompatibility molecules. *J.Cell Biol. 112*: 1099.

22. **Degen, E., M. F. Cohen-Doyle, and D. B. Williams.** 1992. Efficient dissociation of the p88 chaperone from MHC class I molecules requires both ß2-microglobulin and peptide. *J. Exp. Med. 175*: 1653.

23. **Baas, E. J., H.-M. van Santen, M. J. Kleijmeer, H. J. Geuze, P. J. Peters, and H. L. Ploegh.** 1992. Peptide-induced stabilization and intracellular transport of empty HLA class I complexes. *J. Exp. Med. 176*: 147.

24. **Jackson, M. R., M. F. Cohen-Doyle, P. A. Peterson, and D. B. Williams.** 1993 Intracellular transport of MHC class I molecules is regulated by the molecular chaperone, Calnexin. *Science,* in press.

25. **Oritz-Navarrette, V., and G. J. Hämmerling.** 1991. Surface appearance and instability of empty H-2 class I molecules under physiological conditions. *Proc. Natl. Acad. Sci. USA. 88*: 3594.

26. **Benjamin, R. J., J. A. Madrigal, and P. Parham.** 1991. Peptide binding to empty HLA-B27 molecules on viable human cells. *Nature 351*: 74.

27. **Sykulev, Y., A. Brunmark, M. R. Jackson, J. Cohen, P. A. Peterson and H. N. Eisen.** 1993. Affinity and kinetics of reactions between antigen-specific T-cell receptor and peptide-MHC complexes. Submitted.

2

Where Do MHC Class I-Associated Peptides Come From?

Jonathan W. Yewdell, Josephine H. Cox, Heidi Link, Igor Bacik,
and Jack R. Bennink

PEPTIDE GENERATION: THE NEXT FRONTIER

The discovery of MHC class I molecule-restricted recognition of viral proteins in 1974 gave birth to the field of molecular cellular immunology (1). Two decades later, the puzzle of MHC-restriction has been clearly resolved as the consequence of the T cell receptor recognizing an array of residues contributed by both the class I molecule and a peptide of 8 to 10 residues derived from a foreign protein. Indeed, the image of a peptide snug in the antigen binding groove of MHC class I molecules is an icon of modern immunology, even gracing departmental stationery.

The antigenic peptides presented by class I molecules can be derived from proteins localized to membranes, cytosol, or nucleus. No class of protein is known to be spared from the peptide generating machinery. One of the enigmas regarding the class I antigen processing system is how antigenic peptides are generated from this diverse set of proteins. This question is of practical importance, since the peptide generating machinery no doubt influences not only which peptides are produced from a given protein, but also which proteins are favored for peptide production. The purpose of this paper is to review current understanding of the generation of antigenic peptides and to summarize the recent contributions of our laboratory to this problem.

PEPTONS OR PROTEINS AS THE SOURCE OF ANTIGENIC PEPTIDES?

Observations that synthesis of antigenic peptides could be directed by pro-moterless transfected genes, or transfected genes with frame shifts or stop codons upstream from peptide coding regions, spawned the elegant pepton hypothesis (2). In brief, it was proposed that antigenic peptides derive not from proteins, but from short translation products (peptons) made from either mRNA, or most spectacularly, from short RNAs produced expressly for the purpose of peptide biosynthesis. The beauty of the hypothesis is its provision for monitoring genes independently of their levels of expression. This would enable the immune system to detect mutations in both coding and non-coding regions leading to malignant transformation, and to detect the presence of latent viruses. (One of the drawbacks of the hypothesis is that the immune system would also have to deal with an enormous repertoire of self-peptons). The pepton hypothesis receives some support from observations that frame shift mutations do not abolish antigen presentation of peptides from transfected nuclear genes (3,4).

These and other related findings clearly indicate that biochemically detect-able amounts of proteins are not required for sensitization of target cells for lysis by T_{CD8+}. One finding weighs heavily against the pepton hypothesis, however, or at least its relevance *in vivo* to the production of immuno-dominant peptides. All of the antigenic peptides encoded by viral gene products correspond to open reading frames of bona fide viral proteins. Similarly, all of the class I-associated peptides that have been sequenced by chemical or physical methods match open reading frames in cases where genetic information encoding the peptide is known. This strongly suggests that the peptides associated with class I molecules identified to date are derived from translation of standard mRNAs in the proper reading frame. To be fair, however, it is not unreasonable that any pepton-mechanism that might exist would produce peptides in small quantities and be limited to cellular genes, or at least to genes present in the nucleus (many of the viral peptides characterized are derived from viruses that replicate in the cytosol). None the less, evidence has yet to be presented that peptons are produced in sufficient abundance to trigger immune responses *in vivo*.

Several findings indicate that peptides can be derived from proteins. First, the introduction of exogenous proteins into the cytosol has been repeatedly demonstrated to result in presentation of antigenic peptides to T_{CD8+} *in vitro* and to induce T_{CD8+} responses *in vivo*. While it has never been established whether the antigenic peptides generated under these conditions originate from intact proteins, or their denatured or fragmented derivatives, this finding establishes that cells have the capacity to produce antigenic peptides independent of protein synthesis. Second, it has been found that alterations of proteins that result in accelerated degradation can enhance antigen presentation in some circumstances (5). This latter finding provides the only direct link between proteolysis of intact proteins and the generation of antigenic peptides.

In summary, there are good reasons to believe that many, if not most immunogenic biosynthesized peptides (which represent the vast majority of physiologically relevant antigenic peptides) require the ribosomal synthesis of polypeptides from open reading frames of standard mRNAs. It is an essentially open question, however, whether antigenic peptides are derived from full length native proteins or from some other ribosomal product.

THE QUESTIONABLE ROLE OF THE PROTEASOME IN ANTIGEN PROCESSING

The efficient presentation of biosynthesized cytosolic and nuclear proteins, the requirement for exogenous proteins to be delivered to the cytosol, and the location of the peptide transporter (TAP) in the membrane of the endoplasmic reticulum (ER) and (perhaps cis-Golgi complex as well) (6), all point to the cytosol as the starting point of the antigen processing pathway for endogenous and exogenous antigens. The TAP-dependence of presentation of most antigens, in conjunction with the apparent preference of TAP for peptides of less than 20 residues (7-9), indicate that cytosolic proteases must play an important role in antigen processing.

The most active cytosolic protease activity is associated with the 20S proteasome particle, a 700 kDa structure that constitutes approximately 1% (!) of the protein present in mammalian cells. The proteasome is composed of more than 10 distinct proteins. A subpopulation of proteasomes contains two subunits (LMP2 and LMP7) encoded by MHC genes intimately interdigitated with those encoding the two subunits of TAP. This suspicious genetic location, and the γ-interferon enhancement of LMP gene transcription prompted speculation that LMP-containing proteasomes contribute to the production of antigenic peptides. These speculative fires were recently fueled by reports that LMP-containing proteasomes demonstrate enhanced ability to cleave indicator peptides on the COOH-terminal side of hydrophobic or basic residues (10,11). This would be consistent with the preference of virtually all characterized class I molecules for peptides containing hydrophobic or basic COOH-terminal residues.

The role of LMP-containing proteasomes in antigen processing must be considered tentative, however, since cells lacking LMP genes were found to present at least two determinants to T_{CD8+} (12), and more generally, express class I molecules containing antigenic peptides quantitatively and qualitatively similar to class I molecules derived from normal cells (13). It remained possible, however, that LMP gene products enhance the rate or efficiency of peptide production or are needed for the production of a sub-population of peptides.

We investigated this question using human cells lacking LMP genes (14). Infection of cells with recombinant vaccinia viruses (rVV) expressing mouse class I molecules enabled us to examine presentation of peptides to virus-specific mouse T_{CD8+} using a standard ^{51}Cr-microcytotoxicity assay. We found that LMP-deficient cells could present 2 virus-derived peptides in association with H-2 D^d, 3 virus-derived peptides in association with H-2 K^d, and 2 virus-derived peptides in association with H-2 K^b. Indeed, we were unable to identify antigens that the cells were incapable of presenting to T_{CD8+}. By adding brefeldin A (BFA) at various times after virus infection to block the delivery of additional class I-peptide complexes to the cell surface, we measured the rate of presentation of two viral peptides to T_{CD8+}. In each case, LMP-deficient cells were sensitized for lysis with similar kinetics as wild-type cells. Taken at face value, these findings do not support a role for LMP-containing proteasomes in the production of antigenic peptides.

It is important to recognize that these experiments do not address the role of non-LMP containing proteasomes in antigen processing. The 20S proteasome likely functions as the active subunit of a larger structure termed the 26S proteasome, which is at least partly responsible for the degradation of proteins that have been covalently attached via selected lysine residues to ubiquitin. Ubiquitin is an abundant protein highly conserved between all eukaryotes that is essential to many cellular functions, including the cytosolic degradation of proteins. Several cell lines selected from mutagenized cells on the basis of growth arrest at temperature $\geq 39°C$ demonstrate a deficiency of an enzyme (termed E1) essential for the conjugation of ubiquitin to proteins. The ability of these cells to degrade short lived or misfolded proteins is inversely proportional to the time and temperature of incubation; complete inhibition generally requires incubation for 1 hour at 43°C.

It was recently reported that following incubation for one hour at 41°C, one such mutant cell line (ts20) was unable to present exogenous ovalbumin loaded into the cytosol by osmotic lysis of pinosomes, while its wild type parent demonstrate only a slight decrease in antigen presentation (15). Based on this finding, it was concluded that E1 is necessary for the generation of antigenic peptides from exogenously delivered ovalbumin.

To examine the importance of E1 in the presentation of viral antigens, we have used ts1AS9 cells, a L929 cell-derivative with a thermolabile E1 (16). It was previously shown that a mutation in the E1 gene (which is located on the X chromosome, probably accounting for the high frequency that cell lines with mutations in the gene are isolated) was responsible for the temperature-sensitive growth of the cells. Since the function of the E1 in ts1AS9 cells has not been characterized, we first determined that the effect of elevated temperatures on protein ubiquitination is similar in magnitude to that observed with ts20 cells, as measured by Western blotting of ubiquitinated cellular proteins with an anti-ubiquitin antibody. Further, we found that ubiquitination at elevated temperature is restored by expression of wild type human E1 following infection of cells with a rVV expressing the gene. Despite the failure of cells to ubiquitinate proteins following incubation at 43°C for 1 hour, the same treatment does not interfere with their ability to present any of 4 influenza virus proteins biosynthesized following infection with influenza virus. Presentation of a minor viral structural protein delivered in small quantities to the cytosol from non-infectious virus following fusion of viral and cellular membranes is also not compromised by heat-inactivation of E1. Incubation of cells with BFA at various times after exposure to infectious virus revealed that presentation of determinants occurred at least as rapidly in heated ts A1S9 cells as in wild-type cells, strongly suggesting that the cell lines generated and presented peptides with similar efficiency.

These findings suggest that E1 is not generally required for the generation of antigenic peptides from either exogenously provided or endogenously synthesized antigens. As E1 is thought to be essential for the conjugation of ubiquitin to proteins, this implies that generation of peptides via a ubiquitin-independent pathway occurs at least as efficiently as any ubiquitin-dependent pathway that may function, and indicates that the involvement of other cytosolic proteases in antigen processing must be entertained.

TAP AS A MIRROR OF CYTOSOLIC PROTEASE SPECIFICITY

TAP is an essential part of the processing pathway for all but a minor subset of peptides that are either derived from ER-insertion sequences, or from proteins delivered to the ER by NH_2-terminal insertion sequences (more about this latter pathway below). Although we and others previously demonstrated the TAP-independent presentation of peptides derived from cytosolic proteins, this pathway is far less efficient than the standard route. The extremely low levels of T_{CD8+} in mice lacking TAP, presumably due to the involvement of TAP in thymic positive selection of T_{CD8+}, is probably the most compelling evidence for the critical nature of TAP in antigen presentation *in vivo* (17).

Since most antigenic peptides must interact with TAP, the specificity of TAP for peptide length and sequence provides a critical clue towards the nature of the cytosolic proteases that function in antigen processing. Three groups recently demonstrated TAP-mediated transport of peptides in cell free or semi-intact cell systems (7-9). Although the details of TAP-specificity remain to be fleshed out, the present results suggest that TAP is able to transport peptides of 8 to 10 residues, and that somewhat longer peptides can be transported at a similar efficiency.

Whatever the *in vitro* transport assays reveal, they will have to be correlated with antigen presentation in cultured cells and *in vivo*. To investigate the potential for TAP to transport small peptides, we produced a number of rVVs expressing an initiating Met followed by the 8 or 9 residues corresponding to naturally processed viral determinants. To minimize synthesis of COOH-extended peptides by ribosomal read through, the inserted genes contained two stop codons after the coding region. Normal cells infected with each of the rVVs presented the biosynthesized peptides to the appropriate T_{CD8+}, while TAP-deficient cells presented the peptides poorly, if at all (18). These data indicate that if cytosolic proteases produce the properly-sized peptide (or a peptide one residue longer, since we are uncertain as to the efficiency of removal of the NH_2-terminal Met), these peptides would be conveyed by TAP to MHC class I molecules. Obviously, they do not rule out the possibility that longer peptides produced in the cytosol are transported into the ER where they are trimmed before or after associating with MHC class I molecules. Indeed, as described in the next section, results from two different experimental strategies we have employed suggest that such trimming can occur in the secretory pathway.

PEPTIDE PRODUCTION AND TRIMMING IN THE SECRETORY PATHWAY

In the course of examining the presentation of short biosynthesized peptides, we found that extension of a nonameric biosynthesized determinant by two residues greatly diminished its presentation (19). Based on the finding that the serum dipeptidyl peptidase angiotensin converting enzyme (ACE) efficiently removed the dipeptide from a synthetic equivalent of the peptide (20), we generated a rVV expressing ACE. Co-expression of ACE with the longer peptide enabled efficient TAP-dependent presentation of the peptide (21). ACE appeared to act exclusively in the secretory pathway, since we were unable to detect enyzmatically active non-glycosylated ACE in detergent extracts. These findings were most consistent with the idea that the COOH-terminally extended peptide is transported via TAP into the secretory pathway where ACE converts it into the antigenically active nonamer. We cannot exclude the possibility, however, that the extended peptide is cleaved in the cytosol by ACE that is either re-imported from the secretory pathway, or not exported from the cytosol following biosynthesis.

Despite this mechanistic ambiguity, these findings provide the first direct demonstration that protease expression can affect the types of antigenic peptides produced by cells. The practical implications are that cells expressing specialized proteases will present a different spectrum of peptides to the immune system, and that proteases expressed by viruses like human immunodeficiency virus or poliovirus could alter the types of viral *and* cellular antigenic peptides produced by infected cells. This latter consideration predicts that some peptides recognized by "virus-specific" T_{CD8+} will actually be encoded by cellular genes (of course, such peptides could also result from virus induced alterations in cellular gene expression or metabolism).

To more directly explore the potential of antigenic peptide trimming in the secretory pathway, we examined the capacity of TAP-deficient cells to present peptides delivered to the secretory compartment by ER-insertion sequences. It was originally shown that addition of an ER-insertion sequence derived from an adenovirus glycoprotein to an eleven residue peptide from influenza virus M1 protein allowed for its HLA A2.1-restricted presentation by TAP-deficient cells (22). Subsequently, using rVVs, we found that this ER-insertion sequence, or another ER-insertion sequence from α-interferon, could facilitate the presentation of numerous different peptides in association with H-2 K^d, D^b, K^b, or K^k (18,21). We next created rVVs that encode a NH_2-terminal ER-insertion sequence followed by a K^b-restricted peptide (K^bRP) and ending with a K^d-restricted peptide (K^dRP), or the ER-insertion sequence followed by the K^dRP and ending with the K^bRP. Infection of TAP-expressing cells with either of the rVVs led to their lysis by T_{CD8+} specific for either determinant. Following infection of TAP-deficient cells, however, only the determinant located at the COOH-terminus was efficiently presented in each case. Other rVVs expressing the tandem peptides without the ER-insertion sequence did not sensitize TAP-deficient cells for lysis (in contrast to TAP expressing cells), demonstrating that antigen processing in these circumstances depends on a functional ER-insertion sequence, and by inference, must be occurring in the secretory compartment. Since rVVs expressing either the K^bRP or the K^dRP alone COOH-terminal to the ER-insertion sequence efficiently sensitize TAP-deficient cells for lysis by the appropriate T_{CD8+} (18), we know that the middle peptide in the tandem construct can be efficiently liberated from the ER-insertion sequence, presumably due to the action of signal peptidase.

Based on findings with synthetic peptides, extension of natural peptides by even a single residue at the COOH-terminus almost always results in a greatly diminished affinity, and consequent decrease in antigenicity. Extensions at the NH_2-terminus have less of an effect, but addition of more than approximately three residues again results in greatly diminished affinity (the lesser effect at the NH_2-terminus might, however, be due to trimming by aminopeptidases, which are abundantly expressed at the cell surface). Thus, the presentation of the COOH-terminal peptide from the tandem peptide almost certainly represents the action of a protease present in the secretory compartment that cleaves within 2 or 3 residues of the K^bRP-K^dRP junction. This protease could be signal peptidases itself, cleaving the gene product 10 residues or so downstream of the expected cleavage site, although the efficiency of such activity would be expected to be very low. Alternatively, the COOH-terminal peptide might be liberated via two cleavages, the first being signal peptidase cleavage at the predicted junctional cleavage site, the second being mediated by signal peptidase or another peptidase in the ER (or Golgi complex, for that matter). In this case, the proteolytic activity (an amino peptidase?) must also efficiently destroy the amino terminal determinant in the process of liberating the COOH-terminal determinant, since it is presented or at low or undetectable levels.

These findings extend prior observations that peptides derived from ER-insertion sequences are present on HLA A2.1 molecules expressed by TAP-deficient cells (23,24). The protease responsible for the production of these signal-derived peptides was not identified, but it is possibly a mammalian homolog of a bacterial signal peptide peptidase, which functions to degrade ER-insertion sequences.

It was recently reported that a determinant(s) in the lumenal domain of HIV gp160 was efficiently presented to T_{CD8+} by TAP-deficient cells (25). Peptides were apparently generated in the secretory pathway, since removal of the ER-insertion sequence reduced presentation to background levels. Retention of the protein in the ER by expression of an ER-retained version of CD4 (which binds GP160), did not diminish presentation, consistent with the idea that peptides are generated within the ER itself, perhaps through the action of proteases thought to exist in the ER for the express purpose of degrading aberrantly folded proteins.

We have examined the presentation of full length proteins delivered to the ER by either natural or genetically engineered NH_2-terminal ER-insertion sequences. A K^d -restricted determinant from the influenza virus A/PR/8 HA is presented at low levels by TAP-deficient cells. The same determinant is not presented from A/Japan/305 HA. In the case of the PR8 HA, presentation is likely due to determinants produced in the cytosol, since identical levels of presentation are observed with a cytosolic form of HA lacking an ER-insertion sequence. In additional experiments we found that TAP-deficient cells do not present any of 3 determinants present in full length NP delivered to the ER by addition of a NH_2-terminal ER-insertion sequence derived from α-interferon. Biochemical studies confirmed that this protein is efficiently delivered to the ER, since all immunoprecipitated material contains N-linked oligosaccharides. At least some NP is exported from the ER, since NP with complex-type oligosacchrides resistant to digestion with endoglycosidase H is secreted from cells.

Together with a previous report that a measles virus glycoprotein is not presented by TAP-deficient cells (26), these findings suggest that the ability of the ER (and the rest of secretory pathway) to generate antigenic peptides from intact proteins is rather limited. By contrast, it appears that proteases in the secretory compartment may have a more general capacity to trim peptides to the proper size for class I association. Important questions for future studies are the relevance of such trimming to the production of TAP-dependent antigenic peptides, the precise location(s) of the proteases in the various compartments of the secretory pathway, and the possible involvement of class I molecules in tethering peptides during trimming.

PEPTIDE CHAPERONES

However and wherever peptides are hatched, they must eventually come to roost in class I molecules. It was recently reported that antigenic peptides co-purify with either a cytosolic heat shock protein (HSP) or a resident ER HSP (27,28). In both cases, peptides were released by addition of ATP to the protein-peptide complex. These findings inspired the proposal that antigenic peptides are ferried about the cell by these HSPs, and perhaps other molecular chaperones as well.

Such a mechanism would presumably have evolved in response to limitations in the ability of peptides to freely diffuse in cells. On their own, peptides of 8 to 20 residues in length would diffuse at a rate of between 1 and 0.1 μ^2/sec (29) (the higher rate is based on a radius of gyration of 10Å, the lower on a radius of gyration of 100Å; these values very likely encompass the size range of peptides of these lengths). If TAP is active at all portions of the ER, then the presence of the ER network in virtually all regions of the cytoplasm would ensure that peptides would never have to travel greater than 1 μ, and usually much less, to encounter TAP. This would occur within one second (peptides generated in the nucleus would be an exception, having to travel perhaps as much as 5 to 10 μ). This distance is perhaps 10 to 100 fold greater than the peptide would have to travel to reach a chaperone present at 10^6 copies/cell. However rapidly the peptide could bind to the chaperone, be carried to TAP and elute, the price paid for this increased complexity is almost certainly too great to warrant savings of less than a second (the same argument holds for transport of peptides to class I molecules on the other side of the ER membrane). Thus, if the binding of peptides to specific molecular chaperones is relevant to antigen processing, it likely functions not to speed delivery of peptides to TAP, but as a means of preventing peptides from binding non-productively to other proteins in the cytosol or ER. This would minimize sequestration of antigenic peptides, but perhaps its real biological significance would be to prevent the inadvertent inactivation of the unsuspecting object of the peptide's affection.

COMING TO GRIPS WITH *DRIPS*?

The fact of the matter is that we don't really know where or how most antigenic peptides are generated. It is reasonably certain that the unadulterated pepton hypothesis applies, at most, to the generation of small quantities of antigenic peptides encoded by nuclear genes, and not the antigenic peptides studied under most circumstances. It is not at all clear, however, whether under physiological circumstances these more abundant cell-encoded peptides or virus-encoded peptides are derived from intact biosynthesized proteins or from the defective products of protein synthesis (premature translation products, misfolded full length forms), which for lack of a better term, we will call Defective RIbosomal Products (DRIPs). Indeed, it is plausible that antigenic peptides are primarily produced by a specialized subpopulation of ribosomes dedicated to DRIP production.

If DRIPs were a major source of antigenic peptides, then the rate of antigenic peptide production would be dependent largely on the rate of translation, and not the total amounts of protein synthesized, or the degradation rate of full length proteins. This would explain many of the puzzling characteristics of class I antigen presentation. The rapid and efficient presentation of viral peptides following infection in the midst of a vast excess of cellular proteins would be expected, since viral mRNAs reach levels of cellular mRNAs within minutes of infection, while several or more hours are needed for viral proteins to achieve levels comparable to cellular proteins. The efficient generation of peptides from viral proteins with NH_2-terminal ER-insertion sequences or from cytosolic or nuclear proteins with low or negligible degradation rates would make sense. The DRIP hypothesis shares several attractive features with its progenitor, the pepton hypothesis, in explaining why peptides recovered from class I molecules don't simply reflect the abundance of proteins in the cell, and allowing the immune system to sample self and foreign proteins in a manner independent of the properties of the native protein.

It is important to stress that the production of antigenic peptides from DRIPs (or from peptons, for that matter) would still require the action of proteases. As it appears that TAP prefers peptides of no more than 20 residues (and perhaps much smaller), at least some, and possibly all of the proteolytic events must occur in the cytosol. In searching for the relevant proteases, the broad specificities, abundance, and sheer known and potential variety of proteasomes isolated from cells make them prime suspects. As in jurisprudence, however, a justly rendered guilty verdict requires conclusive evidence. We believe that our findings that antigen processing requires neither ubiquitin targeted proteolysis nor LMP-containing proteasomes provides reasonable doubt regarding the role of proteasomes in antigen processing. While at the same time these findings do not prove the innocence of proteasomes, they do suggest that either there is a proteasome-independent pathway capable of producing antigenic peptides, or that antigenic peptides are produced by non-LMP containing proteasomes from non-ubiquitinated proteins.

In considering that proteasomes may not be essential to antigen processing, it is worth noting that other proteases are known to exist in the cytosol (ribosomes even possess their own serine proteinase [EC number 3.4.21.], whose biological relevance is unknown). Cell biologists are just now beginning to consider myriad roles of known proteases in cells. Given the obvious importance of proteolysis in cellular regulation, there can be little doubt that many more proteases remain to be discovered. There is every reason to believe that we will have many happy years searching for the sources of antigenic peptides and characterizing the various cytosolic and secretory proteases that create the now familiar denizen of the class I binding groove.

REFERENCES

1. **Zinkernagel, R. M. and P. C. Doherty**. 1974. Restriction of in vitro T cell-mediated cytotoxicity in lymphocytic choriomeningitis within a syngeneic or semiallogeneic system. *Nature 248*:701.

2. **Van Pel, A. and T. Boon**. 1989. T cell-recognized antigenic peptides derived from the cellular genome are not protein degradation products but can be generated directly by transcription and translation of short subgenic regions. A hypothesis. *Immunogenetics 29*:75.

3. **Fetten, J. V., N. Roy, and E. Giboa**. 1991. A frameshift mutation at the NH_2 terminus of the nucleoprotein gene does not affect generation of cytotoxic T lymphocyte epitopes. *J. Immunol. 147*:2697.

4. **Shastri, N. and F. Gonzalez**. 1993. Endogenous generation and presentaiton of the ova peptide/K^b complex to T-cells. *J. Immunol. 150:2724*.

5. Townsend, A. , J. Bastin, K. Gould, G. Brownlee, M. Andrew, B. Coupar, D. Boyle, S. Chan, and G. Smith. 1988. Defective presentation to class I-restricted cytotoxic T lymphocytes in vaccinia-infected cells is overcome by enhanced degradation of antigen. *J. Exp. Med. 168*:1211.

6. Kleijmeer, M. J., A. Kelly, H. J. Geuze, J. W. Slot, A. Townsend, and J. Trowsdale. 1992. Location of MHC-encoded transporters in the endoplasmic reticulum and cis-Golgi. *Nature 357*:342.

7. Shepherd, J. C., T. N. M. Schumacher, P. G. Ashton-Rickardt, S. Imaeda, H. L. Ploegh, C. A. Janeway,.Jr., and S. Tonegawa. 1993. TAP-1-dependent peptide translocation in vitro is ATP dependent and peptide selective. *Cell 74*:577.

8. Neefjes, J. J., F. Momburg, and G. J. Hämmerling. 1993. Selective and ATP-dependent translocation of peptides by the MHC-encoded transporter. *Science 261*:769.

9. Adrolewicz, M. J., K. S. Anderson, and P. Cresswell. 1993. Evidence that transporters associated with antigen processing translocate a major histocompatibility complex class I-binding peptide into the endoplasmic reticulum in an ATP-dependent manner. *Proc. Natl. Acad. Sci. USA 90*:9130.

10. Gaczynska, M. , K. L. Rock, and A. L. Goldberg. 1993. Gamma-interferon and expression of MHC genes regulate peptide hydrolysis by proteasomes. *Nature 365*:264.

11. Driscoll, J. , M. G. Brown, D. Finley, and J. J. Monaco. 1993. MHC-linked LMP gene products specifically alter peptidase activities of the proteasome. *Nature 365*:262.

12. Momburg, F. , V. Ortiz-Navarrete, J. Neefjes, E. Goulmy, Y. van der Wal, H. Spits, S. J. Powis, G. W. Butcher, J. C. Howard, P. Walden, and G. J. Hämmerling. 1992. Proteasome subunits encoded by the major histocommpatibility complex are not essential for antigen presentation. *Nature 360*:174.

13. Arnold, D. , J. Driscoll, M. Androlewicz, E. Hughes, P. Cresswell, and T. Spies. 1992. Proteasome subunits encoded in the MHC are not

generally required for the processing of peptides bound by MHC class I molecules. *Nature 360*:171.

14. **Yewdell, J. W., C. K. Lapham, I. Bacik, T. Spies, and J. R. Bennink**. 1994. MHC-encoded proteasome subunits LMP2 and LMP7 are not required for efficient MHC class I restricted-presentation of viral antigens. *J. Immunol. in press*:

15. **Michalek, M. T., E. P. Grant, C. Gramm, A. L. Goldberg, and K. L. Rock**. 1993. A role for the ubiquitin-dependent proteolytic pathway in MHC class I-restricted antigen presentation. *Nature 363*:552.

16. **Zacksenhaus, E. and R. Sheinin**. 1990. Molecular cloning, primary structure and expression of the human X linked A1S9 gene cDNA which complements the ts A1S9 mouse L cell defect in DNA replication. *EMBO J. 9*:2923.

17. **Van Kaer, L. , P. G. Ashton-Rickardt, H. L. Ploegh, and S. Tonegawa**. 1992. TAP1 mutant mice are deficient in antigen presentation, surface class I molecules, and CD4-8+ T cells. *Cell 71*:1205.

18. **Bacik, I. , J. H. Cox, R. Anderson, J. W. Yewdell, and J. R. Bennink**. 1994. TAP-independent presentation of endogenously synthesized peptides is enhanced by endoplasmic reticulum insertion sequences located at the amino but not carboxy terminus of the peptide. *J. Immunol. in press*:

19. **Eisenlohr, L. C., J. W. Yewdell, and J. R. Bennink**. 1992. Flanking sequences influence the presentation of an endogenously synthesized peptide to cytotoxic T lymphocytes. *J. Exp. Med. 175*:481.

20.. **Sherman, L. A., T. A. Burke, and J. A. Biggs**. 1992. Extracellular processing of antigens that bind class I major histocompatibility molecules. *J. Exp. Med. 175*:1221.

21. **Eisenlohr, L. C., I. Bacik, J. R. Bennink, K. Bernstein, and J. W. Yewdell**. 1992. Expression of a membrane protease enhances presentation of endogenous antigens to MHC class I-restricted T lymphocytes. *Cell 71*:963.

22. **Anderson, K. , P. Cresswell, M. Gammon, J. Hermes, A. Williamson, and H. Zweerink**. 1991. Endogenously synthesized peptide

with an endoplasmic reticulum signal sequence sensitizes antigen process-ing mutant cells to class I-restricted cell-mediated lysis. *J. Exp. Med.* *174*:489.

23. **Henderson, R. A., H. Michel, K. Sakaguchi, J. Shabanowitz, E. Apella, D. F. Hunt, and V. H. Engelhard**. 1992. HLA-A2.1-associated peptides from a mutant line: A second pathway of antigen presentation. *Science 255*:1264.

24. **Wei, M. L. and P. Cresswell**. 1992. HLA-A2 molecules in an antigen-processing mutant cell contain signal sequence-derived peptides. *Nature 356*:443.

25. **Hammond, S. A., R. C. Bollinger, T. W. Tobery, and R. F. Siliciano**. 1993. Transporter-independent processing of HIV-1 evnelope protein for recognition by CD8+ T cells. *Nature 364*:158.

26. **van Binnendijk, R. S., C. A. van Baalen, M. C. M. Poelen, P. de Vries, J. Boes, V. Cerundolo, A. D. M. E. Osterhaus, and F. G. C. M. UytdeHaag**. 1992. Measles virus transmembrane fusion protein synthe-sized de novo or presented in immunostimulating complexes is en-dogenously processed for HLA class I- and class II-restricted cytotoxic T cell recognition. *J. Exp. Med. 1765*:119.

27. **Udono, H. and P. K. Srivastava**. 1993. Heat shock protein 70-associated peptides elicit specific cancer immunity. *J. Exp. Med. 178*:1391.

28. **Li, Z. and P. K. Srivastava**. 1993. Tumor rejection antigen gp96/grp94 is an ATPase: implications for protein folding and antigen presentation. *EMBO J. 12*:3143.

29. **Paine, P. L.**. 1984. Diffusive and nondiffusive proteins in vivo. *J. Cell Biology 99*:188s.

3

ASSEMBLY AND TRANSPORT OF CLASS I MHC GLYCOPROTEINS

Peter Cresswell, Matthew Androlewicz, Lisa Denzin, Bodo Ortmann, and Bhanu Sadasivan

The successful expression of a functional class I MHC molecule on the surface of a cell demands that the MHC-encoded transmembrane glycoprotein associates with two additional components. These are ß$_2$ microglobulin (ß$_2$m) and a short peptide of 8-10 amino acids. The association of these components occurs in the endoplasmic reticulum (ER) of the cell, and is required for efficient transport out of the ER, through the Golgi apparatus, and to the plasma membrane. The biochemical mechanisms involved in the generation of these complexes, and the molecular explanation for apparent occasional exceptions to the above requirements for successful transport, have been major interests of this laboratory for a number of years. In this paper, we will summarize work performed during the past few years which has both enhanced our understanding of the problem and, as always in any scientific endeavor, raised new questions.

CLASS I MHC STRUCTURE

Class I MHC genes (HLA-A, B and C in humans and H-2K, D and L in the mouse) encode transmembrane glycoproteins of 44-49kDa, containing one (human), two (most mouse) or three (some mouse) N-linked glycans. The lumenal or extracellular N-terminal portion of the molecule can be divided into three distinct domains (α_1, α_2 and α_3).

The membrane proximal domain (α_3) is homologous to constant region domains of immunoglobulins, with a typical intradomain disulfide bond, and it folds into a similar structure. The α_1 and α_2 domains fold together to form a structure characteristic of both class I and class II MHC molecules (1, 2) which consists of eight antiparallel ß-strands overlaid by two α-helices which form a groove or cleft. The short peptide characteristically associated with class I molecules is bound in the groove, with the N- and C-termini and a limited number of its amino acid side chains buried in discrete pockets formed by amino acids present in the floor of the groove or in the flanking α-helices (3, 4). ß$_2$m, which is non-covalently associated with the extracellular region of the class I glycoprotein, is a protein of approximately 12kDa which, like the α_3 domain of the glycoprotein chain, resembles a single immunoglobulin constant region domain. It interacts with residues in the α_3 domain and with the underside of the ß-sheet which forms the floor of the peptide binding groove (1).

ASSEMBLY OF CLASS I MOLECULES

General Scheme of Assembly

The class I heavy chain and ß$_2$m are synthesized on membrane-associated ribosomes with classical N-terminal signal sequences and translocated into the ER where the signal sequences are removed. Addition of the N-linked glycan(s) to the heavy chain occurs co-translationally. Two routes have been defined by which class I-associated peptides enter the ER. A minority of peptides are introduced by signal sequence dependent translocation, and in fact are fragments of signal sequences. The majority enter the ER from the cytosol by a specific translocation mechanism. These three components, the class I heavy chain, ß$_2$m, and peptide, then associate in the ER to form a tripartite complex which is transported to the cell surface.

Generation and Translocation of the Cytosolic Peptides

The majority of the peptides which associate with class I

molecules are derived from cytosolic proteins (5). The proteolysis of these proteins occurs in the cytosol, and is thought to be mediated by a multisubunit proteinase called the proteasome (6). The twenty or so subunits which make up the proteasome assemble into a cylindrical structure capable of mediating the ATP-dependent proteolysis of cytosolic proteins covalently associated with ubiquitin, a small protein the binding of which is believed to mediate their turnover. A cell line with a temperature-sensitive defect in the ubiquitin pathway is impaired in its ability to perform class I-restricted antigen processing at the non-permissive temperature (7). Further circumstantial evidence that the proteasome is involved in the processing of cytosolic components comes from the finding that two proteasome subunits (LMP.2 and LMP.7, where LMP is an abbreviation for "low molecular weight protein", (8)) are encoded by genes in the MHC (reviewed in 9). Proteasomes which include LMP.2 and LMP.7 have an altered substrate specificity, showing a more rapid cleavage after basic or hydrophobic amino acids which may favor the generation of MHC-binding peptides (10, 11).

Cytosolically generated peptides are translocated into the ER by the action of a specific transporter molecule. This transporter is probably a heterodimer, with the genes encoding the two subunits closely linked in the MHC. These genes are called *TAP.1* and *TAP.2*, where TAP stands for "Transporter associated with Antigen Processing" (reviewed in 9). The TAP molecule is a member of a family of molecules which share the function of translocating small molecules across biological membranes using ATP hydrolysis to drive translocation (12). Each TAP gene encodes a protein with an N-terminal region which contains six hydrophobic transmembrane segments connected by short hydrophilic loops and a putatively cytoplasmic C-terminal domain with sequences characteristic of ATP-binding proteins. Thus, the presumed dimer contains twelve transmembrane regions and two ATP-binding domains. The precise structure of the TAP molecule, and the mechanism by which it uses ATP hydrolysis to translocate peptides, is unknown. Until recently its translocation function was only inferred from the defects in the intracellular transport of class I MHC molecules apparent in cell lines with mutations in one or both TAP genes (see below). Recently, however, work from this laboratory and two others has established with a high degree of confidence that the TAP proteins genuinely translocate peptides across the ER membrane (13-15).

In our own work we used the toxin Streptolysin O to

permeabilize human B-lymphoblastoid cell lines (13). Cells so treated acquire pores in the plasma membranes through which soluble molecules can be introduced into the cytoplasmic compartment of the cell. These pores are sufficiently large that antibody molecules (150 kDa) can pass through them. In our experiments, radiolabelled class I-binding peptides were introduced into the cell. In CIR.A3 cells, which have functional TAP genes and also express HLA-A3 molecules, we were able to show that a radio-iodinated 10 amino acid peptide from the HIV nef protein (nef 7B) would bind to HLA-A3 if the cells were permeabilized with Streptolysin O. In these experiments, the cells were incubated on ice for 10 min. with the toxin, washed, and then warmed to 37° in the presence of the peptide. Pores form rapidly in the plasma membrane at 37°, allowing the peptide to enter the cell. After various intervals the cells were extracted with detergent, and an anti HLA-A3 monoclonal antibody was used to immunoprecipitate HLA-A3 molecules, which were counted in a gamma counter to determine the level of associated ^{125}I-labeled peptide. Control non-permeabilized cells exhibited no HLA-A3 association of the nef 7B peptide, arguing that surface HLA-A3 molecules were not responsible for the binding observed. HLA-A3-expressing transfectants of the TAP.1 and TAP.2-negative T2 cell line were unable to efficiently generate HLA-A3-peptide complexes, arguing that the TAP proteins were required for the association. In addition, the mouse TAP.2 mutant cell line, RMA-S, when transfected with the HLA-A3 gene and the human ß$_2$m gene, was unable to generate HLA-A3-nef 7B complexes when permeabilized in the presence of ^{125}I-nef 7B, while the wild-type parent, RMA, similarly transfected with HLA-A3 and ß$_2$m, efficiently generated such complexes.

Binding of ^{125}I-nef 7B to the HLA-A3 molecules was rapid, reaching a plateau within 10 min. of permeabilization. Binding was also dependent on the presence of ATP, since treatment of the permeabilized cells with apyrase, a specific ATPase, eliminated the ability of the cells to generate HLA-A3-nef 7B complexes. Binding was restored by the addition of ATP and an ATP-regenerating system. These experiments have been repeated using CIR and T2 cells expressing a transfected HLA-B27 gene and a radiolabelled B27-binding nonamer peptide derived from a ribosomal protein (5), with virtually identical results.

The most likely explanation for the results is that the radiolabelled peptides, upon entering the cytoplasmic compartment through the Streptolysin O pores, are translocated across the ER membrane by the TAP proteins. Association with class I MHC molecules then occurs. The evidence that the association occurs in the ER is that the introduction of the peptide induces the transport of a pool of class I molecules out of the ER, as measured by the acquisition of endoglycosidase H (endo H) resistance (13). This transport step, in agreement with the observations of others, requires both ATP and cytosolic components which must be added back to the permeabilized cells. Currently, we cannot be certain that the ATP dependence of peptide association with class I MHC molecules specifically reflects its requirement for TAP function. However, a low level of class I MHC-peptide association occurs in T2.A3 cells regardless of the presence of ATP. Thus association following translocation does not require ATP, and the most likely explanation for the ATP requirement in wild-type cells is that TAP-mediated translocation of peptides is ATP-dependent.

The two peptides which we have successfully used in this system were a nonamer and decamer. An additional important question is whether TAP translocation is restricted to peptides of a certain length. We, and others, have approached this question by a competition technique, in which unlabelled peptides are added to the permeabilized cells to determine if they can effectively compete with a radiolabelled peptide for translocation (13-15). Initially we used the ^{125}I-nef 7B peptide for these experiments, measuring inhibition of translocation by assaying for a reduction in HLA-A3 association in CIR.A3 cells in the presence of competitor peptides. This approach is limited in the sense that only non-HLA-A3 binding competitor peptides can be used, because otherwise a reduction in HLA-A3 association could also result from direct competition for HLA-A3 binding. To circumvent this problem, we adopted the approach pioneered by Neefjes et al. (14), in which the radiolabelled peptide incorporates a glycosylation acceptor sequence (asn-X-ser or asn-X-thr), which is a substrate for N-linked glycan addition following translocation of the peptide into the ER. This modification can be detected by binding the radiolabelled glycopeptide to the lectin concanavalin A covalently coupled to Sepharose beads. In our experiments we modified the sequence of the ribosomal protein-derived B27-binding peptide (5) described earlier, using the nonamer RRYQNSTEL. Using this approach we have been able to draw the following general conclusions (M. Androlewicz, manuscript in preparation). First, peptides shorter than eight amino acids are not

TAP translocation substrates by this criterion, i.e., they do not effectively compete for the translocation of the glycosylation acceptor peptide. Second, most peptides of between eight and twelve amino acids are good to excellent translocation substrates. Third, in general, longer peptides (16-30 amino acids) are poor substrates, with some exceptions. The longest peptide tested to date which is a good substrate has 24 amino acids. Thus, the TAP translocation system has a minimal length constraint, but so far we have been unable to define a maximum length requirement.

It is clear from our experiments that some peptides with an optimal length (8-12 amino acids) are nevertheless poor competitors in the translocation system. The most likely explanation is that they are not good translocation substrates. We have found that such peptides fail to inhibit the binding of radiolabelled peptides, derivatized with a photoreactive group, to the TAP proteins themselves (M. Androlewicz, manuscript in preparation), which is good evidence for this hypothesis. Thus, the TAP molecules have some restriction in the sequences of peptides which can be translocated. To date we have been unable to precisely define the components of the peptide sequence, in terms of position or specific amino acid residue at any position, which are responsible for the observed TAP specificity. A current favorite approach to the design of specific vaccines involves the preselection of class I binding epitopes from pathogen encoded proteins of defined sequence based on the known "anchor residues" (16), favored by particular class I alleles. Obviously, it will be important to understand the restrictions imposed by the TAP proteins in translocating peptides into the ER before this approach can be truly effective.

Class I MHC-Binding Signal Sequence Peptides

A characteristic of cell lines, or animals, with defective TAP function, is that surface expression of class I MHC molecules is low (see below). An example running counter to this rule is the product of the human HLA-A2 allele, which is expressed at substantial levels (20-40% of wild-type) on mutant B-cell lines defective in TAP.1 expression (.134, ref. 17) or in both TAP.1 and TAP.2 expression (.174 and T2, refs. 17, 18). HLA-A2 molecules isolated from T2 cells were found to contain a restricted number of peptides, which when sequenced proved to be derived from the signal sequences of known proteins. These were calreticulin (19), an interferon-inducible protein called IP30 (19, 20) and a component of the signal sequence receptor complex, SSRα (20, 21). SSRα and calreticulin are high abundance ER proteins (22-24), while a fraction of IP30 is secreted and the remainder localizes in intracellular vesicles, possibly lysosomes (25).

All of the signal sequence peptides had characteristics of HLA-A2 binding peptides, namely a leucine residue at the second position and a valine at the C-terminal position. Only one of the peptides, derived from IP30, proved to be a classical nonamer. The other peptides varied between 10 and 13 residues. Crystallographic evidence has shown that, for at least one example, a longer peptide can be integrated in the class I binding groove by insertion of the N- and C-termini into the pockets which they normally occupy, with the excess length accommodated by "bulging" in the middle (26). How the longer signal sequence peptides physically fit into the HLA-A2 binding site is unknown.

Initially it seemed likely that HLA-A2 was unusual in its ability to bind signal sequence derived peptides, perhaps because its binding site is unusually hydrophobic. However, signal sequence peptides, derived in fact from MHC molecules themselves, have been found in association with HLA-B7 molecules in wild-type, TAP-expressing cells (27), and one example of a class I-restricted viral epitope which is a signal sequence peptide has been reported in the mouse (28). Thus it appears that this source of peptides may be important in generating cytotoxic T-cell epitopes for a number of class I alleles.

Are there specialized mechanisms involved in generating class I-restricted epitopes from signal sequences, or is the association of such peptides with class I molecules a fortuitous by-product of signal sequence cleavage? The answer to this question is unknown but we are attempting to approach it in a variety of ways. It is clear from previous experiments, performed by this laboratory in collaboration with Zweerink and co-workers and later independently extended by the latter group, that when a minigene encoding a signal sequence followed by a defined HLA-A2-restricted T-cell epitope is expressed in TAP-deficient T2 cells they can be recognized by cytotoxic T-cells specific for the epitope (29). Thus, in the absence of the normal TAP translocation system, signal sequences can drive an epitope into the ER, where it is presumably cleaved from the signal sequence in a conventional manner by signal peptidase prior to its association with the class I molecule. The peptides naturally associated with HLA-A2 molecules in T2 cells are, however, fragments of the signal sequence itself. This implies that further, internal cleavage of the signal sequence must take place to generate the epitope. Is this also performed by the signal peptidase, or is another protease involved? In one case, that of the SSRα signal sequence, we have found that the thirteen amino acid peptide which binds to HLA-A2 is actually a dispensable part of the signal sequence (L. Denzin, unpublished results). The SSRα signal sequence is unusually long (32 amino acids), and the N-terminal 18 amino acids, which exclude the HLA-A2 binding region, function as a perfectly adequate signal sequence when fused to an immunoglobulin λ chain lacking its endogenous signal sequence. In the case of SSRα, then, the initial signal peptidase cleavage may generate an intermediate protein with the HLA-A2-binding sequence still attached, and a subsequent cleavage may release this peptide for HLA-A2 binding. We are continuing to investigate this model.

A characteristic of the class I MHC-associated signal sequence-derived peptides defined to date is that they contain one or more hydrophilic or charged residues. This may be important in preserving solubility once the peptides are generated, favoring binding to class I molecules rather than aggregation, association with heat shock proteins, or segregation into the lipid bilayer.

The Role of Chaperones in the Assembly of Class I MHC Molecules

Considerable evidence has accumulated in recent years which suggests that the folding of newly-synthesized proteins into a native conformation and their assembly into multi-subunit complexes does not occur spontaneously *in vivo*. A number of proteins (collectively described as molecular chaperones) have been described which facilitate the process (reviewed in 30). In bacteria, for example, the proteins groEL and groES, in conjunction with DnaK and DnaJ, have been shown to facilitate protein folding. In the mammalian ER the immunoglobulin binding protein, Bip/GRP78, a member of the Hsp70 family, has been found to play a role in immunoglobulin assembly by binding immunoglobulin heavy chains, and releasing them in an ATP-dependent fashion coincident with light chain association (31). The stress protein GRP94 (endoplasmin, ERP99, gp96, see below) an Hsp 90 homologue, has also been implicated in immunoglobulin assembly, and it and Bip/GRP78 appear to form a trimeric intermediate with the heavy chain prior to light chain association (32). It is generally thought that such proteins function by reversibly binding to exposed hydrophobic regions in misfolded or partly assembled proteins, preventing them from aggregating until a native conformation, with no exposed hydrophobic surfaces, is acquired.

The major chaperone implicated in class I assembly is calnexin (IP90, p88) (33-35). Calnexin is a 65kDa transmembrane phosphoprotein which is localized in the ER by a retention signal consisting of a sequence of basic amino acids at its C-terminus, which protrudes into the cytoplasm of the cell. It has been found to associate with a number of transmembrane proteins during their assembly into multimeric complexes. These include components of membrane immunoglobulins (34), T-cell receptors (34), and class II molecules (36). It has also been isolated in apparently stable association with the signal sequence receptor (including SSRα) in canine pancreatic microsomes (22). One of the first descriptions of the protein which ultimately proved to be calnexin was by Williams and co-workers, who found it to be associated with assembling mouse class I MHC molecules (35). Dissociation of calnexin from class I molecules occurred in parallel with the acquisition of endo H resistance by the heavy chain N-linked glycans, suggesting it played a role in class I assembly. Subsequently, it was found that class I-β_2m dimers in the TAP.2 mutant cell line RMA-S also associated with calnexin, implying that release from calnexin might require the completion of the tripartite class I-β_2m-peptide complex (37).

In human B-cell lines free class I heavy chains, for example in the β_2m-deficient cell line Daudi, associate strongly with calnexin (34, 37, and B. Ortmann, unpublished results). However, HLA class I-β_2m dimers in the TAP-deficient cell line T2 are not strongly associated with calnexin (K. Anderson, unpublished results). Thus, in the human system the primary role of calnexin may be as a class I heavy chain binding protein, facilitating its association with β_2m. Its potential role in facilitating peptide association may be restricted to, or at least more important in, the mouse system.

Experiments by Williams and co-workers have suggested that the primary interaction of class I molecules with calnexin is with the transmembrane region (38). Thus mutant H-2D[b] molecules which are membrane associated by a phosphatidyl inositol glycolipid-linked tail cannot be found in association with calnexin. Other workers have suggested that calnexin primarily associates with assembling glycoproteins by their N-linked glycans (39), having an affinity for the residual glucose residue present on the N-linked glycan following its initial attachment to the protein backbone and the subsequent enzymatic removal of the penultimate and terminal glucose moieties (40). How these two apparently opposing views are to be resolved is not clear.

An additional protein which may play a role in class I assembly is gp96 (GRP94). Srivastava and colleagues have shown that gp96 isolated from mouse tumor cells can elicit a class I-MHC restricted CTL response when used to immunize mice syngeneic to the tumor (41). This has been found to depend on the association of a tumor-specific peptide with the gp96 molecules, which like many stress proteins, can bind peptides (42). The implication of these findings is that in the immunized mice the gp96 is "donating" peptides in some unknown fashion to class I molecules, perhaps in a macrophage. Based on this, and evidence that mouse class I MHC molecules can be detected in association with gp96 (P. Srivastava, personal communication), Srivastava and colleagues have suggested that gp96 may play a key role in class I assembly in normal cells, perhaps serving as a peptide-binding intermediate (42).

A final question is whether the TAP molecules themselves play any role in class I assembly, other than their essential role in providing the major supply of peptides to the ER. Are class I molecules simply waiting in the ER for peptides to enter their binding sites by simple diffusion, or is the transfer of peptides from the TAP proteins to the class I molecules coordinated in some fashion? While the answer to this is not definitively known, we have evidence that class I MHC molecules in the ER are physically associated with TAP proteins prior to their being loaded with peptides (B. Ortmann, manuscript in preparation). The class I MHC molecules are associated with the TAP proteins only in their endo H-sensitive form, arguing again for a role in assembly. Exactly how calnexin, possibly gp96, and the TAP proteins coordinate their activities to assist in the class I assembly process is unclear, but is obviously a major area for future investigation.

Transport of Class I MHC Molecules

With the exception of unusual mouse molecules, such as H-2D^b, which have three N-linked glycans, free class I heavy chains are not expressed on the cell surface. In common with single subunits of many heteromultimeric membrane molecules, they normally associate with various chaperones in the ER, and are ultimately degraded in the absence of the partner subunits required to generate a stable, native conformation. In general, the same is true of peptide-free class I-ß$_2$m dimers, at least when one is discussing the products of human alleles. Human TAP-deficient cell lines express very low levels of class I molecules at the cell surface (17, 18), with the exception of the HLA-A2 molecules which, as described earlier, have a propensity to bind signal sequence-derived peptides. Low amounts of other alleles clearly are expressed at the plasma membrane, since although antibodies may not detect them, antigen-specific CTL can readily kill such mutant cells in an MHC-restricted fashion if the appropriate epitopic peptide is added (43). Presumably, these peptides bind to the low levels of class I molecules expressed on the surface. Whether these molecules arrive at the cell surface "empty", or whether they become available for binding added peptides upon dissociation of an endogenous peptide, perhaps derived from a signal sequence, is unknown.

Mouse class I alleles, at least most of them, are different from human alleles in that they do not appear to absolutely require associated peptides to leave the ER. H-2K^k appears to be an exception. In the TAP.2-negative RMA-S cell line, the endogenous K^b and D^b alleles can be detected at the cell surface at approximately 10% of wild-type levels, and if the cells are cultured at temperatures lower than 37°, this level increases (44). At 26°, the level of surface expression is virtually that of wild-type RMA cells. Ploegh and co-workers have suggested that this reflects a general instability of class I-ß$_2$m dimers in the absence of peptides and that at the lower temperatures the class I-ß$_2$m association is better maintained (44). In fact, if RMA-S cells which have been incubated at 26° are transferred to 37°, the surface expression of K^b and D^b rapidly falls. This decline can be prevented by adding specific K^b- or D^b-binding peptides, respectively (44). These same mouse alleles when expressed in the human TAP-deficient cell line T2 are well surface-expressed in comparison to non-HLA-A2 human class I alleles expressed in the same cells (20, 45). This is so even though they are demonstrably virtually peptide-free (shown for K^b and D^p, ref. 20). Why this should be so is still unclear. The surface expression levels of K^b and D^b in T2 transfectants can be further upregulated by 26° incubation, and their basic properties seem to be the same as in RMA-S cells, particularly when compared to RMA-S cells transfected with the human ß$_2$m gene (46). Human class I molecules in RMA-S, with or without human ß$_2$m, behave similarly to the way they behave in T2 cells (46). In both cell types it appears that the mouse class I molecules can assemble in the absence of peptides and leave the ER, while the products of human class I alleles do this very poorly. Thus the transport characteristics of the class I molecules are inherent, and not a function of the species of origin of the cell type they inhabit.

A possible explanation for the difference in behavior described above is that mouse class I molecules can more closely approximate a native structure in the absence of associated peptide than human class I molecules. One can then say that either the requisite chaperones do a "better job" in assembling mouse class I-ß$_2$m dimers than in assembling human class I-ß$_2$m dimers, allowing them to be transported, or that the requisite chaperones do a poorer job, in that they fail to efficiently retain mouse class I-ß$_2$m dimers in the ER. The precise structural differences between K^b, D^b and D^p and the large number of human molecules studied which are responsible for the difference in behavior are unknown although "exon-shuffling" experiments have implicated the α_1 and α_2 domains in determining the transport phenotype (46). In light of the suggestion that N-linked glycans may be a ligand for calnexin binding, the presence of two or three N-linked glycans on mouse class I molecules compared to only one on human class I molecules might be a potentially important difference. However, the addition by site-directed mutagenesis of a second N-linked glycan to the HLA-B7 heavy chain at the precise position where the second N-linked glycan is present in mouse class I molecules does not improve its cell surface expression in T2 cells (46). The true answer awaits additional experiments.

ACKNOWLEDGMENTS

The authors are indebted to Nancy Dometios for patiently preparing the manuscript. The work from this laboratory was supported by awards from the National Institutes of Health, RO1AI23081, (to P. Cresswell), from Merck Research Laboratories (to M.J. Androlewicz), from the German Cancer Research Center (DKFZ), (to B. Ortmann) and by the Howard Hughes Medical Institute.

REFERENCES

1. **Bjorkman P. J., M. A. Saper, B. Samraoui, W. S. Bennett, J. L. Strominger, and D. C. Wiley.** 1987. Structure of the human class I histocompatibility antigen, HLA-A2. *Nature 329*:506.

2. **Brown, J. H., T. S. Jardetsky, J. C. Gorga, L. J. Stern, R. G. Urban, J. L. Strominger, and D. C. Wiley.** 1993. Three-dimensional structure of the human class II histocompatibility antigen HLA-DR1. *Nature 364*:33.

3. **Fremont, D. H., M. Matsumura, E. A. Stura, P. A. Peterson, I. A. Wilson.** 1992. Crystal structures of two viral peptides in complex with murine MHC class I H-2K^b. *Science 257*:919.

4. **Saper, M. A., P. J. Bjorkman, and D. C. Wiley.** 1991. Refined structure of the human histocompatibility antigen HLA-A2 at 2.6 A resolution. *J. Mol. Biol. 219*:277.

5. **Jardetzky, T. S., W. S. Lane, R. A. Robinson, D. R. Madden and D. C. Wiley.** 1991. Identification of self peptides bound to purified HLA-B27. *Nature 353*:326.

6. **Goldberg, A.L., and K. L. Rock.** 1992. Proteolysis, proteasomes and antigen presentation. *Nature 357*:375.

7. **Michalek, M. T., E. P. Grant, C. Gramm, A. L. Goldberg, and K. L. Rock.** 1993. A role for the ubiquitin-dependent proteolytic pathway in MHC class I-restricted antigen presentation. *Nature 363*:552.

8. **Monaco, J. J., and H. O. McDevitt.** 1982. Identification of a fourth class of proteins linked to the murine major histocompatibility complex. *Proc. Natl. Acad. Sci. USA 79*:3001.

Monaco, J. J. 1992. A molecular model of MHC class-I-restricted antigen processing. *Immunol. Today. 13*:173.

10. **Driscoll, J., M. G. Brown, D. Finley, and J. J. Monaco.** 1993. MHC-linked LMP gene products specifically alter peptidase activities of the proteasome. *Nature 365*:262.

11. **Gaczynska, M., K. L. Rock, and A. L. Goldberg.** 1993. Γ-Interferon and expression of MHC genes regulate peptide hydrolysis by proteasomes. *Nature 365*:264.

12. **Higgins, C. F.** 1992. ABC transporters: From microorganisms to man. *Ann. Rev. Cell Biol. 8*:67.

13. **Androlewicz, M. J., K. S. Anderson, and P. Cresswell.** 1993. Evidence that transporters associated with antigen processing translocate a major histocompatibility complex class I-binding peptide into the endoplasmic reticulum in an ATP-dependent manner. *Proc. Natl. Acad. Sci. USA 90*:9130.

14. **Neefjes, J. J., F. Momburg, G. J. Hammerling.** 1993. Selective and ATP-dependent translocation of peptides by the MHC-encoded transporter. *Science 261*:769.

15. **Shepherd, J. C., T. N. M. Schumacher, P. G. Ashton-Rickardt, S. Imaeda, H. L. Ploegh, C. A. Janeway, Jr., and S. Tonegawa.** 1993. TAP1-dependent peptide translocation *in vitro* is ATP dependent and peptide selective. *Cell 74*:577.

16. **Falk, K., O. Rotzschke, S. Stevanovic, G. Jung, and H - G. Rammensee.** 1991. Allele-specific motifs revealed by sequencing of self-peptides eluted from MHC molecules. *Nature 351*:290.

17. **Spies, T., and R. DeMars.** 1991. Restored expression of major histocompatibility class I molecules by gene transfer of a putative peptide transporter. *Nature 351*:323.

18. **Salter R. D., and P. Cresswell.** 1986. Impaired assembly and transport of HLA-A and -B antigens in a mutant TxB cell hybrid. *EMBO J. 5*:943.

19.   Henderson, R. A., H. Michel, K. Sakaguchi, J. Shabanowitz, E.
       Appella, D. F. Hunt, and V. H. Engelhard. 1992.
       HLA-A2.1-associated peptides from a mutant cell line: a second
       pathway of antigen presentation. *Science 255*:1264.

20.   Wei, M. L. and Cresswell, P. 1992. HLA-A2 molecules in an
       antigen processing mutant cell contain signal sequence-derived
       peptides. *Nature 356*:443.

21.   Hartmann, E., T. A Rapoport, and S. Prehn. 1992. Signal
       sequence identified. *Nature 358*:198.

22.   Wada, I., D. Rindress, P. H. Cameron, W - J. Ou, J. J. Doherty,
       D. Louvard, A. Bell, D. Dignard, D. Y. Thomas, and J. J. M.
       Bergeron. 1991. SSRα and associated calnexin are major
       calcium binding proteins of the endoplasmic reticulum
       membrane. *J. Biol. Chem. 266*:19599.

23.   Prehn, S., J. Herz, E. Hartmann, T. V. Kurzchalia, R. Frank, K.
       Roemisch, B. Dobberstein, and T. A. Rapoport. 1990. Structure
       and biosynthesis of the signal-sequence receptor. *Eur. J.
       Biochem. 188*:439.

24.   Michalak, M., R. E. Milner, K. Burns, and M. Opas. 1992.
       Calreticulin. *Biochem. J. 285*:681.

25.   Luster, A. D., R. L. Weinshank, R. Feinman, and J. V. Ravetch.
       1988. Molecular and biochemical characterization of a novel
       Γ-Interferon-inducible protein. *J. Biol. Chem. 263*:12036.

26.   Guo, H - C., T. S. Jardetzky, T. P. J. Garrett, W. S. Lane, J. L.
       Strominger, and D. C. Wiley. 1992. Different length peptides
       bind to HLA-Aw68 similarly at their ends but bulge out in the
       middle. *Nature 360*:364.

27.   Huczko, E. L., W. M. Bodnar, D. Benjamin, K. Sakaguchi, N. Z.
       Zhu, J. Shabanowitz, R. A. Henderson, E. Appella, D. F. Hunt,
       and V. H. Engelhard. 1993. Characteristics of endogenous
       peptides eluted from the class I MHC molecule HLA-B7
       determined by mass spectrometry and computer modeling. *J. I.*

151:2572.

28.   **Buchmeier, M. J., and R. M. Zinkernagel**. 1992. Immunodominant T cell epitope from signal sequence. *Science* *257*:1142.

29.   **Anderson, K. A., P. Cresswell, M. Gammon, J. Hermes, A. Williamson, and H. Zweerink**. 1991. Endogenously synthesized peptide with an endoplasmic reticulum signal sequence sensitizes antigen processing mutant cells to class I-restricted cell mediated lysis. *J. Exp. Med. 174*:489.

30.   **Kelley, W. L. and C. Georgopoulos**. 1992. Chaperones and protein folding. *Curr. Opin. in Cell Biol. 4*:984.

31.   **Bole, D. G., L. M. Hendershot, and J. F. Kearney**. 1986. Posttranslational association of immunoglobulin heavy chain binding protein with nascent heavy chains in nonsecreting and secreting hybridomas. *J. Cell. Biol. 102*:1558.

32.   **Melnick, J., S. Aviel, and Y. Argon**. 1992. The endoplasmic reticulum stress protein GRP94, in addition to BiP, associates with unassembled immunoglobulin chains. *J. Biol. Chem. 267*:21303.

33.   **Ahluwalia N., J. J. M. Bergeron, I. Wada, E. Degen, and D. B. Williams**. 1992. The p88 molecular chaperone is identical to the endoplasmic reticulum membrane protein, calnexin. *J. Biol. Chem. 267*:10914.

34.   **Hochstenbach, F., V. David, S. Watkins, and M. B. Brenner**. 1992. Endoplasmic reticulum resident protein of 90 kilodaltons associates with the T-and B-cell antigen receptors and major histocompatibility complex antigens during their assembly. *Proc. Natl. Acad. Sci. USA 89*:4734.

35.   **Degen, E., and D. B. Williams**. 1991. Participation of a novel 88-kD protein in the biogenesis of murine class I histocompatibility molecules. *J. Cell Biol. 112*:1099.

36. **K. S. Anderson, and P. Cresswell**. 1993. A role for calnexin (IP90) in the assembly of class II MHC molecules. *EMBO J.*, in press.

37. **Degen, E., M. F. Cohen-Doyle, and D. B. Williams**. 1992. Efficient dissociation of the p88 chaperone from major histocompatibility complex class I molecules requires both ß$_2$-microglobulin and peptide. *J. Exp. Med. 175:*1653.

38. **Margolese, L., G. L. Waneck, C. K. Suzuki, E. Degen, R. A. Flavell, and D. B. Williams**. 1993. Identification of the region on the Class I histocompatibility molecule that interacts with the molecular chaperone, p88 (Calnexin, IP90). *J. Biol. Chem*. *268*:17959.

39. **Ou, W - J., P. H. Cameron, D. Y. Thomas, and J. J. M. Bergeron**. 1993. Association of folding intermediates of glycoproteins with calnexin during protein maturation. *Nature 364*:771.

40. **Hammond, C., I. Braakman, and A. Helenius**. 1993. Role of N-linked oligosaccharide recognition, glucose trimming and calnexin during glycoprotein folding in the endoplasmic reticulum. *Proc. Natl. Acad. Sci. USA*, in press.

41. **Srivastava, P. K., A. B. DeLeo, and L. J. Old**. 1986. Tumor rejection antigens of chemically induced sarcomas of inbred mice. *Proc. Natl. Acad. Sci. USA 83*:3407.

42. **Li, Z., and P. K. Srivastava**. 1993. Tumor rejection antigen gp96/grp94 is an ATPase: implications for protein folding and antigen presentation. *EMBO J. 12*:3143.

43. **Cerundolo, V., J. Alexander, K. Anderson, C. Lamb, P. Cresswell, A. McMichael, F. Gotch, and A. Townsend**. 1990. Presentation of viral antigen controlled by a gene in the major histocompatibility complex. *Nature 345*:449.

44. **Ljunggren H - G., N. J. Stam, C. Ohlen, J. J. Neefjes, P. Hoglund, M - T. Heemels, J. Bastin, T. N. M. Schumacher, A.**

Townsend, K. Karre, and H. Ploegh. 1990. Empty class I molecules come out in the cold. *Nature 346*:476.

45.   Alexander, J., J. A. Payne, R. Murray, J. A. Frelinger, and P. Cresswell. 1989. Differential transport requirements of HLA and H-2 class I glycoproteins. *Immunogenetics 29*:380.

46.   Anderson, K. S., J. Alexander, M. Wei, and P. Cresswell. 1993. Intracellular transport of class I MHC molecules in antigen processing mutant cell lines. J. *Immunol. 151*:3407.

47.   Alexander, J., J. A. Payne, B. Shigekawa, J. A. Frelinger, and P. Cresswell. 1990. The transport of class I major histocompatibility complex antigens is determined by sequences in the α_1 and α_2 protein domains. *Immunogenetics 31*:169.

4

USE OF HUMAN B CELL MUTANTS TO LOCATE AND STUDY GENES FOR ANTIGEN PROCESSING

Robert DeMars , Stephanie Ceman , Richard Rudersdorf , Jean Petersen , Yoji Shimizu , and Roberta Greenwood

INTRODUCTION: ISOLATION OF ANTIGEN PROCESSING MUTANTS

In the great majority of cases, foreign antigenic proteins appear within antigen presenting cells by being synthesized in the cytosol or by entering cells from the cell surface or external milieu via invaginations of the plasma membrane that become endosomes. The two modes of origin are followed by cleavage of the proteins into peptides in separate compartments of the cell. Those peptides that can be antigenic bind to a groove formed by folding of the extracellular domains of class I and class II proteins that are encoded in the major histocompatibility complex (=MHC) on the short arm of human chromosome 6 (=Chr. 6p). The class I-peptide and class II-peptide complexes then move to and become anchored in the plasma membrane, where they can activate T cells by binding their antigen receptors. There are several highly polymorphic class I (HLA-A, -B and -C) and class II (HLA-DP, -DQ and -DR) proteins in humans; the amino acid sequences of their groove-forming segments and the sequences of available peptides determine which combinations can form complexes that can bind T cell antigen receptors. The processes that generate the peptides and their complexes and then display them at the cell surface are collectively referred to below as 'AP' for antigen processing; 'AP genes' encode the molecules that execute AP.

According to the summary above, the MHC could be regarded as an assemblage of genes that encode peptide-presenting class I and class II molecules. One of the exciting discoveries unexpectedly made with the expermentation approach described below is that diverse genes needed for the intracellular production of class I-peptide and class II-peptide complexes closely neighbor the class II genes, i.e., evolution has assembled AP genes along with genes for peptide presentation in the MHC. This article emphasizes studies used to locate these genes in humans and is not intended to be a detailed, general review about AP genes [e.g., (1-4), which contain references for background information in this chapter]. Issues concerning functions of the genes and, biological effects of their deficiencies and polymorphisms are discussed elsewhere.[1]

Our research on the MHC originally required the isolation of cell mutants in which mutations in class I or class II genes reduced cell surface expression of HLA molecules. Mutants were isolated on the basis of reduced ability to bind antibodies that recognized specified kinds of HLA molecules. Two unanticipated phenomena resulted in the coincidental isolation among such mutants of other mutants that had AP defects:

(i) The great majority of class I and class II molecules purified from cells bear diverse self peptides that are processed from self proteins and, like foreign antigenic peptides, have sequence motifs that define groove-binding ability. Thus, the intracellular formation of complexes with self peptides normally occurs during the production of class I and class II molecules. We didn't know that newly synthesized HLA-A, -B and -C molecules that fail to form such complexes are degraded within the cell. This phenomenon permitted the unsuspected isolation of mutants in which reduced cell surface display of class I molecules results from AP defects. Class II HLA molecules that do not form the normally predominant varieties of complexes with self peptides are displayed on the cell surface but are structurally abnormal; in some cases, these molecules do not bind antibodies that recognize conformational epitopes on the normal molecules, again permitting the isolation of AP mutants.

(ii) Several AP genes are located in the MHC; when gamma rays were used to delete class I or class II loci, linked AP genes were sometimes deleted, too. This first resulted in the unsuspected isolation of mutants that had heterozygous deletions of AP loci, which was the key to then isolating homozygous AP mutants.

The great majority of mutations are phenotypically recessive and isolation of phenotypically null mutants requires that both copies of autosomal genes be mutated. Thus, the frequencies ($<10^{-9}$) of spontaneous homozygous mutants among homozygous wild type cells are ordinarily too low to permit their confident isolation. A solution is to isolate null mutants from cells that are already heterozygous, so that only one gene copy need be mutated. This is the key to facile isolation of HLA antigen-loss mutants: most humans are heterozygous at their HLA loci and antibodies that recognize the molecules encoded by one allele can be used to isolate the relatively frequent ($\approx10^{-6}$) spontaneous mutants that have mutations in just that allele. The frequency of such mutants can be increased to $\approx10^{-4}$-10^{-3} by mutagenesis of the parental cells with mutagens that predominantly cause base substitutions (e.g., EMS, MNNG), frame shifts (e.g., ICR-191) or deletions (e.g., gamma rays).

Only one copy of any gene on the single X and Y chromosomes in XY

[1] DeMars, R. 1994. Antigen processing genes in the major histocompatibility complex: functions and biological effects. BioEssays, in preparation.

cells and only the single transcribed copy of most X-chromosomal genes on the active (i.e., non-Barr body) X chromosome in XX cells need be mutated in order to generate a mutant phenotype. Both copies of some X chromosomal loci are expressed in XX cells; excepting these loci, if AP genes were located on the X or Y chromosomes, mutants affected in non-vital functions would be readily selected as HLA antigen-loss mutants of cells derived from virtually any normal human. In fact, human cell AP mutants have, so far, been isolated only from cells that were initially hemizygous for certain parts of the MHC.

LOCATION AND ISOLATION OF GENES NEEDED FOR CLASS I-ASSOCIATED AP: THE TAP1 AND TAP2 GENES

The AP mutants described below were descendants of an Epstein-Barr virus-transformed lymphoblastoid cell line (=LCL), LCL 721 (Fig. 1) derived from a normal human female. LCLs constitutively express large amounts of class I and class II HLA molecules on their surfaces, perform AP and are effective antigen-presenting cells. Mutant LCLs in which cell surface expression of a specified HLA molecule is reduced can be isolated by exposing the parental cell population to complement plus a complement-binding antibody that recognizes the HLA molecule: cells that express the HLA molecule bind the antibody and are killed by subsequent binding of complement but HLA loss mutants bind little or no antibody plus complement and survive.

Isolation of AP mutants began by making LCL 721 hemizygous for the MHC: gamma ray mutagenesis followed by selection against HLA-B8 yielded diverse B8-loss mutants (5). Mutant .45 (Fig. 1) proved to completely lack one Chr. 6p. Gamma irradiation of mutant .45 followed by selection for mutants lacking HLA-A2 expression yielded mutants .61 and .134, while selection for loss of HLA-DR expression led to the isolation of mutant .174 (Fig. 1). In mutants .61, .134 and .174: (i) Cell surface expression of HLA-B5 is absent and that of HLA-A2 is reduced by at least 50 percent. A2 and B5 genes and normal amounts of their transcripts are present. Fusion with appropriate LCLs restores normal cell surface expressions of A2 and B5 (6). (ii) Cytolytic T cells that recognize an epitope of influenza virus matrix protein in association with HLA-A2 do not lyse mutant cells that produce the protein in the cytosol. The same T cells lyse mutant cells that have been exposed to a synthetic peptide that has the epitope. Therefore, mutants .61,.134 and .174 are unable to form intracellular complexes between HLA-A2 and peptides derived from antigenic protein produced in the cytosol. (iii) The HLA-A2/beta-2m heterodimers in lysates of the mutant cells are unstable at 37° but are stabilized in the presence of an antigenic peptide that associates with HLA-A2. Enzymatic analysis indicates that class I molecules produced in .174 are degraded before they leave the endoplasmic reticulum (=ER).

Stabilization of HLA-A2 from the mutants by antigenic peptides suggested that association with self-peptides normally is necessary for departure of class

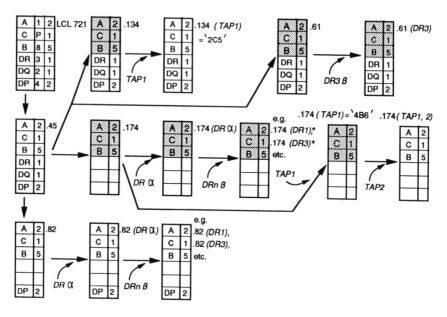

Figure 1. Origins of cell lines discussed in this chapter. LCL 721 is a cloned lymphoblastoid cell line created by infecting a B lymphocyte from a normal human female with Epstein-Barr virus. Boxes show the HLA loci and alleles. Shading indicates cell surface expression of the indicated HLA molecules is reduced as a result of antigen processing mutations. Empty boxes designate deleted genes. HLA-loss mutants were isolated after gamma ray mutagenesis of parental cells followed by treatment with complement plus antibodies that bound to HLA molecules against which selection was imposed. Curved arrows indicate cloned genes that were transferred into LCLs by means of electroporation. Molecules encoded by transgenes expressed in transferent cell lines are shown in parentheses following the names of the recipient cell lines. DR1* and DR3* indicate that the transgene-encoded DR molecules are abnormally conformed and unstable as a result of an antigen processing mutation in .174 recipient cells.

I molecules from the ER and their display at the cell surface. The presence of some HLA-A2 on .61, .134 and .174 was puzzling in this respect but was explained by analysis of peptides bound to HLA-A2 produced in a .174-derived cell hybrid, 'T2', which has abnormalities (i) - (iii) above, and in control hybrid 'T1': the great majority of T1-derived A2 is associated with many different, 8-10 amino-acid-long self-peptides, but most T2-derived A2 is associated with longer peptides that are derived from leader sequences that are cleaved from certain proteins within the ER. This observation and other evidence suggest that abnormalities (i) -(iii) result from failure of mutants .61, .134 and .174 to form complexes between newly synthesized class I molecules and self- or antigenic peptides that are produced in the cytosol and introduced into the ER. This defect is now known to result from lost expression of the TAP1 and TAP2 genes; discovery of these genes illustrates how cell mutants and basic considerations concerning mutagens and mutation rates can be used to locate AP genes.

The normal amounts of HLA-A2 and -B5 on hybrids made by fusing mutants .61, .134 and .174 with normal cells suggested that their mutations were phenotypically recessive. Absence of similar mutants among hundreds of HLA antigen-loss mutants isolated from normal LCLs indicated that the mutations probably didn't occur on the X or Y chromosomes. In contrast, isolation of several independent mutants from mutant .45, which was already hemizygous for Chr. 6p, suggested that the mutations occurred on that chromosome arm. Gamma ray mutagenesis defined a region in which the mutations could have occurred by inducing an $\approx$ 1 mb deletion on the single copy of Chr. 6p in .45, creating a homozygous deletion in the MHC of .174 (Fig. 2). The region that could contain the AP genes of interest was further defined by mapping endpoints of homozygous deletions that are in mutants .82 and 5.2.4 and overlap the .174 deletion (Fig. 2). Since .82 has a normal class I phenotype and 5.2.4 has the mutant AP phenotype, the AP genes were thought to be in DNA present in .82 and absent in .174 and 5.2.4 (8).

Cosmid clone probes spanning the DNA region containing the AP gene(s) detected several mRNAs and cDNA clones; deletion of all of them from .174 and 5.2.4 prevented identification of the specific gene(s) that caused the mutant phenotype. However, mutant .134 lacked just the newly discovered mRNA encoded by the gene now named TAP1. Functionality of this gene was demonstrated by showing that transfer of a cDNA clone restored normal AP and class I phenotypes to .134 (e.g., in 2C5, Fig. 1) (9, 10). This result proved that the abnormal AP and class I phenotypes of .61, .134 and .174 could be engendered by losing expression of just one gene at a time, but it was quickly learned that mutations in different genes could have the same effect.

Expression of a TAP1 transgene in .174 (e.g., in 4B6, Fig. 1) did not restore normal class I phenotypes, indicating that additional genes that had been deleted were needed. The TAP1 nucleotide sequence has characteristics of the family of ATP-binding ('ABC') transporter genes: in organisms ranging from bacteria to humans, ABC-transporter proteins are anchored in membranes and transport one kind or another of many kinds of molecules across them. The sizes of the transported entities range from those of small ions to those of

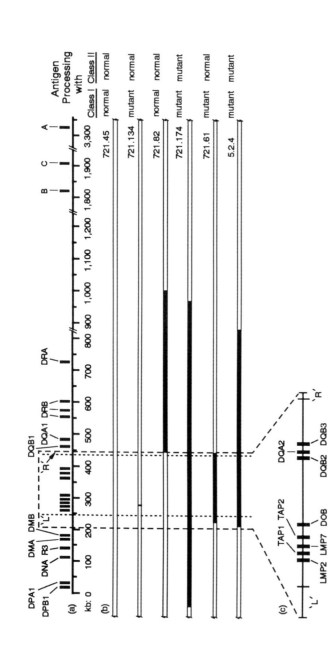

Figure 2. Homozygous MHC deletions in LCL mutants (Fig. 1) used to locate antigen processing genes in the region enclosed by heavy dashed lines. (a) MHC map adapted from (7). (b) Deletions (solid lines) were mapped with Southern blotting and probes derived from known loci or cosmids. A subgenic TAP1 mutation is shown in .134 because TAP1 DNA is present but the mRNA is absent. (c) MHC region containing antigen processing genes. A gene (not shown) for class II-associated antigen processing is in 'L' or 'R'; 'L' was assumed (16) in depicting the .61 and 5.2.4 deletions.

peptides. Thus, the TAP1 protein was an excellent candidate for being involved in transport of peptides from the cytosol into the ER, where they could form complexes with class I molecules. But functional ABC transporters are encoded by 1-4 genes (depending on the transporter) that provide for two ATP-binding sites and two membrane-spanning domains; the TAP1 gene provided for only one of each, suggesting that additional genes deleted from .174 might encode a 'half-transporter' missing in TAP1-expressing transferents of .174. The gene now named TAP2, ≈ 10 kb away from TAP1, proved to have just such a sequence. Several kinds of evidence (10-12) prove that the TAP1 and TAP2 proteins cooperate in AP and in evoking normal class I phenotypes: (i) Antibodies that bind only to the TAP1 protein co-precipitate the TAP2 protein. (ii) Loss of TAP2 expression evokes mutant phenotypes closely resembling those caused by TAP1 deficiency. Transfer of a TAP2 cDNA gene restores normal phenotypes to TAP2-deficient mutants. (iii) Transfer of a TAP2 cDNA gene into a TAP1-expressing, phenotypically mutant derivative of .174 (4B6 in Fig. 1) restored surface expression of class I molecules.

The TAP1 and TAP2 genes very closely neighbor the LMP2 and LMP7 loci (Fig. 2). These genes encode components of 'proteasomes', which are abundant in the cytosol and capable of cleaving proteins into diverse peptides. Transcription of LMP2 and LMP7 and of the TAP, class I and class II genes is co-induced by interferon γ. The location, functional context and transcription regulation of the LMP2 and LMP7 genes identified them as interesting candidates for participating in the generation or trafficking of cytosolic peptides that then associate with class I molecules in the ER (1, 2). This possibility is challenged because transfer of just the TAP1 and TAP2 genes into LCL .174 restored cell surface expression of class I molecules (11, 12). Nevertheless, it remains possible that LMP2 and LMP7 are AP genes.

LOCATION OF MHC GENES NEEDED FOR CLASS II-ASSOCIATED AP: AP II GENES

Two independent lines of evidence indicate that the MHC region containing the TAP genes also contains at least one gene that is needed for class II-mediated antigen presentation. (A). Mutagenesis of an LCL that was hemizygous for the MHC (13) followed by selection for HLA-DR3-loss mutants resulted in the isolation of some mutants that no longer bound mAb 16.23, which binds to a conformational epitope on normal DR3 molecules. However, the mutants still bound a mAb that recognizes any DR allotype and shared several abnormal traits: (i) Loss of the 16.23 epitope indicated that the DR3 molecules were abnormally conformed. (ii) The DR3 alpha and beta chains dissociated in the presence of sodium dodecyl sulfate (=SDS) at 4° C. (iii) When given exogenous antigenic proteins, the mutants did not activate antigen-specific T cell clones that were restricted by either HLA-DR, -DQ or -DP; activation occurred if the mutants were given exogenous antigenic peptides. Traits (i) -(iii) indicate that the mutations impair the formation of intracellular complexes between peptides and class II molecules in the exogenous AP pathway and that association with self-peptides has an important effect on the properties of class II molecules that are produced in normal cells (13, 14)

(B). Mutant 721.174, (Fig. 1) from which all class II genes known to be expressed were deleted, (Fig. 2) offered the opportunity to use gene transfer to create a collection of LCLs, each of which expresses just one kind of class II molecule. This was initiated by transferring the DR alpha gene into .174; the DR alpha gene is not polymorphic and the resultant cell line, .174(DRα) (Fig. 1), could be used as a 'founder' LCL for receiving different DRβ alleles.

Transferent LCLs [.174(DR1) and .174 (DR3), Fig. 1].with normal amounts of HLA-DR1 and -DR3 on their surfaces, as judged by mAb L243 binding, were produced in this way. However, the .174(DR3) cells did not bind mAb 16.23, despite the fact that transferent cells [.82(DR3), Fig. 1] made by transferring the same DRα and DR3β cDNA genes into deletion LCL .82 bound mAb 16.23. DR molecules produced in .174-derived transferents were unstable in SDS at 4°. Transferent .174(DR1) cells were not lysed by DR1-restricted T cells when given intact influenza proteins but were lysed when given exogenous T cell-recognized antigenic peptides. Thus, AP abnormalities (i)-(iii) observed with mutants described under (A) above also were caused by the homozygous deletion in the completely unrelated LCL .174 (15).

The presence of abnormalities (i)-(iii) in mutant 5.2.4 and their absence in .82(DR3) suggested that at least one gene needed for class II-mediated antigen presentation is located in ≈ 230 kb that is present in .82 and deleted from .174 and 5.2.4 (Fig. 2). We provisionally refer to such genes as 'APII genes' (antigen processing associated with class II). Recent analysis of deletion mutant 721.61 (Fig. 1) has narrowed the DNA interval that contains an APII gene to ≈ 40 kb (16). Mutant .61 has one deletion breakpoint between the DQB1 and DQB3 loci (Fig. 2). Cosmid clone probes spanning the DQB1-DQB3 interval have not detected mRNA in normal cells and indicate that the .61 and .82 deletion breakpoints either overlap or are separated by ≤ 3 kb of DNA ('R', Fig. 2) that must be rich in highly repeated sequences. It is unlikely that 'R' contains an APII gene. The other .61 deletion breakpoint is located in ≈ 40 kb of DNA ('L', Fig. 2) that has been difficult to span with cloned overlapping DNA segments. A 5.2.4 deletion breakpoint is also located in 'L'. Since .61 does not have abnormalities (i)-(iii) above, at least one APII gene is located in DNA ('R' and 'L') that is present in .61 and .82 and absent in .174 and 5.2.4. We have cloned 8 kb of DNA extending into 'L' from its centromereward side by screening a lambda vector library of genomic DNA with a cosmid probe that bounds that side of 'L'. An interesting probe was used to clone a second piece of 'L' DNA. Endonuclease Bam H1 releases from genomic DNA of .61 a fragment in which DNA segments that flank its MHC deletion are joined. The 'L' part of this joint fragment was used as a probe to detect an ≈ 12 kb fragment of LCL 721 (AP normal) DNA cloned in a phage. The 12 kb and 8 kb fragments do not overlap and account for about one half of 'L'; they are being used to probe mRNA, screen cDNA clones and to detect clones of additional DNA from 'L'.

Isolation of candidate APII genes is imminent. Functionality of the clones will be evaluated by transferring them into APII-deficient mutants and observing if they restore normal class II properties and AP function. As with the TAP genes, if at least two neighboring genes are required, transfer of only one of them into transferent cells such as .174(DR3), from which all genes in the region have been deleted, will not restore normal APII phenotypes. It may be necessary to isolate and transfer at least one additional APII gene into .174(DR3) cells that already express the first APII gene isolated. It should be easy to restore normal phenotypes to mutants that have mutations in just one APII gene by transferring a single APII gene, but it's possible that the cloned gene and mutated gene will be different; while fusions between three independent mutants did not result in complementation (13), the number of mutants tested does not make the probability that a second locus exists negligible. Clues to function may be provided by the nucleotide sequences of APII genes being isolated.

The inferred impairment in formation of intracellular class II-peptide complexes implied that DR molecules produced in APII-defective mutants might lack peptides. Instead, while about 30 percent of the DR3 molecules isolated from .174(DR3) were 'empty', 60-70 percent of them bore peptides that were 21-24 amino acids long and were derived from residues 80-104 of the invariant chain, Ii (17). These peptides are longer than the great majority of peptides that are associated with class II molecules in normal cells and are 11-18 amino acids long. Synthetic peptide Ii 82-106 effectively inhibited binding of an unrelated groove-binding peptide to purified DR3, suggesting that the 25-mer can bind to DR3 in solution. But essentially all of the DR3 isolated from .174(DR3) was SDS-unstable, despite association of most of it with the nested 21-24-mers derived from Ii 80-103. Similar observations (18) were made with DR3 molecules isolated from the .174-derived cell hybrid, T2, which has abnormalities (i)-(iii) described above (19). Furthermore, those DR3 molecules in solution became stabilized to SDS in the presence of antigenic peptides that are presented with DR3 to T cells: again, the Ii-derived peptides blocked binding of the antigenic peptides to the DR3 molecules but did not stabilize them.

Do the DR-associated Ii 80-103 peptides provide clues to the APII defect in mutant .174 and phenotypically similar mutants? Ii associates with newly synthesized class II molecules within the ER and is thought to block binding of other peptides until it is cleaved from class II molecules after they have traversed the Golgi apparatus. The abundant Ii-derived peptides on DR3 produced in .174 show that extensive cleavage of Ii occurs, but they are long peptides that can effectively inhibit binding of other peptides in the class II groove. The presence of the long Ii-peptides on most .174-derived DR3 cannot, by itself, account for the virtual absence of the normally observed 11-to-18 mers from such DR3: substantial fractions of the DR3 molecules from .174(DR3) and T2 cells bind the shorter antigenic peptides (17, 18). Perhaps attachment of the shorter peptides to these DR3 molecules was blocked by the

Ii 80-103-derived peptides within the cells and the Ii peptides became separated from the DR3 during its purification. Thus, the great majority of DR molecules might normally bear Ii 80-103-derived peptides. Further cleavage of these peptides would be required for normal peptide loading and might be defective in .174. Otherwise, at least several possible kinds of defects could cause the abnormal APII phenotype of .174. (i) Conjunction of compartments that contain groove-binding processed antigenic and self-peptides with other compartments that contain class II molecules and the nested Ii 80-103-derived peptides may be defective. (ii) Transport of processed peptides from vesicles in which they are generated to class II -containing vesicles may be impaired. (iii) While binding of peptides to class II molecules in solution occurs in the absence of other proteins, efficient peptide loading onto class II molecules in cells may require the participation of proteins that are absent in .174. (iv) Antigen-processing proteases or conditions (e.g., acidification) that affect their activities may be defective in .174. Are normal amounts of ordinarily groove-binding peptides produced in .174?

How do Ii peptides apparently bind to class II molecules without stabilizing them? The explanation proposed below takes into account the following points (20): (a) Ii 80-94 neither stabilizes DR1 in solution nor competes with antigenic peptides for binding to DR1 in solution; Ii 89-103 does both. (b) Ii 89-103 has sequence motifs proposed for DR1-binding antigenic peptides; Ii 80-94 lacks the motifs. (c) the nested 21-to-24-mers derived from Ii 80-103 are found on DR1 isolated from APII-normal cells and more abundantly on DR3 isolated from APII-defective .174 and T2 cells. Ii 89-103 is also found on DR1 (20), but from APII-normal cells only (17). An Ii peptide having a size usually associated with groove-binding peptides is found on DR3 from APII-normal cells but does not share sequence with Ii 89-103 (21).

The basic statements of the hypothesis are: (i). The sequence of Ii 89-103 allows it to bind in the DR1 groove and, thereby, stabilize DR1 and block binding of other DR1-binding peptides. (ii). Sequence included in Ii 80-94 binds to DR1 (and other class II molecules) outside of but adjacent to the groove; Ii 80-94 bound at that position neither stabilizes DR1 nor competes with other DR1-binding peptides. (iii). Ii 80-103 very preferentially binds to DR1 and DR3 by means of the Ii 80-94 sequence, allowing its Ii 89-103 segment to obstruct the groove but not enter it in a way that results in normal conformation and stability of LCL-derived DR1 and DR3. (iv). Ii 80-103 can sometimes bind promiscuously in the DR groove by means of its Ii 89-103 segment, resulting in occasional stabilization of DR1 and DR3, as has been reported (18, 20). The degree to which Ii 80-103 stabilized insect cell-grown, truncated DR1 was not reported (20). Therefore, one can't at this time determine how strongly such stabilization contradicts statement (iii), but note that the insect cell-grown, truncated DR1 that was used behaves differently from normal human cell-derived DR1 (22).

A fuller discussion of this hypothesis is precluded here but a few experiments should indicate if the hypothesis is at all useful. (a) A radiolabeled version of Ii 80-94 should be used to directly measure binding of the peptide to DR1 and other class II molecules isolated from normal cells. (b) Unlabeled

Ii 80-94 should compete with labeled Ii 80-103 for most binding to DR1 from APII-normal cells. (c) Synthetic peptides in which Ii 80-94 (C-terminus) is joined to antigenic, DR-stabilizing peptides should bind to DR1 and inhibit binding of unconjugated antigenic peptide but should poorly stabilize DR1. More generally, peptides in which Ii 80-94 is joined to sufficiently long antigenic or non-antigenic sequences might be interesting tools for manipulating immune reactions.

OTHER AP PATHWAYS, MUTANTS, AND GENES

Disruption of the familiar endogenous and exogenous AP pathways in the mutants discussed in preceding sections has facilitated detection of alternative AP pathways. For example, certain influenza virus epitopes that are synthesized in the cytosol are presented to T cells in association with HLA-DR1 molecules. Thus, peptides derived from cytosolic proteins associate not only with class I molecules in the ER but also associate with class II molecules in an unidentified compartment of the cell (23). The presentation of cytosolic epitopes with DR1 is greatly impaired in .174(DR1) cells, in which the exogenous pathway is defective, as described above (24). Does deletion of the same APII gene from .174 disrupt class II-mediated presentation of both cytosolically and exogenously derived antigens? In addition, presentation of a minigene-encoded influenza virus-derived peptide that is synthesized in the cytosol and presented with DR1 to T cells is impaired in the specifically TAP1⁻mutant .134 and is restored in .134(TAP1) transferent cells (2C5 in Fig. 1) (22). Perhaps this peptide (or a processed form) bypasses the invariant chain-imposed inhibition of peptide binding to class II molecules in the ER. But other interesting explanations exist. E. Long, et al, discuss alternative AP pathways elsewhere in this volume.

The experimentation approach used to find the TAP and APII genes may also provide clues to still other AP genes and processes. LCL mutant .220 (not shown) was isolated from HLA-A, -B-null mutant .184 (not shown) after successive gamma irradiations starting with .45 (Fig. 1) had first sequentially deleted the HLA-A and -B loci (25). Transfer of HLA-A and -B genes into .220 shows it to be a new kind of mutant, in which the amounts of transgene-encoded class I molecules on transferent cells depend on the alleles that are transferred: HLA-A1 and -B8 are reduced by 90-95 percent, while six other A and B alleles are reduced little, if at all. At least normal amounts of A1 and B8 mRNA are present in phenotypically deficient transferent cells. Normal A1 cell surface expression is restored in cell hybrids made by fusing .220 (A1) transferent cells with mutants .61 and .174: non-expression of TAP1, TAP2, LMP2, LMP7 and of any other genes that are deleted from .174 (Fig. 2) cannot explain the .220 defect. Thus, the location of the genetic lesion in .220 must be sought elsewhere on Chr. 6 and, especially in microscopically detected hemizygous regions of several other autosomes.

There are several testable predictions regarding the nature of the genetic defect in .220. (i) While apparently normal sizes and amounts of A1 and B8 mRNA are observed in .220(A1) and .220(B8) transferent cells, there may be an allele-specific abnormality that results in secretion of these class I molecules. (ii) Dysfunction of a protein involved in post-translational modification of class I proteins might differentially affect the amounts of diverse class I molecules that appear at the .220 cell surface. Proteins, such as calnexin, that are associated with class I proteins during their production are one possibility. (iii) An AP defect may impair the production or loading of peptides on A1 and B8 molecules in .220. Although deficiency of the TAP1 and -2 and LMP-2 and -7 genes is excluded, deficiency in other proteasome components might reduce the production of peptides that have A1- and B8-binding motifs. The small amounts of A1 and B8 expressed on .220 transferent cells might be unstable due to association with atypical peptides, a phenomenon also observed with HLA-A2 molecules produced in TAP-deficient mutants. Such defective formation of intracellular complexes between A1 and B8 and antigenic peptides might manifest itself functionally, for example as a lack of response of antigen-restricted A1 or B8-restricted T cells to .220 cells that express the antigenic protein in the cytosol.

Additional AP genes, if they exist, might be found by isolating candidate mutants on the basis of what has been learned about the TAP and APII mutants. Mutations that generally impede formation of class I-peptide complexes should eliminate superficial HLA-B5. Viable HLA-DR3+ mutants in which formation of class II-peptide complexes is impaired should not bind mAb 16.23. Therefore, it might be feasible to detect non-lethal mutations affecting AP genes, especially in chromosome regions that are already hemizygous. Cell lines with diverse hemizygous regions exist and gene transfer could be used to make them HLA-B5+ or HLA-DR3+ if necessary. Complementation tests with existing TAP and APII point mutants could be used to detect new kinds of AP mutants, as was done with .220. Mutagenesis would be needed to make such mutants frequent enough to isolate if homozygous mutations must occur in diploid chromosome regions. Base change mutagenesis would allow the isolation of viable AP mutants by avoiding dominant lethal effects of extensive heterozygous chromosome rearrangement resulting from use of chromosome breaking agents. However, such mutants would probably lack the deletions that were so useful in locating the TAP and APII genes. One of several ways of compensating for this deficiency would be to combine mutagenesis with a base changer and DNA breaker, using each at a dose that yields single gene mutants at frequencies of $\approx 10^3$. Some of the rare ($\approx 10^{-6}$) homozygous mutants might have a visible heterozygous deletion marking a chromosome region containing an AP gene.

ACKNOWLEDGMENTS

*Laboratory of Genetics, University of Wisconsin, Madison, WI 53706, and †Department of Microbiology and Immunology, University of Michigan Medical School, Ann Arbor, MI, 48109. Supported by grants AI15486 and GM0713317 from the National Institutes of Health. S.C. was a Cremer Foundation Scholar. Paper no. 3387 from the Laboratory of Genetics.

REFERENCES

1. **DeMars, R. and T. Spies.** 1992. New genes in the MHC that encode proteins for antigen processing. *Trends in Cell Biology 2:*81.

2. **Monaco, J. J.,** 1992 Pathways of antigen processing. *Immunol. Today 13:*173.

3. **Townsend, A. and J. Trowsdale.** 1993. The transporters associated with antigen presentation. *Seminars in Cell Biol. 4:*53.

4. **Monaco, J.** 1992. Genes in the MHC that may affect antigen processing. *Current Opinion in Immunol. 4:*70.

5. **Kavathas, P., F. .H. Bach and R. DeMars,** 1980, Gamma ray-induced loss of expression of HLA and glyoxylase I alleles in lymphoblastoid cells. *Proc. Natl. Acad. Sci., USA 77:*4251.

6. **DeMars, R., R. Rudersdorf, C. Chang, J. Petersen, J. Strandtmann, N. Korn, B. Sidwell and H. T. Orr.** 1985. Mutations that impair a post-transcriptional step in expression of HLA-A and -B antigens. *Proc. Natl. Acad. Sci., USA 82:*8183.

7. **Campbell, R. D. and J. Trowsdale.** 1993. Map of the human MHC. *Immunol. Today 14:*349.

8. **Spies, T., M. Bresnahan, S. Bahram, D. Arnold, G. Blanck, E. Mellins, D. Pious and R. DeMars.** 1990. A gene in the major histocompatibility complex class II region controlling the class I antigen presentation pathway. *Nature 348:*744.

9. **Spies, T. and R. DeMars,** 1991. Restored expression of major histocompatibility class I molecules by gene transfer of a putative peptide transporter. *Nature 351:*323.

10. **Spies, T. V. Cerundolo, M. Colonna, P. Cresswell, A. Townsend and R. DeMars.** 1992. Presentation of endogenous viral antigen by major histocompatibility class I molecules dependent on putative peptide transporter heterodimer. *Nature 355:*644.

11. **Arnold, D., J. Driscoll, M. Androlewicz, E. Hughes, P. Cresswell and T. Spies.** 1992. Proteasome subunits encoded in the MHC are not generally required for the processing of peptides bound by MHC class I molecules. *Nature 360:*171.

12. **Momburg, F., V. Ortiz-Navarette, J. Neefjes, E. Gouling, Y. van de Wal, H. Spits, S. J. Powis, G. W. Butcher, J. C. Howard, P. Walden and G. J. Hammerling.** 1992. Proteasome subunits encoded by the major histocompatibility complex are not essential for antigen presentation. *Nature 360:*174.

13. **Mellins, E. S. Kempin, L. Smith, T. Monji and D. Pious.** 1991. A gene required for class II-restricted antigen presentation maps to the major histocompatibility complex. *J. Exp. Med. 174:*1607.

14. **Mellins, E., L. Smith, B. Arp, T. Cotner, E. Celis and D. Pious.** 1990. Defective processing and presentation of exogenous antigens in mutants with normal HLA class II genes. *Nature 343:*71.

15. **Ceman, S., R. Rudersdorf, E. O. Long and R. DeMars.** 1992. MHC class II deletion mutant expresses normal levels of transgene encoded class II molecules that have abnormal conformation and impaired antigen presentation ability. *J. Immunol. 49:*754.

16. **Ceman, S., J. W. Petersen, V. Pinet, E. O. Long and R. DeMars.** 1993. A gene required for normal MHC class II expression and function is localized to ≈ 40 kb of DNA in the class II region of the MHC. *J. Immunol.,* accepted for publication.

17. **Sette, A., S. Ceman, R. T. Kubo, K. Sakaguchi, E. Appella, D. F. Hunt, T. A Davis, H. Michel, J. Shabanowitz, R. Rudersdorf, H. M. Grey and R. DeMars.** 1992. Invariant chain peptides in most HLA-DR molecules of an antigen-processing mutant. *Science 258:*1801.

18. **Riberdy, J. M., J. R. Newcomb, M. J. Surman, J. A. Barbosa and P. Cresswell.** 1992. HLA-DR molecules from an antigen-processing mutant cell line are associated with invariant chain peptides. *Nature 360:*474.

19. **Riberdy, J. M. and P. Cresswell.** 1992. The antigen processing mutant T2 suggests a role for MHC-linked genes in class II antigen presentation. *J. Immunol. 148:*2586.

20. **Chicz, R. M., R. G. Urban, W. S. Lane, J. C. Gorga, L. J. Stern, D. A. A. Vignali and J. L. Strominger.** 1992. Predominant naturally processed peptides bound to HLA-DR1 are derived from MHC-related molecules and are heterogeneous in size. *Nature 358:*764.

21. **Chicz, R. M., R. G. Urban, J. C. Gorga, D. A. A. Vignali, W. S. Lane and J. L. Strominger.** 1993. Specificity and promiscuity among naturally processed peptides bound to HLA-DR alleles. *J. Exp. Med. 178:*27.

22. **Stern, L. J. and D. C. Wiley.** 1992. The human class II MHC protein HLA-DR1 assembles as empty heterodimers in the absence of antigenic peptide. *Cell 68:*465.

23. **Malnati. M. S., M. Marti, T. LaVaute, D. Jaraquemada, W. Biddison, R. DeMars and E. O. Long.** 1992. Processing pathways for presentation of cytosolic antigen to MHC class II-restricted T cells. *Nature 357:*702.

24. **Malnati, M. S., S. Ceman, M. Weston, R. DeMars and E. O. Long.** 1993. Presentation of cytosolic antigen by HLA-DR requires a function encoded in the class II region of the MHC. *J. Immunol.,* in press.

25. **Shimizu, Y.,** 1987. Use of DNA-mediated gene transfer and HLA mutant B-lymphoblastoid cells to study the expression of human class I major histocompatibility complex genes. Ph. D. dissertation, University of Wisconsin-Madison.

5

MULTIPLE PATHWAYS OF ANTIGEN PROCESSING FOR MHC CLASS II-RESTRICTED PRESENTATION

Eric O. Long, Paul A. Roche, Mauro S. Malnati, Timothy LaVaute,
and Valérie Pinet

ANTIGEN PRESENTATION BY MHC CLASS II MOLECULES

MHC class II molecules present antigens to CD4+ T cells. The purpose of this presentation is to trigger immune responses to various pathogens. Most immune responses depend on the initial recognition of antigen by CD4+ helper T cells. Thus, class II molecules serve an essential function in immune surveillance. Presentation of antigen by class II molecules (1,2) occurs after antigen has been taken up by antigen presenting cells (APC) and processed intracellularly into peptides that can bind to class II molecules, and after class II-peptide complexes have returned to the cell surface. An effective immune surveillance will be achieved only if class II molecules can capture a wide range of peptides. Not only is it necessary for each class II allele to bind many different peptide sequences, but it would also be advantageous for class II molecules to bind peptides that have been generated in several subcellular sites. Since processing requirements for antigens with different chemical properties must vary, different repertoires of peptides are likely to be found among subcellular compartments. Although certain specialized processing compartments may be useful, the constraint of an exclusive one may be detrimental to the immune system. The term "processing pathway" will be used here to describe the various intracellular events that lead to cell surface presentation of antigenic peptides bound to class II molecules. This review will describe how a multiplicity of class II processing pathways may be achieved.

In contrast to antigen presentation by MHC class I molecules, no constraint on the source of peptides presented by class II molecules need be applied. It is important that class I molecules present peptides derived only from endogenous antigens in order to target CD8+ cytotoxic T cell responses to infected or transformed cells and to avoid destruction of bystander cells. This is achieved by the loading of peptides onto class I molecules in the early exocytic pathway. Thus, class I-restricted T cell responses are directed at cells that produce foreign antigens, whereas class II-restricted T cell responses are triggered by cells presenting any foreign antigen that has gained access to endosomes. As described here, antigens from different subcellular sources may reach class II processing compartments.

Newly synthesized class II $\alpha\beta$ heterodimers are targeted to endosomes by their association with the invariant chain (Ii). As demonstrated for antigen internalized into B cells by Ig, processed antigen is presented by newly synthesized class II molecules (3). The protease-sensitive Ii is degraded once $\alpha\beta$Ii complexes arrive in endosomes, and thereafter Ii-free class II molecules bind peptides and reach the cell surface. Ii serves three functions: it promotes efficient folding of newly synthesized $\alpha\beta$ heterodimers and their transport out of the endoplasmic reticulum, it provides a sorting signal for endosomal targeting, and it prevents peptide binding to $\alpha\beta$Ii complexes until it is removed in a proteolytic compartment. Spleen cells from Ii-deficient mice were defective in the processing and presentation of several antigens (4,5), presumably because newly synthesized class II molecules failed to be targeted to processing compartments.

Current evidence suggests that late endocytic compartments are most efficient for antigen presentation. Liposome-encapsulated antigens delivered specifically to macrophage lysosomes were more efficiently presented than those delivered to earlier endosomal compartments (6). Furthermore, intracellular class II molecules at steady state, as analyzed by immunoelectron microscopy in a human B cell line, are in a late endosomal compartment (termed MIIC) where newly synthesized class II molecules accumulate (7). It is not known whether MIIC is a processing compartment, a peptide loading compartment, or a sorting compartment for peptide-loaded class II molecules on their way to the cell surface. A clue to these questions may be provided once the product of the C2P gene is identified. C2P is a gene in the class II region of the MHC that controls antigen presentation by class II molecules (8). The C2P product acts most likely late in the class II processing pathway since class II molecules isolated from cells missing C2P (and the rest of the MHC class II region) were loaded with peptides derived from Ii rather than peptides derived from the normal complement of endosomal proteins (9,10).

We will review here data from our laboratory suggesting that class II molecules are not limited to the presentation of antigen encountered in a single specialized compartment. In addition, endocytosis of antigen is not always required since cytosolic antigen can also be presented. This multiplicity in the pathways used by class II molecules may result both from different intracellular transport routes for class II molecules, and from different mechanisms for antigen delivery to processing compartments.

TRANSPORT OF CLASS II MOLECULES TO ENDOSOMES VIA THE CELL SURFACE

Proteins transported to endosomes from the exocytic pathway can be targeted either directly from the trans-Golgi network (TGN) or by internalization from the cell surface. It is not known to what extent either of these pathways is used by newly synthesized $\alpha\beta$Ii complexes. It has been inferred from immunoelectron microscopy studies that class II molecules are targeted directly to endosomes from the TGN (7). However, it is difficult to deduce transport pathways from the intracellular distribution of class II molecules at steady state. Targeting of class II molecules to endosomes from the TGN may well occur, but this possibility has not been directly tested. On the other hand, we have demonstrated that a large number of $\alpha\beta$Ii complexes is targeted to endosomes by rapid internalization from the surface of human B cell lines (11). Careful quantitation of the number of cell surface $\alpha\beta$Ii complexes, and of their internalization rate, led to the determination that 3000 such complexes are targeted to endosomes by internalization every minute. Considering that only a few hundred class II-peptide complexes are sufficient to trigger a T cell response, this large number of $\alpha\beta$Ii complexes internalized from the cell surface is likely to contribute to functional antigen presentation.

The importance of a targeting pathway via the cell surface was underscored by the finding that the same amino-terminal end of the Ii cytoplasmic tail previously shown to target $\alpha\beta$Ii complexes to endosomes (12,13), also contained signals for internalization from the cell surface (11). Ii dissociated from internalized $\alpha\beta$Ii complexes with a half-life of about 1 h, releasing $\alpha\beta$

heterodimers that are presumably capable of binding immunogenic peptides. Recent experiments have ruled out the possibility that Ii dissociation from αβIi complexes occurred at the cell surface: Ii dissociation was inhibited by chloroquine and by blocking energy metabolism, implying that it took place in an acidic vesicle after internalization of the surface αβIi complexes. Furthermore, since the dissociation increased gradually over time, this process may take place in several classes of endosomes. Consequently, this transport pathway may provide a mechanism for the binding of peptides to class II molecules in several endocytic compartments.

DIFFERENT PROCESSING PATHWAYS FOR EXOGENOUS ANTIGEN

The sensitivity of internalized antigens to low pH and proteases must vary greatly. While some class II-restricted T cell epitopes require reduction of disulfide bonds and proteolytic cleavages in very late endosomal compartments, others may be accessible for presentation within unprocessed native proteins. Protease-sensitive epitopes that are produced in mildly acidic early endosomes may even be completely degraded before antigen delivery to a late compartment. How can such peptides be captured by class II molecules? One possible mechanism was suggested by the finding that αβIi complexes internalize from the cell surface, thereby traversing several endosomal compartments. On the other hand, peptide binding to mature cell surface class II molecules during their recycling through early endosomes also represents a potential mechanism. Constitutive recycling of surface class II molecules has been measured on a human B cell line (14). Using a different assay, we have confirmed extensive recycling of surface class II molecules in an EBV-B cell line. Even though the half-life of certain peptide-class II complexes in human APC was as long as that of class II molecules, indicating that no peptide exchange occurred (15), other peptide-class II complexes in mouse APC have a shorter half-life than that of class II molecules, implying that the association with class II molecules is reversible (16,17). Peptide competition experiments in mouse APC showed that peptide exchange can occur with pre-existing class II-peptide complexes (18). However, there is no evidence that peptide loading onto recycling mature cell surface class II molecules provides a way for APC to capture peptides that are generated from natural antigens in early endosomes.

Presentation of exogenous antigen typically requires expression of Ii in the APC, but certain antigens can be presented by cells devoid of Ii (4,19). Two interpretations can be proposed to explain Ii-independent presentation. First, newly synthesized class II may gain access to class II processing compartments independently of Ii. Second, cell surface class II molecules may acquire peptides during internalization from and recycling back to the cell surface.

Data from our laboratory support the second possibility. The natural antigen influenza virus A was used to determine the processing requirements of two HLA-DR1-restricted epitopes, one in the matrix protein M1, the other in the hemagglutinin protein H3. Virus particles bind to sialic acid moieties at the surface of APC and are efficiently taken up for processing. Very low doses of inactivated virus particles were sufficient to detect presentation of either epitope by a B cell line to specific DR1-restricted cytotoxic T cell clones. Presentation of M1 by DR1-transfected human fibroblasts was achieved only by cells that co-expressed Ii, suggesting that targeting of newly synthesized class II molecules to late endosomes by association with Ii is required for M1 presentation. In contrast, presentation of H3 by DR1-transfected fibroblasts was independent of Ii. Furthermore, presentation of H3 persisted in the absence of the C2P gene, as demonstrated with the transfected mutant cell .174(DR1). Most of the MHC class II region, including the C2P gene, is deleted in mutant .174 (20). These data suggested that H3 processing was not limited to late endocytic compartments to which newly synthesized class II molecules are targeted by Ii and where C2P is thought to act.

Cells were treated with protein synthesis inhibitors prior to the addition of antigen to test whether H3 presentation was dependent on newly synthesized class II molecules. Presentation of H3 by transfected fibroblasts and by the B cell line was not inhibited even after a 9 h inhibition of protein synthesis. In contrast, M1 presentation was completely dependent on newly synthesized proteins. Since depletion of class II molecules from MIIC after inhibition of protein synthesis takes 6 - 9 h (7), our data strongly suggest that pre-existing mature class II $\alpha\beta$ heterodimers at the cell surface can still participate in the presentation of some exogenous antigens. Most likely, such molecules bind peptides during recycling through endosomal compartments. The low doses of virus at which these responses to the immunodominant DR1-restricted epitope in H3 were observed also suggest that this processing pathway is used during normal immune responses to influenza virus.

Therefore, two distinct processing pathways exist for presentation of exogenous antigen. The "classical pathway", used for M1 presentation, requires expression of Ii, ongoing protein synthesis, and at least one function encoded in the MHC class II region. In contrast, the alternative pathway revealed by H3 presentation is independent of Ii expression, independent of newly synthesized class II molecules, and independent of C2P. The ability of mature class II molecules to bind peptides for presentation to CD4+ T cells may be a useful mechanism for the capture of epitopes derived from antigens that are rapidly degraded upon uptake into APC.

PROCESSING OF CYTOSOLIC ANTIGEN

Evidence that cytosolic antigen could be processed endogenously for presentation by class II molecules in untransfected cells was obtained with a recombinant vaccinia virus expressing the influenza virus matrix protein M1 (21). It was possible to distinguish between the class I and class II-mediated presentation of M1 in the same cells by use of different inhibitors: the HLA-A2-restricted presentation was completely blocked by brefeldin A and unaffected by chloroquine, whereas HLA-DR1-restricted presentation was completely blocked by chloroquine and only partially inhibited by brefeldin A. Similar findings were reported for the presentation of influenza virus neuraminidase introduced into the cytosol by fusion with inactivated virus particles (22). However, another study reported chloroquine-insensitive and brefeldin A-sensitive class II-restricted presentation of M1 in cells fused with acid treated virus (23). Since the interpretation of data obtained with inhibitors that have profound effects on subcellular morphology and metabolism can be misleading, a genetic approach was chosen to test whether cytosolic antigen presentation by class II molecules involved the class I processing pathway (i.e. TAP-mediated peptide delivery into an early exocytic compartment).

Two mutant B cell lines (described by DeMars et al., Chapter 4 in this volume) defective in the class I processing pathway provided evidence for an independent pathway for the class II-restricted presentation of endogenously processed cytosolic antigen. Mutant .134 is specifically deficient in the TAP1 subunit of TAP, and mutant .61 carries a homozygous deletion that eliminates the TAP1, TAP2, LMP2 and LMP7 genes. Presentation of the natural cytosolic M1 and of an engineered cytosolic form of H3 was independent of these four genes (24,25). These results meant either that the processing pathway for class II-restricted presentation was independent of peptides translocated into the exocytic pathway, or that newly synthesized class II molecules can be loaded with peptides that entered the endoplasmic reticulum independently of TAP. The second interpretation is highly unlikely since a short cytosolic peptide, corresponding to the DR1-restricted epitope of H3, required TAP for presentation (24). Therefore, endogenous processing of cytosolic H3 is unlikely to produce short peptides that bind to newly synthesized class II molecules in the absence of TAP. Together with the chloroquine sensitivity of cytosolic antigen presentation, these data suggest that antigen is delivered from the cytosol to a class II processing compartment.

A testable prediction and an important question are raised by this hypothesis. The prediction is that presentation of cytosolic antigen should depend on C2P, as does the presentation of most exogenous antigen. The mutant cell .174(DR1) was used to test it. Mutant .174(DR1) was defective in the presentation of both endogenous antigens M1 and cytosolic H3 (25). These results, and those with mutant .61, demonstrated that presentation of exogenous and cytosolic antigen each requires at least one function encoded in the MHC class II region that is not TAP1, TAP2, LMP2 and LMP7. It is likely that C2P is required for effective presentation in both cases. The definitive answer awaits the isolation of an HLA-DR1 positive C2P mutant.

The important question raised by our data is whether cytosolic protein presentation is constitutive in uninfected cells. The published evidence for TAP-independent class II-restricted presentation of cytosolic antigen was obtained with cells infected by recombinant vaccinia viruses. These experiments were performed with short infections (≤ 4 h) to avoid cytopathic effects. Nevertheless, the processing pathway responsible for the presentation of cytosolic antigen may have been induced by the virus infection. Results from two independent approaches suggest that cytosolic proteins can be presented constitutively by class II molecules. First, three recent studies have identified peptides derived from cytosolic proteins among the peptides bound to class II molecules of mouse and human B cells (26-28). Although the bulk of peptides bound to class II molecules is clearly derived from internalized proteins, some peptides may derive from proteins that were processed by an alternative endogenous pathway. Second, stably transfected cells expressing high levels of a cytosolic form of lysozyme were recognized by a specific class II-restricted T cell hybridoma (29). We have also observed constitutive class II-restricted presentation of cytosolic H3 in a B cell line stably transfected with an episomal expression vector. Similar transfections in TAP-defective mutants will be produced to test for the role of TAP in the constitutive presentation of cytosolic antigen. Although the unlikely possibility of endogenous antigen re-uptake via endocytosis has not been excluded, these data suggest that cytosolic self proteins can be presented by class II molecules.

An endogenous pathway for processing and class II-restricted presentation of cytosolic proteins to helper T cells can be useful in several ways. First, a helper T cell response may be greatly accelerated by the ability of the immune system to detect peptides from cytosolic proteins on class II molecules of infected cells. In the many cases of viruses that infect class II-expressing cells, peptide presentation could occur more rapidly than if it depended on antigen re-uptake by an APC. Second, constitutive presentation of peptides derived from cytosolic proteins expands the range of self proteins to which tolerance can be established, thus avoiding potential autoimmune reactions.

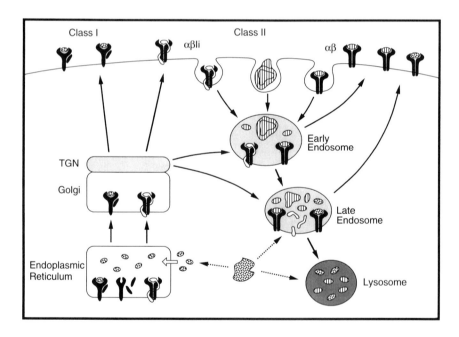

Fig. 1. Processing pathways for class II-restricted antigen presentation. Stippled peptides represent peptides derived from cytosolic antigen, hatched peptides those derived from endocytosed antigen.

THE NEXT FRONTIER

The next frontier in class II-restricted antigen presentation lies beyond the safe description of pathways (illustrated in Figure 1) in the unexplored territory of mechanisms. Targeting of newly synthesized class II molecules to endosomes mediated by the cytoplasmic tail of Ii must involve signals that interact with components of the internalization machinery or with carrier vesicles that bud from the TGN. Interesting problems arise once αβIi complexes reach endosomal compartments. Ii-derived peptides have to be replaced by other peptides. The defect in C2P mutants suggests that this step is mediated by a specialized gene product. Newly formed class II-peptide complexes must escape lysosomal degradation and eventually find their way to the cell surface. Is there a specialized rescue mechanism for intracellular class II molecules or do they follow beaten trafficking paths? The C2P-independent presentation of some antigen by mature class II molecules raises many questions. Can peptide exchange take place during rapid transit through recycling endosomes? If not, some internalized class II molecules would have to be recycled by another route. Are peptides loaded onto class II molecules by this alternative pathway similar to the bulk of peptides bound to class II? Finally, several existing cellular mechanisms may account for the TAP-independent presentation of cytosolic antigen: macroautophagy, microautophagy and translocation from the cytosol to a proteolytic endosomal compartment. Does the immune system exploit one of those or has it developed its own?

A multiplicity of processing pathways for antigen presentation by class II molecules makes eminent sense. Given the importance of helper T cell responses, the immune system simply cannot afford to take a blinded look at foreign antigens.

REFERENCES

1. **Long, E. O.** 1992. Antigen processing for presentation to CD4+ T cells. *New Biol.* 4:274.

2. **Germain, R. N. and D. H. Margulies.** 1993. The biochemistry and cell biology of antigen processing and presentation. *Annu. Rev. Immunol. 11*:403.

3. **Davidson, H. W., P. A. Reid, A. Lanzavecchia, and C. Watts.** 1991. Processed antigen binds to newly synthesized MHC class II molecules in antigen-specific B lymphocytes. *Cell 67*:105.

4. **Viville, S., J. Neefjes, V. Lotteau, A. Dierich, M. Lemeur, H. Ploegh, C. Benoist, and D. Mathis.** 1993. Mice lacking the MHC class II-associated invariant chain. *Cell 72*:635.

5. **Bikoff, E. K., L. Y. Huang, V. Episkopou, J. van Meerwijk, R. N. Germain, and E. J. Robertson.** 1993. Defective major histocompatibility complex class II assembly, transport, peptide acquisition, and CD4+ T cell selection in mice lacking invariant chain expression. *J. Exp. Med. 177*:1699.

6. **Harding, C. V., D. S. Collins, J. W. Slot, H. J. Geuze, and E. R. Unanue.** 1991. Liposome-encapsulated antigens are processed in lysosomes, recycled, and presented to T cells. *Cell 64*:393.

7. **Peters, P. J., J. J. Neefjes, V. Oorschot, H. L. Ploegh, and H. J. Geuze.** 1991. Segregation of MHC class II molecules from MHC class I molecules in the Golgi complex for transport to lysosomal compartments. *Nature 349*:669.

8. **Mellins, E., S. Kempin, L. Smith, T. Monji, and D. Pious.** 1991. A gene required for class II-restricted antigen presentation maps to the major histocompatibility complex. *J. Exp. Med. 174*:1607.

9. **Riberdy, J. M., J. R. Newcomb, M. J. Surman, J. A. Barbosa, and P. Cresswell.** 1992. HLA-DR molecules from an antigen-processing mutant cell line are associated with invariant chain peptides. *Nature 360*:474.

10. **Sette, A., S. Ceman, R. T. Kubo, K. Sakaguchi, E. Appella, D. F. Hunt, T. A. Davis, H. Michel, J. Shabanowitz, R.**

Rudersdorf, H. M. Grey, and R. DeMars. 1992. Invariant chain peptides in most HLA-DR molecules of an antigen-processing mutant. *Science 258*:1801.

11. **Roche, P. A., C. L. Teletski, E. Stang, O. Bakke, and E. O. Long.** 1993. Cell surface HLA-DR-invariant chain complexes are targeted to endosomes by rapid internalization. *Proc. Natl. Acad. Sci. USA 90*:8581.

12. **Bakke, O. and B. Dobberstein.** 1990. MHC class II-associated invariant chain contains a sorting signal for endosomal compartments. *Cell 63*:707.

13. **Roche, P. A., C. L. Teletski, D. R. Karp, V. Pinet, O. Bakke, and E. O. Long.** 1992. Stable surface expression of invariant chain prevents peptide presentation by HLA-DR. *EMBO J. 11*:2841.

14. **Reid, P. A. and C. Watts.** 1992. Constitutive endocytosis and recycling of MHC glycoproteins in human B-lymphoblastoid cells. *Immunol. 77*:539.

15. **Lanzavecchia, A., P. A. Reid, and C. Watts.** 1992. Irreversible association of peptides with class II MHC molecules in living cells. *Nature 357*:249.

16. **Harding, C. V., R. W. Roof, and E. R. Unanue.** 1989. Turnover of Ia-peptide complexes is facilitated in viable antigen-presenting cells: biosynthetic turnover of Ia vs. peptide exchange. *Proc. Natl. Acad. Sci. U. S. A. 86*:4230.

17. **Marsh, E. W., D. P. Dalke, and S. K. Pierce.** 1992. Biochemical evidence for the rapid assembly and disassembly of processed antigen-major histocompatibility complex class II complexes in acidic vesicles of B cells. *J. Exp. Med. 175*:425.

18. **Adorini, L., E. Appella, G. Doria, F. Cardinaux, and Z. A. Nagy.** 1989. Competition for antigen presentation in living cells involves exchange of peptides bound by class II MHC molecules. *Nature 342*:800.

19. **Nadimi, F., J. Moreno, F. Momburg, A. Heuser, S. Fuchs, L. Adorini, and G. J. Hammerling.** 1991. Antigen presentation of hen egg-white lysozyme but not of ribonuclease A is augmented by the major histocompatibility complex class II-associated invariant chain. *Eur. J. Immunol. 21*:1255.

20. **Ceman, S., R. Rudersdorf, E. O. Long, and R. DeMars.** 1992. MHC class II deletion mutant expresses normal levels of transgene encoded class II molecules that have abnormal conformation and impaired antigen presentation ability. *J. Immunol. 149*:754.

21. **Jaraquemada, D., M. Marti, and E. O. Long.** 1990. An endogenous processing pathway in vaccinia virus-infected cells for presentation of cytoplasmic antigens to class II-restricted T cells. *J. Exp. Med. 172*:947.

22. **Hackett, C. J., J. W. Yewdell, J. R. Bennink, and M. Wysocka.** 1991. Class II MHC-restricted T cell determinants processed from either endosomes or the cytosol show similar requirements for host protein transport but different kinetics of presentation. *J. Immunol. 146*:2944.

23. **Nuchtern, J. G., W. E. Biddison, and R. D. Klausner.** 1990. Class II MHC molecules can use the endogenous pathway of antigen presentation. *Nature 343*:74.

24. **Malnati, M. S., M. Marti, T. LaVaute, D. Jaraquemada, W. Biddison, R. DeMars, and E. O. Long.** 1992. Processing pathways for presentation of cytosolic antigen to MHC class II-restricted T cells. *Nature 357*:702.

25. **Malnati, M. S., S. Ceman, M. Weston, R. DeMars, and E. O. Long.** 1993. Presentation of cytosolic antigen by HLA-DR requires a function encoded in the class II region of the MHC. *J. Immunol.*, in press.

26. **Nelson, C. A., R. W. Roof, D. W. McCourt, and E. R. Unanue.** 1992. Identification of the naturally processed form of hen egg white lysozyme bound to the murine major histocompatibility complex

class II molecule I-Ak. *Proc. Natl. Acad. Sci. U. S. A. 89*:7380.

27. **Newcomb, J. R. and P. Cresswell.** 1993. Characterization of endogenous peptides bound to purified HLA-DR molecules and their absence from invariant chain-associated alpha beta dimers. *J. Immunol. 150*:499.

28. **Chicz, R. M., R. G. Urban, J. C. Gorga, D. A. Vignali, W. S. Lane, and J. L. Strominger.** 1993. Specificity and promiscuity among naturally processed peptides bound to HLA-DR alleles. *J. Exp. Med. 178*:27.

29. **Brooks, A. G. and J. McCluskey.** 1993. Class II-restricted presentation of a hen egg lysozyme determinant derived from endogenous antigen sequestered in the cytoplasm or endoplasmic reticulum of the antigen presenting cells. *J. Immunol. 150*:3690.

6

PROCESSING OF EXOGENOUS ANTIGENS FOR PRESENTATION BY CLASS I AND CLASS II MHC MOLECULES: A DISSECTION OF INTRACELLULAR PATHWAYS

Clifford V. Harding

T lymphocytes recognize immunogenic peptides bound to major histocompatibility complex (MHC) molecules. Antigen processing comprises the conversion of protein antigens to peptide-MHC complexes. Class I MHC (MHC-I) molecules bind "endogenous" peptides derived from cytosolic proteins; these peptides are transported into the ER to bind nascent MHC-I molecules. Class II MHC (MHC-II) molecules target to endocytic compartments, where they bind peptides catabolically derived from "exogenous" proteins internalized by endocytosis or phagocytosis (or endogenous proteins that target to endocytic compartments). Subsequently, both classes of MHC molecules are displayed on the plasma membrane for T cell recognition.

We are dissecting the mechanisms of antigen processing and the roles of various endocytic compartments therein. A compartment with lysosomal properties ("early lysosomes") appears to be important in the class II antigen processing pathway. Liposome-encapsulated antigens can be targeted for release in lysosomes, and this results in efficient antigen processing (1, 2). Antigens expressed in the cytoplasm of bacteria can also be efficiently processed, apparently in phagolysosomes, for presentation by MHC-II molecules (3). High levels of MHC-II molecules are present in some lysosomes and phagolysosomes (4), where they can bind immunogenic peptides (5). In addition, we have characterized an alternative processing pathway whereby exogenous particulate antigens, such as extracellular or vacuolar bacteria, may be processed for MHC-I as well as MHC-II presentation (6), contrary to the "endogenous/MHC-I vs. exogenous/ MHC-II" rule; the mechanisms of this are a focus of current research.

Cell Biology of the MHC-II Antigen Processing Pathway

Many topics of antigen processing and MHC function have been reviewed (7-9) and are discussed elsewhere in this volume; a comprehensive review cannot be presented here. Our focus is a dissection of functions of the endocytic pathway in antigen processing. MHC molecules bind peptides for presentation to T cells, and the endocytic pathway contains the proteolytic mechanisms that degrade internalized proteins to produce immunogenic peptides. Thus, endocytic organelles play a central role in the processing of exogenous antigens.

Proteins and even small particulate antigens (e.g. viruses and liposomes) can be internalized via endosomes, which can fuse with each other and also generate recycling vesicles. In endosomes internalized proteins are sorted and targeted for recycling (also transcytosis in polarized cells) or transport to lysosomes for degradation. These mechanisms of sorting and transport lead to the "maturation" of endosomes, wherein certain proteins are recycled and depleted from later endosomes, while others are retained and concentrated prior to fusion with lysosomes. Phagocytosis represents an additional mechanism of internalization, whereby macrophages can engulf larger particles, such as bacteria, into phagosomes. Phagocytosis represents a separate pathway that does not follow the precise mechanisms of endosomal maturation. However, phagosomes fuse with endosomes, generate recycling vesicles that convey materials back to the plasma membrane (10), and also fuse with lysosomes.

The first question is where within the endocytic pathway the immunogenic peptides are proteolytically generated, to be subsequently recycled for presentation. The endocytic compartments potentially involved include earlier endosomal compartments and later lysosomal compartments (we will discuss phagocytic processing below). Each of these represents a heterogeneous group of vesicles with differences in maturity and composition. Lysosomes may be defined as late endocytic vesicular structures that have a high physical density (revealed by subcellular fractionation), high levels of lysosomal proteases, high levels of characteristic membrane glycoproteins such as lysosome associated membrane protein 1 (Lamp-1), and an absence of mannose-6-phosphate receptor (MPR), which is found in late endosomes (this is best defined for

the 215 kD MPR, but also appears to hold for the 46kD MPR) (11).
Endosomes also contain lower levels of "lysosomal" proteases and
membrane proteins, such as cathepsin D, and LAMP-1 (12). Thus,
immunogenic peptides could be produced by endosomal antigen
proteolysis, and these peptides could then bind to MHC-II molecules in the
same endosomal compartment. This model is appealing, since there is
ample evidence of recycling pathways from endosomes to the plasma
membrane, and recycling from late endocytic structures, especially
lysosomes, is not well appreciated. However, recent evidence indicates an
important role for late endocytic compartments with lysosomal properties
in generating immunogenic peptides for presentation to T cells. This
information also has implications for previously unappreciated aspects of
transport pathways involving late endocytic compartments.

Early evidence for the involvement of lysosomes in antigen
processing came from a comparison of the processing of antigens
encapsulated in two different types of liposomes (1, 2). Acid-resistant
liposomes (Table I) sequestered their contents within the liposome
membrane during endocytosis and intracellular transport until their

Table I. Liposomes Used in Antigen Processing Studies

Liposome characteristic	Acid-sensitive	Acid-resistant
Composition[a]	PE/PHC PE/CHEMS PE/SG	PC/PS
Site of antigen release	early endosomes	lysosomes
Cytosolic delivery	yes[b]	no
MHC-II processing in vitro (HEL)	++	+++
MHC-I processing in vitro (OVA)	++	(+)[c]
MHC-I processing in vivo (OVA)	+++	+++

[a]PE, phosphatidylethanolamine; PHC, palmitoylhomocysteine; CHEMS,
cholesterol hemisuccinate; PC, phosphatidylcholine; PS,
phosphatidylserine; SG, succinylglycerol.
[b]Cytosolic delivery is approximately 0.1-1% for PE/CHEMS, i.e., the
vast majority of antigen remains within endocytic vesicles.
[c]Requires higher concentration than with acid-sensitive.

delivery to lysosomes (defined by density upon subcellular fractionation), where the antigens were released (1). Acid-sensitive liposomes contained a proton-titratable stabilizing lipid; in acidified endosomes, protonation of this lipid induced a phase transition that released the encapsulated material. Thus, the two types of liposomes were vectors for the release of antigens directly into either the endosomal or the lysosomal environment. To study their processing, liposome-encapsulated antigens were incubated with macrophages, and the cells were subsequently fixed with paraformaldehyde. Peptide-MHC-II complexes that were generated during the processing incubation were detected using antigen-specific T hybridoma cells. Antigens encapsulated in the acid-resistant liposomes were efficiently processed and presented to T cells after their release directly into lysosomes (1, 2), indicating that lysosomes, defined by density, had important antigen processing capabilities. In other studies, transferrin receptor-directed delivery of antigen to early endosomes was insufficient for processing (13), also implicating late endocytic compartments, but different results have also been obtained (14).

Several additional approaches have indicated important roles for late endocytic, lysosomal or phagolysosomal compartments in antigen processing. In macrophages, the reduction of disulfide bonds in internalized substrates is mediated specifically within dense lysosomes (15). Thus, antigenic epitopes that require disulfide reduction (16, 17) must be processed in this compartment. In addition, antigen processing is inhibited at 18°C (18), a condition that preferentially blocks lysosomal function, although certain late endosomal functions may also be blocked at this temperature. The phagocytic processing of bacteria results in the presentation of peptide epitopes derived from proteins expressed throughout the bacteria (including the bacterial cytoplasm), suggesting that extensive degradation is involved, consistent with the functional capabilities of phagolysosomal compartments (3). Together these studies suggest that lysosomal compartments can play an important role in antigen processing, although they do not exclude endosomal contribution to the processing of some epitopes. In fact, the compartments involved may vary between different antigens or different antigen processing cells.

A related question is to specifically identify the intracellular sites that contain MHC-II molecules and where peptides bind to MHC-II molecules. Most MHC-II molecules are expressed on the cell surface, but immunoelectron microscopy studies have reported MHC-II within endosomes (12) and an electron dense endocytic compartment with lysosomal characteristics (4, 5, 19). In interferon-γ activated peritoneal macrophages, abundant MHC-II labeling was present in a late endocytic compartment (retaining endocytic tracers after 1-2 hrs) that contained the lysosomal markers cathepsin D and Lamp-1, and was devoid of the 46 kD MPR. We termed this compartment tubulo-vesicular or early lysosomes (4). MHC-II was also found in MPR-positive endosomes, though at lower levels. The early lysosomal compartment was distinct from more mature vacuolar lysosomes, both morphologically and kinetically (mature vacuolar lysosomes have a more homogenously dense appearance by electron microscopy and accumulate endocytic tracers after an overnight chase). Electron microscopy of dense lysosomal subcellular fractions from Percoll gradients showed vesicular structures similar to the early lysosomes demonstrated in whole cells, indicating that these vesicles possessed the high density characteristic of lysosomes (5). In addition, a large proportion of biosynthetically labeled MHC-II molecules were immunoprecipitated from lysosomal fractions 1-2 hr after biosynthesis. MHC-II was up to 10-fold more concentrated in the lysosomal membranes than in the light density membranes (5), indicating that MHC-II in these fractions did not arise simply by contamination of lysosomal fractions with lighter density organelles. Early lysosomes may be the site where intra-cellular MHC-II lingers before transport to the plasma membrane (20). In summary, MHC-II molecules reside in a late endocytic compartment with lysosomal properties, although the exact relationship of this compartment to other late endocytic compartments requires further analysis.

In order to identify the site of peptide binding to MHC-II molecules, macrophages were incubated with a model antigen, hen egg lysozyme (HEL), followed by subcellular fractionation on Percoll density gradients. The individual fractions were then disrupted and the resulting membranes were incubated with T hybridoma cells specific for HEL(52-61)-I-A^k. After 120 min of exposure to HEL, macrophage lysosomes contained peptide-MHC-II complexes that were recognized by T cells (Fig. 1A). Furthermore, at earlier time points after the addition of antigen (e.g. 15 min, when antigen presentation is not yet detectable on the surface of intact cells) peptide-MHC-II complexes were detected abundantly in the lyso-somal fractions (Fig. 1B), but were absent or minimally detected in light density fractions (containing endosomes plus plasma membrane). Thus, peptide-MHC-II complexes were first formed in lysosomal fractions and only later appeared on the plasma membrane. This suggests that the complexes were formed first in lysosomes and then transported to the plasma membrane (Fig. 2), but further study of MHC-II transport between lysosomes and plasma membrane is required to test this hypothesis.

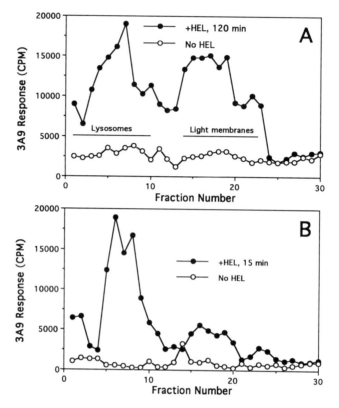

Figure 1. Formation of specific peptide-MHC-II complexes in lysosomes (modified from Ref. 5). Macrophages were incubated with or without HEL (5 mg/ml), washed, homogenized and fractionated on 27% Percoll gradients. Individual fractions were subjected to hypotonic shock and then partially lyophilized to disrupt membrane orientation (the peptide binding domain of MHC-II would be expected to be lumenal in endocytic compartments). 3A9 T hybridoma cells were used to determine the level of HEL(52-61)-I-A^k complexes present. IL-2 secretion by the T cells, a measure of peptide-MHC-II recognition, was measured by the proliferation of IL-2 dependent CTLL cells incubated with supernatants from the T cell assay. Each point represents a single fraction assayed in a single well. A. 120 min HEL incubation, revealing complexes in both plasma membrane and lysosomal fractions. B. 15 min HEL incubation. At this time point delivery of antigen to lysosomes is just beginning, and presentation at the cell surface is absent or minimal. Here specific peptide-MHC-II complexes are found primarily in lysosomes.

An Alternative MHC-I Antigen Processing Pathway for Exogenous
Antigens: Processing of Liposomes and Vacuolar Bacteria

"Exogenous" antigens are internalized by endocytosis and processed
within endosomes and lysosomes to produce peptides that then generally
bind to MHC-II molecules. On the other hand, "endogenous" antigens are
those that are synthesized within the presenting cell (e.g. viral antigens)
and processed, at least in many cases, within the cytosol to produce
peptides that then bind to MHC-I molecules. CD4 and CD8 also bind
MHC-II and MHC-I molecules, respectively, such that peptide-MHC-II
complexes are generally recognized by CD4+ T cells, while peptide-
MHC-I complexes are recognized by CD8+ T cells. Accordingly,
immunization with exogenous proteins elicits primarily a CD4 T cell
response. Exogenous antigens can enter the class I processing pathway if
they can penetrate from the membrane-bound endocytic vesicular system
into the cytosol (2, 21-23). In vitro, endosomes can be breached by an
osmotic shock (21) or by fusion of virion and endosomal membranes (23)

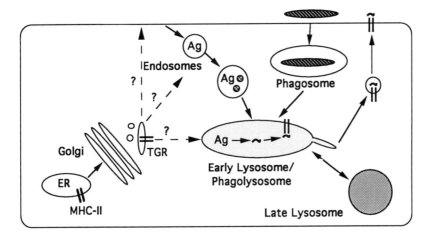

*Fig. 2. A model for the class II MHC antigen processing pathway
(modified from Ref. 5). Nascent MHC-II molecules are transported from
the endoplasmic reticulum (ER) to the trans Golgi reticulum (TGR) with
subsequent targeting to early lysosomes/phagolysosomes (the exact route
for this remains uncertain). Immunogenic peptides are proteolytically
generated and bind to MHC-II molecules within the early lysosomal/
phagolysosomal compartment. Peptide-MHC-II complexes may then be
recycled to the plasma membrane.*

to release exogenous antigen into the cytosol for class I processing. However, in some cases class I processing of exogenous particulate antigens occurs without an identified cytosolic penetration step. Exogenous cell-associated antigens (24), vacuolar bacteria, (e.g. Salmonella, 25, 26) and vacuolar parasites (e.g. Leishmania) can elicit MHC-I restricted T cell responses. In these cases, the antigens or immunogenic peptides may either transit through the cytosol by undefined means or follow a novel intravesicular pathway to a compartment where they bind to MHC-I molecules. As indicated in Table I, typical acid-resistant liposomes do not mediate cytosolic delivery, yet i.v. immunization with antigen encapsulated in acid-resistant liposomes efficiently induces MHC-I restricted T cell responses (27). These responses may involve processing mechanisms similar to those involved in generating responses to other particulate antigens, such as bacteria (6) or synthetic particles (28).

In order to study the processing of bacterial antigens, E. coli and Salmonella typhimurium were transfected to express fusion proteins consisting of a bacterial protein with defined cellular localization and function fused with a small antigenic sequence derived from OVA. Several OVA fusion proteins expressed in either E. coli or Salmonella typhimurium were processed efficiently for presentation by K^b, regardless of varied localization (bacterial cytoplasm, periplasm or outer surface) of the epitope within the bacteria (6). We eliminated the possibility that MHC-I presentation of bacterial antigen was an artifact due to release from the bacteria of peptide fragments containing the OVA epitope that could bind directly to surface MHC-I molecules without further processing. Bacterial OVA fusion protein was not presented by fixed macrophages or non-phagocytic EL4 cells, both of which present the synthetic OVA(257-264) peptide efficiently. Furthermore, presentation of bacterial antigen by macrophages was blocked by cytochalasin D, which disrupts actin microfilaments and inhibits phagocytosis. These results suggest the existence of a phagocytic pathway by which exogenous bacterial antigens can be processed and then presented by MHC-I molecules.

Both Salmonella and E. coli strain HB101 have no known mechanism to escape the vacuolar phagosomal/phagolysosomal system, and electron microscopy studies indicated an exclusively intravacuolar localization of intracellular bacteria (6). Furthermore additional functional evidence supports a processing mechanism distinct from the classical MHC-I pathway (i.e., the pathway of cytosolic antigen catabolism and loading of peptide onto MHC-I in the ER). In these experiments macrophages were incubated with cycloheximide or Brefeldin A to block protein synthesis or

ER/Golgi transport. Both of these drugs block the classical MHC-I processing pathway, but they had little or no effect on the processing and presentation of the OVA epitope expressed in E. coli. This suggests that the processing pathway used pre-existing MHC-I molecules derived from a post-Golgi compartment. While the classical MHC-I processing model best explains the major pathway for MHC-I molecules, certain phagocytic APC may express an accessory pathway to allow presentation of exogenous particulate antigens.

The previous studies suggest a number of questions for future research, which will include the following efforts.

1. Further localization of MHC-II within endocytic compartments in antigen presenting cells. Important differences may exist in the processing compartments expressed by different types of antigen presenting cells, e.g. macrophages vs. B cells.

2. Localizing the formation of peptide-MHC-II complexes. Preliminary data with a single antigen have localized the earliest formation of peptide-MHC-II complexes to a dense lysosomal compartment. This hypothesis should be further tested and extended to additional antigens, which could differ in their site(s) of processing. Other antigen presenting cell types should also be examined.

3. Dissecting the pathway for processing of exogenous antigens for presentation by MHC-I molecules.

REFERENCES

1. Harding, C. V., D. S. Collins, J. W. Slot, H. J. Geuze, and E. R. Unanue. 1991. Liposome-encapsulated antigens are processed in lysosomes, recycled and presented to T cells. *Cell 64*:393.

2. Harding, C. V., D. S. Collins, O. Kanagawa, and E. R. Unanue. 1991. Liposome-encapsulated antigens engender lysosomal processing for class II MHC presentation and cytosolic processing for class I presentation. *J. Immunol. 147*:2860.

3. **Pfeifer, J., M. J. Wick, D. Russell, S. Normark, and C. V. Harding.** 1992. Recombinant E. coli express a defined, cytoplasmic epitope that is efficiently processed in macrophage phagolysosomes for class II MHC presentation to T lymphocytes. *J. Immunol.* *149*:2576.

4. **Harding, C. V., and H. J. Geuze.** 1992. Class II MHC molecules are present in macrophage lysosomes and phagolysosomes that function in the phagocytic processing of Listeria monocytogenes for presentation to T cells. *J. Cell Biol. 119*:531.

5. **Harding, C. V., and H. J. Geuze.** 1993. Immunogenic peptides bind to class II MHC molecules in an early lysosomal compartment. *J. Immunol. 151*:3988.

6. **Pfeifer, J. D., M. J. Wick, R. L. Roberts, K. F. Findlay, S. J. Normark, and C. V. Harding.** 1993. Phagocytic processing of bacterial antigens for class I MHC presentation to T cells. *Nature 361*:359.

7. **Harding, C. V.** 1993. Cellular and molecular aspects of antigen processing and the function of class II MHC molecules. *Am. J. Resp. Cell Mol. Biol. 8*:461.

8. **Harding, C. V., and H. J. Geuze.** 1993. Antigen processing and intracellular traffic of antigens and MHC molecules. *Curr. Opin. Cell Biol. 5*:596.

9. **Germain, R. N., and D. H. Margulies.** 1993. The biochemistry and cell biology of antigen processing and presentation. *Annu. Rev. Immunol. 11*:403.

10. **Muller, W. A., R. M. Steinman, and Z. A. Cohn.** 1983. Membrane proteins of the vacuolar system. III. Further studies on the composition and recycling of endocytic vacuole membrane in cultured macrophages. *J. Cell Biol. 96*:29.

11. Bleekemolen, J. E., M. Stein, K. von Figura, J. W. Slot, and H. J. Geuze. 1988. The two mannose 6-phosphate receptors have almost identical subcellular distributions in U937 monocytes.*Eur. J. Cell Biol. 47*:366.

12. Guagliardi, L. E., B. Koppelman, J. S. Blum, M. S. Marks, P. Cresswell, and F. M. Brodsky. 1989. Co-localization of molecules involved in antigen processing and presentation in an early endocytic compartment. *Nature 343*:133.

13. Niebling, W. L., and S. K. Pierce. 1993. Antigen entry into early endosomes is insuffient for MHC class II processing. *J. Immunol. 150*:2687.

14. McCoy, K. L., M. Noone, J. K. Inman, and R. Stutzman. 1993. Exogenous antigens internalized through transferrin receptors activate CD4+ T cells. *J. Immunol. 150*:1691.

15. Collins, D., E. R. Unanue, and C. V. Harding. 1991. Reduction of disulfide bonds within lysosomes is a key step in antigen processing. *J. Immunol. 147*:4054.

16. Jensen, P. E. 1991. Reduction of disulfide bonds during antigen processing. Evidence from a thiol-dependent insulin determinant. *J. Exp. Med. 174*:1121.

17. Hampl, J., B. Gradehandt, H. Kalbacher, and E. Rude. 1992. In vitro processing of insulin for recognition by murine T cells results in the generation of A chains with free CysSH. *J. Immunol. 148*:2664.

18. Harding, C. V. and E. R. Unanue. 1990. Low-temperature inhibition of antigen processing and iron uptake from transferrin: deficits in endosome functions at 18°C. *Eur. J. Immunol. 20*:323.

19. Peters, P. J., J. J. Neefjes, V. Oorschot, H. L. Ploegh, and H. J. Geuze. 1991. Segregation of MHC class II molecules from MHC class I molecules in the Golgi complex for transport to lysosomal compartments. *Nature 349*:669.

20. **Neefjes, J. J., V. Stollorz, P. J. Peters, H. J. Geuze, H. L. Ploegh.** 1990. The biosynthetic pathway of MHC class II but not class I molecules intersects the endocytic route. *Cell 61*:171.

21. **Moore, M. W., F. R. Carbone, and M. J. Bevan.** 1988. Introduction of soluble protein into the class I pathway of antigen processing and presentation. *Cell 54*:777.

22. **Reddy, R., F. Zhou, L. Huang, F. Carbone, M. Bevan, and B. T. Rouse.** 1991. pH sensitive liposomes provide an efficient means of sensitizing target cells to class I restricted CTL recognition of a soluble protein. *J. Immunol. Meth. 141*:157.

23. **Yewdell, J. W., J. R. Bennink, and Y. Hosaka.** 1988. Cells process exogenous proteins for recognition by cytotoxic T lymphocytes. *Science 239*:637.

24. **Carbone, F. R., and M. J. Bevan.** 1990. Class I-restricted processing and presentation of exogenous cell-associated antigen in vivo. *J. Exp. Med. 171*:377.

25. **Flynn, J. L., W. R. Weiss, K. A. Norris, H. S. Siefert, S. Kumar, and M. So.** 1990. Generation of a cytotoxic T-lymphocyte response using a Salmonella antigen-delivery system. *Mol. Microbiol. 4*:2111.

26. **Aggarwal, A., S. Kumar, R. Jaffe, D. Hone, M. Gross, and J. Sadoff.** 1990. Oral Salmonella: Malaria circumsporozoite recombinants induce specific CD8+ cytotoxic T cells. *J. Exp. Med. 172*:1083.

27. **Collins, D. S., K. Findlay, and C. V. Harding.** 1992. Processing of exogenous liposome-encapsulated antigens in vivo generates class I MHC-restricted T cell responses. *J. Immunol. 148*:3336.

28. **Kovacsovics-Bankowski, M., D. Clark, B. Benacerraf, and K. L. Rock.** 1993. Efficient major histocompatibility complex class I presentation of exogenous antigen upon phagocytosis by macrophages. *Proc. Natl. Acad. Sci. USA 90*:4942.

7

GENETIC APPROACHES TO CLASS II-RESTRICTED ANTIGEN PROCESSING AND PRESENTATION

Donald Pious, Steven Fling, and Tatsue Monji

INTRODUCTION

In this report we describe somatic genetic approaches which have produced the first class II-restricted antigen processing/presentation mutants, and what these mutants have revealed about antigen processing and presentation. We have focussed principally on work from our lab, although important contributions in this area have been made by other labs, a number of which are represented in this volume.

In prokaryotes and lower eukaryotes, the isolation and characterization of mutants which are defective in biological functions have been powerful means of investigating these functions and of identifying the genes encoding them. In higher eukaryotes, the mutant approach has not been as widely used, primarily because of several obstacles to genetic studies normally associated with whole animal experiments. These obstacles include long generation times and small number of progeny relative to mutation frequencies, which are typically low, and the lack of efficient alternatives to mating for obtaining recessive mutants. However, somatic cell cultures, which have relatively short generation times and grow exponentially, address the first two of these problems in obtaining mutants. As will become evident below, the system we have used involving immunoselection against HLA antigens in lymphoid cells takes advantage of some unique aspects of the HLA system, including high heterozygote frequency and codominant expression of HLA alleles, to circumvent the problem of the low frequency of recessive mutants. We believe our experience in these studies provides a useful example of the promise (and some of the problems) of this approach to identifying genes involved in immune functions.

Because somatic mutants are present at low frequency in unselected cell populations, even if the cells have been mutagenized, a means of selection is usually required for their isolation. In the case of the class II antigen processing mutants, the discovery of a selective system, which was critical for the isolation of these mutants, was in part serendipitous. This experience highlights an important feature of mutant selection, the fact that mutants selected to address one biological question not uncommonly also yield mutants which address a different question; the latter may prove more interesting than the former. Once such mutants are isolated, however, the selection which yielded them can be used in an intentional way, as described in this chapter, to obtain additional mutants which reveal more about the function and identify genes encoding related functions.

PHENOTYPIC FEATURES OF CLASS II ANTIGEN PROCESSING/PRESENTATION MUTANTS

In the course of characterizing mutants of a B lymphoblastoid cell line which had lost the ability to bind an HLA-DR3 specific, conformationally dependent antibody, Mab 16.23, we discovered a set of mutants which had essentially lost the ability to present class II (DR, DQ and DP) restricted exogenous proteins to the appropriate T cells (1). The distinctive phenotypic features manifested by these mutants and now a second set of mutants are:

1. conformational alterations in class II molecules manifested by:
   a. loss of the DR3 epitope recognized by Mab 16.23 and other conformation sensitive anti-class II antibodies (1);
   b. instability of extracted class II dimers in SDS (1);
   c. reduced Staphylococcal enterotoxin A binding (2);
2. loss of the ability to present class II restricted whole antigens, but retention of the ability to present class II restricted peptides (1);
3. a distinctive abnormality in the kinds of peptides bound to class II molecules (3);
4. normal cell surface levels of class II molecules (1);
5. reduced stimulation of some class II alloreactive clones (4).

Although not every independently arising mutant has been analyzed for all of these features, most of the mutants have, and of those, all features are present. These and other data, discussed below, support the conclusion that in a given mutant, all these abnormalities result from a mutation in a single gene. However, mutations in more than one gene can result in this phenotype.

THE GENETICS OF CLASS II ANTIGEN PROCESSING/PRESENTATION MUTANTS

We have isolated two genetically distinct sets of mutants with similar phenotypic features. The progenitors of the two sets are two different MHC hemizygous deletion mutants of the same B lymphoid cell line (Fig. 1). The first class II antigen processing mutants, designated "c2p-1" (class II presentation- 1), were isolated from 8.1.6, which has a hemizygous deletion in the HLA class II region of about 750kb (ref. 2 and Fig. 2). A second set of mutants was isolated from hemizygous deletion mutant 3.1.0; 3.1.0 has a much larger deletion, of about 40Mb, extending roughly from 10Mb centromeric of the HLA region to the telomere of chromosome 6p. Although the choice of the hemizygous deletion mutant 8.1.6 as progenitor of the first of these mutants was fortuitous, the subsequent choice of 3.1.0 as progenitor of the second set was intentional based on its very large hemizygous deletion flanking the HLA region, as discussed below.

A Gene Mapping to The MHC Class II Region Is Required for Antigen Processing/Presentation

Approximately 30% of the mutants isolated by 16.23 selection of mutagenized 8.1.6 clones had the processing/presentation phenotype. Their frequency, approximately 1×10^{-5}, seemed much too high for recessive mutations in a progenitor cell which is dizygous for the gene(s) affected in the mutants. Because progenitor 8.1.6 was known to have an approximately 750kb hemizygous deletion within the HLA class II/class III

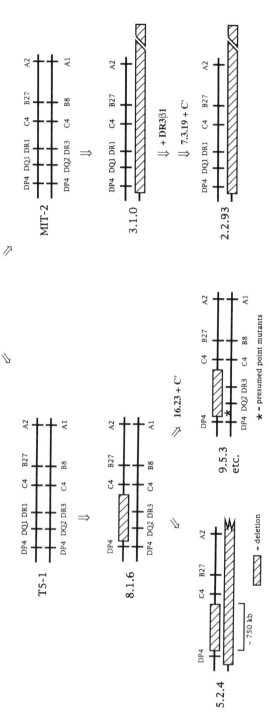

Fig. 1. *Derivation of c2p-1 and c2p-2 mutants from PGLC33H.* Deletions and relative breakpoints are designated by striped bars, within the indicated HLA haplotypes, oriented with the centromere to the left. The telomeric breakpoint of 5.2.4 is undetermined but minimally extends telomeric of HLA-A. Mutant 9.5.3 is representative of the set of c2p-1 mutants. The 3.1.0 deletion is estimated to be 40 Mb, extending from about 10 Mb centromeric of the MHC to 6pter.

region, it seemed plausible that the processing/presentation gene(s) might map to this region. We addressed this question using another mutant, 5.2.4, which was also derived from 8.1.6. Mutant 5.2.4 had sustained a deletion of the complete HLA haplotype not affected in 8.1.6, rendering it homozygously deleted for 750kb in the class II/III region (Figs 1 and 2). Because 5.2.4 retained one expressed DPA and -B gene pair, we were able to test it for the processing/presentation phenotype. 5.2.4, unlike progenitor 8.1.6, was unable to present the DP4 restricted antigens Hepatitis B surface antigen (HBsAg) and Tetanus toxin (TT) to DP4 restricted T cells specific for these antigens (6). However it effectively presented the relevant synthetic HBsAg and TT peptides. In addition, Western blots of detergent extracts of 5.2.4 cells with a Mab to DP showed essentially complete DP dimer dissociation compared to the controls. These results strongly suggested that the gene(s) affected in the 8.1.6-derived c2p-1 mutants mapped to the 750kb region defined by the hemizygous deletion in 8.1.6. It was possible to refine the mapping further based on two additional independently derived mutants, 721.82 (7,8) and 2.4.93 (S. Fling and D. Pious, unpublished data), which have overlapping but not identical homozygous deletions in the class II region (Fig. 2). These mutants are normal with respect to the phenotypic features of the c2p-1 mutants, suggesting that the c2p-1 gene resides in the approximately 230kb segment homozygously deleted in 5.2.4 but not in 721.82 or 2.4.93.

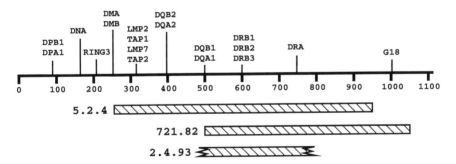

Fig. 2. Portions of the HLA class II region lost in homozygous deletion mutants indicate the region containing the c2p-1 locus. Deletions are designated by striped bar. 5.2.4 is defective for c2p-1 function and 721.82 and 2.4.93 are wildtype. 2.4.93 was derived from 3.1.0 by immunoselection against DR and its deletion extends minimally from HLA-DRA to HLA-DQ1 but does not extend into TAP2.

The Phenotype Of The c2p-1 Mutants Is Recessive

We investigated the genetic basis for the c2p-1 defect further by making somatic cell hybrids between several c2p-1 mutants and another mutant, 3.1.3; 3.1.3 expresses no DRβ mRNA but is wild type with respect to class II processing/presentation (6). For selection of hybrids, we introduced different dominant selectable markers into the cells to be hybridized. The resulting hybrid cells, unlike either input cell, bound Mab 16.23 and presented intact antigens normally to DR3 restricted T cells, and extracted class II dimers from the hybrid had normal stability in SDS gels. Thus the c2p-1 mutation is recessive.

Three Independent c2p-1 Mutants And 5.2.4 Are All Defective In The Same Gene

We next sought to determine whether the same gene(s) was mutated in different 8.1.6 derived mutants of independent origin and in 5.2.4. All six somatic cell hybrids representing all the possible combinations among these three 8.1.6 derived mutants and between these mutants and 5.2.4 were noncomplementary with respect to 16.23 binding and DR dimer dissociation (6). An independent mutant of a different cell line, T2, which is also homozygously deleted in this region, also failed to complement the 16.23 selected mutants. In addition, a hybrid made between 5.2.4 and one of the 16.23 selected mutants retained the antigen presentation defective phenotype. We concluded from these data that these mutants are all defective in the same gene, which maps to the class II region and is required for class II processing/presentation.

A Second Set Of Mutants, "c2p-2", Are Defective In A Chromosome 6p Gene Different Than c2p-1

In searching for other genes which might play roles in class II restricted antigen processing, we were struck by a remarkable feature of the MHC, the fact that the MHC contains tightly linked non-homologous genes whose products interact. Linkage of homologous genes whose products interact is common in eukaryotes; it has an obvious basis in gene duplication. On the other hand, linkage of genes whose products interact but which are not homologous has rarely been found in eukaryotes. Among the few examples outside the MHC are the cluster of cytokines and their receptors on human chromosome 5q, and the two keratin- retinoic acid receptor- homeobox clusters on human chromosomes 12q and 17q. A search we have made of the human genome data base for linkage among some 70 genes with known interactive non-homologous partners (cytokines and polypeptide hormones and their respective receptors, the coagulation and complement cascades, etc), turned up only one additional linkage of this kind. However, the number of functional partners which have been sequenced and mapped is still small, so it is uncertain how uncommon such linkages are.

The MHC contains at least two such linkages. One consists of the complement C2 and C4 genes, which lie in the class III region only 30kb apart. C2 and C4 are not detectably related in sequence, but their gene products directly interact in the classical complement pathway. The linkage of the TAP 1 and 2 genes to the MHC class I genes, 1.5Mb apart, constitutes a second such linkage. The TAP genes code for subunits of the peptide transporter, which interacts with MHC class I gene products; the TAP genes are unrelated to MHC genes in sequence. An additional interesting possibility is that one or more of the HSP70 genes mapping to the MHC class III region plays a role in antigen processing.

Based on these considerations of linkage of MHC genes whose products interact, we investigated the possibility that there are additional antigen processing genes syntenic to the MHC but outside the 750kb region hemizygously deleted in 8.1.6. We again took advantage of the fact that recessive mutants have a much higher frequency in cells which are hemizygous as opposed to dizygous with respect to the gene of interest; for recessive mutations, the mutant frequency in hemizygous progenitor cells should be the square root of the frequency in the dizygous progenitor. Thus we chose as progenitor 3.1.0 (9), which has an large, hemizygous, terminal deletion of chromosome 6p of about 40Mb (Fig. 1), or about 50 times the size of the deletion in 8.1.6. The centromeric breakpoint in 3.1.0 is about 10Mb centromeric to the DP cluster; 3.1.0 is therefore hemizygously deleted for the complete MHC (about 4Mb), as well as for the flanking 10Mb centromeric and 25 Mb telomeric to the MHC.

A DR3B1 cDNA was first transduced into 3.1.0 via a retroviral expression system to permit selection with anti-DR3 Mab 7.3.19, an antibody whose binding is known to be diminished in the 8.1.6-derived processing mutants. This antibody was used rather than 16.23 because its apparently higher affinity resulted in more efficient selection. The selections made on 3.1.0, as outlined above, have yielded a variety of interesting mutants whose characterization is still in progress. One subset of the 3.1.0-derived mutants, represented by 2.2.93 (Fig. 1), has phenotypic features resembling those of the c2p-1 mutants (5). These newer, 3.1.0 derived mutants and the c2p-1 mutants are complementary in somatic cell hybrids made between the two. We infer from these results that the 3.1.0-derived mutants are affected for a different gene(s) than c2p-1, and therefore have designated one mutant, 2.2.93, as representing a second locus, c2p-2. Based on the mutant frequency argument presented above, it is likely that the c2p-2 gene maps within the region hemizygously deleted in 3.1.0. Whether additional loci are affected within the set of 3.1.0 derived mutants remains to be determined.

Various approaches are available for identifying and/or cloning the genes defective in those mutants. Because there exists a fairly detailed genomic map of this region, "candidate" genes within the affected region can be sequenced in mutant and progenitor to identify mutations, and can be transfected into mutants cells in appropriate expression vectors to determine whether they can correct the mutant phenotype. cDNA libraries in high efficiency episomal expression vectors can be used for the same purpose. These studies are currently underway in our laboratory, and we have now, at the time of this writing, identified a cDNA which complements the defect in 2.2.93. The cDNA, representing an MHC-linked gene, encodes an integral membrane protein of ~30kD. Based on the homology of "c2p-2" to other known genes, the exact mapping of the 5.2.4 deletion breakpoint, and the complementation between c2p-2 and c2p-1 mutants, we believe that the "c2p-2" product interacts with the product of the c2p-1 locus, probably as a heterodimer. How this molecule affects function and how mutations in its components result in the observed phenotype remain to be elucidated.

THE CELL BIOLOGY OF CLASS II ANTIGEN PROCESSING/PRESENTATION MUTANTS

Although the available evidence and considerations of the genesis of the c2p-1 and -2 mutants strongly suggest that each mutant arises from a mutation in a single gene, the mutants all manifest the complex array of phenotypic changes outlined in an earlier section. There are insufficient data as yet to support a single integrated mechanism which would account for all the pleiotropic effects of these single gene mutations. We believe that such a single mechanism exists, however, and there are data which account for some of these phenotypic changes and provide a speculative basis for interpreting others.

The c2p-1 Mutants Have An Abnormality In Class II-Bound Peptides Which Can Explain Their Presentation Defect

When we first characterized the phenotypic changes in the c2p-1 mutants, we proposed, as one possible mechanism consistent with the then-available data, that the mutants might have a defect in generation of peptides in the lyso-endosomal compartment resulting in "empty" class II molecules on the cell surface, i.e., cell surface class II molecules lacking bound peptides. Although this mechanism could account for some of the phenotypic changes in these mutants, recent studies of the peptides eluted from affinity-purified class II molecules of mutant cells indicate that the class II molecules of the mutants are not deficient in bound peptides. However, the peptides eluted from class II molecules of mutant cells differ radically from those of non-mutant cells (10,11). The peptides eluted from non-mutant cells consist of a heterogeneous mixture of peptides, each of low relative abundance (12). On the other hand, peptides eluted from the c2p-1 mutants consist almost exclusively of a set of ragged-ended peptides from a single region (amino acids 81-103) of the Invariant chain (3). The same Invariant chain$_{81-103}$ peptides are present in class II molecules from non-mutant cells, but in trace abundance.

The almost complete absence of any bound peptides other than the Invariant chain peptides from the class II molecules of mutant cells readily accounts for their inability to present class II-restricted antigens after incubation with the exogenous proteins. The ability of the mutants to stimulate class II restricted T cells after incubation with cognate peptides, as opposed to intact proteins, can probably be accounted for by the combined effect of a few empty class II molecules on the cell surface of the mutants and the ability of T cells to respond to APCs with small numbers of class II molecules with bound cognate peptide.

Removal Of Invariant Chain Peptides From And Binding Of Cognate
Peptides To Class II Molecules Of Mutants Restores Their Class II Dimer
Stability And SEA Binding Ability

Two manifestations of the conformational changes in class II
molecules of the mutants, the instability of their class II dimers in SDS and
their reduced ability to bind SEA, can be corrected by manipulating the
peptides bound to the class II molecules (2,3). Treatment of mutant cells at
pH4.5 or lower for 4-8 hours results in the complete removal of the
Invariant peptides from class II molecules; after this treatment, class II
molecules of mutant cells which are exposed to cognate peptide at neutral
pH become stabilized into the compact form found normally in non-mutant
cells, and SEA binding capacity is also restored. Restoration of compact
class II dimers and SEA binding capacity does not occur in mutant cells
incubated with cognate peptide without prior treatment at low pH to remove
the Invariant chain peptides, nor does restoration of class II dimer stability
or SEA binding capacity occur in cells exposed to low pH and then to non-
cognate peptides. These data suggest that a stable class II dimer results
from a class II dimer to which a cognate peptide is bound. Further they
suggest that the presence of the Invariant chain$_{81\text{-}1013}$ peptides bound to
class II molecules of the mutants may account for the fact that these
molecules lack bound cognate peptides.

The binding site for the Invariant chain$_{81\text{-}103}$ peptides to class II
molecules is not known. If these peptides bind in the peptide binding
groove, two of their properties may alter class II conformation in the
mutants: the length of the Invariant chain$_{81\text{-}103}$ peptides, which is at the
upper limit of the lengths of the normal peptides bound to class II
molecules from non-mutant cells, and their unusually large number of
prolines. The most abundant of the Invariant chain$_{81\text{-}103}$ peptides contains
six prolines, and the smallest or "core" peptide, five prolines. Compared to
the predominantly shorter, rod-like peptides found in class II molecules of
non-mutant cells (13), the length of the Invariant chain peptides as well as
the large number of prolines, with their tendency to induce peptide bends,
could alter the conformation of class II molecules by producing bulging of
the peptide binding groove.

Mab 16.23 Binds To A Conformational Determinant On Class II Molecules

Given that Mab 16.23 is able to immunoselect antigen processing/presentation mutants, the nature of the determinant recognized by this antibody is of considerable interest. The conformational alterations in class II molecules of the mutants caused by the presence of the bound Invariant chain peptides and/or the absence of bound cognate peptide could account for the loss of the epitope recognized by Mab 16.23, if the epitope recognized by 16.23 were a conformational determinant on the DR3 β chain. An alternative possibility, that the epitope recognized by Mab 16.23 in non-mutant cells is a cognate self- peptide bound to DR3 molecules, seems to be excluded by the fact that Mab 16.23 binds to at least 70% of DR3 molecules in progenitor 8.1.6 cells (3), far in excess of the abundance of any peptide bound to DR3 molecules in these cells. Thus it seems probable that Mab 16.23 binds to a conformational determinant on DR3 β chains which is either disrupted in DR3 molecules to which Invariant chain$_{81-103}$ peptides are bound, or is altered as a consequence of the absence of bound cognate peptide.

What Is The Primary Functional Defect In The Mutants?

From the foregoing it seems probable that absence of cognate peptides bound to most of the class II molecules in the mutants and/or the presence of the Invariant chain peptides accounts for their phenotypic abnormalities. Additional evidence supporting this conclusion comes from another mutant, 10.24.6 (3,14). 10.24.6, unlike the c2p-1 and -2 mutants, has a DRA structural gene mutation resulting in an additional glycosylation of the DRα chain. Mutant 10.24.6 has the same constellation of defects as the c2p-1 and -2 mutants, including the inability to present DR3 restricted exogenous protein antigens, inability to bind Mab 16.23, instability of DR dimers, reduced SEA binding, predominance of Invariant chain$_{81-103}$ peptides bound to DR3, and the restoration of DR dimer stability and SEA binding by low pH/cognate peptide. However, in 10.24.6 these defects are limited to DR3 molecules. Thus in both c2p and 10.24.6 mutants, which differ in their genetic defects, the presence of the Invariant chain peptides replacing cognate peptides as the predominant peptides bound to class II molecules is associated with the same complex mutant phenotype, and at least two of the abnormalities in both kinds of mutant, dimer instability and defective SEA binding, can be corrected by the replacement of these peptides with cognate peptides. Because the change in DR3 molecules in 10.24.6 is the addition of an extra oligosaccharide chain, we speculate that the presence of this bulky group in DR molecules of 10.24.6 interferes with the contact of another molecule which must interact with class II molecules either for the removal of Invariant chain $_{81-103}$ peptides or the binding of cognate peptides. In this scenario, the c2p mutants could have mutations in the genes encoding this molecule, ablating its function.

If the primary phenotypic defect in the c2p mutants is in fact the presence of Invariant chain$_{81-103}$ peptides bound to most class II molecules in place of cognate peptides, the obvious and intriguing question is: what accounts for the excessive abundance of the class II-bound Invariant chain peptides? Two alternative mechanisms immediately suggest themselves. One is that the binding of the Invariant chain peptides occurs by default, secondary to an absence of cognate peptides in the appropriate compartment which can compete successfully with the Invariant chain$_{81-103}$ peptides for binding sites. The absence of cognate peptides could occur either because of a defect in antigen processing, or because of a failure of class II molecules or of antigenic proteins or processed peptides to traffic to the appropriate cell compartment for assembly into class II-cognate peptide complexes. However, the Invariant chain blocks cognate peptide binding (10), and in competitive binding studies, an Invariant chain peptide is able to block binding to class II molecules of the single cognate peptide studied (11). If this competitive binding advantage of Invariant peptides over cognate peptides obtains for other cognate peptides, the default mechanism would seem unlikely. An alternative possibility is that the primary defect in the mutants is the failure of the Invariant chain$_{81-103}$ peptides to dissociate completely from class II molecules, as they do in non-mutant cells, with the class II-Invariant chain$_{81-103}$ peptide complex being an intermediate in this process which accumulates behind the block in the mutants. The failure of the class II-Invariant peptide complex to dissociate in the mutants could result from mutational loss of an activity required for Invariant peptide removal, as postulated above, or from a failure of the complex to traffic to the appropriate compartment for Invariant peptide removal. The failure of removal of Invariant chain peptides would leave no class II dimers available to bind cognate peptides.

FUTURE DIRECTIONS

One of the important points we hope has become evident in this review is the power of mutational and other somatic genetic approaches to identify new genes and to explore the biological roles of their gene products. A number of unusual features of the system we have developed involving MHC mutants in human B lymphoid lines make it especially well suited for such an approach. Among them are features which have facilitated efficient immunoselection and the subsequent genetic analysis of the mutants, including the availability of cell lines which are heterozygous for closely linked, codominantly expressed genes which are abundantly expressed on the cell surface, and the ability to generate hemizygous and homozygous deletion mutants from these cell lines. Also important for efficient genetic analysis of the mutants is the relatively constant background chromosomal stability and diploidy of human EB virus-transformed B cell lines and their immortality and clonability. Functional analyses of the mutants take advantage of the ability of these cells to process and present class II restricted antigens. And finally, the capacity of these cells to be transfected with reasonable efficiency by electroporation and to be infected with retroviral vectors is an important asset in the cloning of genes identified by mutant analysis.

A major challenge for future studies will be to determine whether it is possible to adapt these approaches or to develop alternative ones which can be applied to the mutational analysis of non-MHC linked genes which are important in antigen processing and presentation. Among the specific questions we think it important to address in the near future are:

1. How do the c2p-1 and -2 gene products function in the generation of cognate peptide-bound class II molecules?
2. In what cell compartment(s) do these functions take place?
3. Are there other genes syntenic to the MHC which are involved in processing/presentation?
4. Can we identify, by mutational or other means, other genes, not syntenic with the MHC, which have roles in class II-restricted antigen processing presentation?

ACKNOWLEDGMENTS

We gratefully acknowledge the important contributions to this work of Dr. Elizabeth Mellins, Lynn Dixon, Sherry Kempin and Laura Smith, former members of our lab, and current members Ben Arp, Dr. Tom Cotner and Sheri Fujihara.

This work was supported by a predoctoral traineeship (to T.M) and research grants from the National Institutes of Health, and by a Cancer Research Institute postdoctoral fellowship (to S.F.).

REFERENCES

1.   **Mellins, E., L. Smith, B. Arp, T. Cotner, E. Celis, and D. Pious.** 1990. Defective processing and presentation of exogenous antigens in mutants with normal HLA class II genes. *Nature.* 343:71.

2.   **Pious, D., L. Smith, T. Monji, T. Cotner, J. Kerner, and S. Dillon.** Superantigen binding to MHC class II molecules is affected by class II-bound peptides, submitted.

3.   **Monji, T., A.L. McCormack, J.R. Yates III, and D. Pious.** Sequential treatment of class II presentation (c2p-1) mutants with acidic pH and peptide reconstitutes class II dimer stability but not 16.23 antibody binding, submitted.

4.   **Cotner, T., E. Mellins, A. Johnson, and D. Pious.** 1991. Mutations affecting antigen processing impair class II-restricted allorecognition. *J. Immunol.* 146:414.

5.   **Fling, S., B. Arp and D. Pious.** Manuscript in preparation.

6.   **Mellins, E., S. Kempin, L. Smith, T. Monj, and D. Pious.** 1991. A gene required for class II restricted antigen presentation maps to the major histocompatibility complex. *J. Exp. Med.* 174:1607.

7.   **Spies, T., M. Bresnahan, S. Bahram, D. Arnold, G. Blanck, E. Mellins, D. Pious, and R. DeMars.** 1990. A gene in the major histocompatibility complex class II region controlling the class I antigen presentation pathway. *Nature* 348:744.

8.   **Ceman, S., R. Rudersdorf, E.O. Long, and R. DeMars.** 1992. MHC class II deletion mutant expresses normal levels of transgene encoded class II molecules that have abnormal conformaton and impaired antigen presentation ability. *J. Immunol.* 149:754.

9.   **Levine, F., H. Erlich, B. Mach, R. Leach, R. White, and D. Pious.** 1985. Deletion mapping of HLA and chromosome 6p genes. *Proc. Natl. Acad. Sci. USA* 82:3741.

10.   **Roche, P.A., and P. Cresswell.** 1990. Invariant chain association with HLA-DR molecules inhibits immunogenic peptide binding. *Nature* 345:615.

11.   **Sette, A., S. Ceman, R.T. Kubo, S. Kazuyasu, E. Appella, D.F. Hunt, T.A. Davis, H. Michel, J. Shabanowitz, R. Rudersdorf, H.M. Grey, and R. DeMars.** 1992. Invariant chain peptides in most HLA-DR molecules of an antigen-processing mutant. *Science* 258:1801.

12.   **Hunt, D.F., M. Hanspeter, T.A. Dickinson, J. Shabanowitz, A.L. Cox, K. Sakaguchi, E. Appella, H.M. Grey, and A. Sette.** 1992. Peptides presented to the immune system by the murine class II major histocompatibility complex molecule 1-A^d. *Science* 256:1817.

13.   **Brown, J.H., T.S. Jardetzky, J.C. Gorga, L.J. Stern, R.G. Urgan, J.L. Strominger, and D.C. Wiley.** 1993. Three-dimensional structure of the human class II histocompatibility atigen HLA-DR1. *Nature* 364:33.

14.   **Mellins, E., P. Cameron, M. Amaya, D. Pious, L. Smith, and B. Arp.** 1993. A mutant HLA-DR molecule associated with invariant chain peptides. *J. Exp. Med.*, in press.

8

REGULATION OF MHC CLASS II INTRACELLULAR TRANSPORT AND PEPTIDE LOADING

Paola Romagnoli, Flora Castellino, and Ronald N. Germain

INTRODUCTION

T cells recognize fragments of antigenic proteins in association with MHC molecules. These antigenic peptides are generated by degradation of proteins in the cytoplasm and in the endocytic pathway, which are the two intracellular compartments in which pathogens reside. To insure the detection of all intracellular pathogens, the two classes of MHC molecules, class I and class II, have differentially evolved to optimize the capture of antigens in these two different compartments. MHC class I molecules are specialized for binding peptides derived from endogenously synthesized proteins or proteins entering the cytosol, while MHC class II molecules are optimized for capture of peptides originated from the processing of proteins present in the endocytic pathway (1).

Upon synthesis and translocation in the ER, MHC class I molecules acquire a transport-competent conformation by loading peptides and associating with β2-microglobulin (β2m) (2). These stable complexes exit the ER and through the default exocytic pathway reach the plasma membrane in 15-30 minutes after synthesis. Unoccupied MHC class I molecules are retained in the ER as incompletely folded proteins (3). This mechanism avoids surface expression of empty class I molecules and favors the acquisition of compartment-specific (ER) peptides.

Because proper folding and subunit assembly is a prerequisite for effective post-ER transport of all membrane and secretory proteins, MHC class II α and β chains, like class I heavy chains and β2m, need to be correctly assembled into heterodimers with a transport-competent conformation following synthesis and translocation into the ER lumen. In contrast to class I molecules that utilize cytosolic peptides imported into the ER to help achieve this state, a transport-competent conformation must be achieved by MHC class II molecules without binding such ER peptides, so that the class II molecules can reach the endocytic pathway with a binding site still available for the acquisition of antigenic peptides produced at that site. The class II molecules must also be in a transport competent state to leave endosomes and arrive at the cell surface. Our recent work has focused on the contributions of a non-MHC encoded protein termed invariant chain to such proper class II assembly and transport, on the pathway followed by class II during transit to endocytic / lysosomal sites of protein degradation and peptide binding, and on the control of class II exit from the peptide loading site(s). A model of the relationship between proper folding and peptide-binding site occupancy of MHC class II molecules in two different intracellular compartments (ER and endocytic pathway) will be presented.

FOLDING AND PEPTIDE BINDING SITE OCCUPANCY OF MHC CLASS II MOLECULES IN THE EARLY BIOSYNTHETIC COMPARTMENT

Invariant Chain Contributes to Initial Assembly and Early Postsynthetic Transport of MHC Class II Proteins

MHC class II molecules assemble in the ER as a stoichiometric complex with trimers of invariant chain (Ii) (4). The rapid and complete coassembly of invariant chain with MHC class II molecules in the ER hinted at two possible functions of invariant chain in the early biosynthetic compartment: 1) by associating with class II molecules invariant chain could promote the proper folding of MHC class II molecules necessary for intracellular transport, and 2) invariant chain could inhibit loading of MHC class II molecules with ER peptides.

Experimental data supporting the first model initially came from studies using injected or transfected cells. In *Xenopus* oocytes injected with class II and Ii mRNA, Ii was found to promote Golgi transit of MHC class II molecules, first suggesting a chaperon-like function of invariant chain (5). This observation has been confirmed and extended in mammalian cells (6, 7). In transfected fibroblasts invariant chain promotes ER egress, Golgi transit, and cell surface expression of MHC class II molecules.

The relevance of these observations was limited by the use of transfected cell lines that normally do not express class II molecules and Ii. To study invariant chain functions in professional antigen presenting cells under physiological conditions, we have generated mice lacking Ii by gene targeting techniques (8). The effects of invariant chain absence have been studied in splenic B cells by confocal microscopy and by SDS-PAGE analysis of labelled proteins. The intracellular localization of MHC class II molecules in spleen B cells of wild type animals and of mutant animals is dramatically different. MHC class II molecules in mice lacking Ii are present mainly in the ER, with only a low level of surface expression, while in normal animals they are found in large amounts at the cell surface and in small vesicles. These results show that Ii makes an important though not absolutely required contribution to ER egress of MHC class II molecules in professional antigen presenting cells. Immunoprecipitation of metabolically labelled proteins from splenic B cells of such Ii-deficient mice also surprisingly demonstrated that in the absence of Ii, class II α and β chains do not assemble properly, remaining largely as free chains rather than progressing to correctly folded heterodimers.

These invariant chain-deficient animals complement gene transfer models for studying the cell biology of retention of non-mutant proteins in the ER. ER-retention of proteins can occur by formation of transport-incompetent aggregates and/or by stable interaction with ER resident proteins (9, 10). In MHC class II-positive, Ii-negative transfected fibroblasts or in spleen cells from Ii-negative mice, sedimentation velocity analysis reveals the presence of high molecular weight aggregates consisting of either single chains (homoaggregates) or αβ complexes (heteroaggregates) (11). The retention in the ER of MHC class II molecules is mediated by the association of these aggregates with the immunoglobulin heavy chain binding protein (Bip), grp94 and p72 (12), all of which are ER resident proteins and members of the heat shock protein (HSP) family. Co-expression of Ii reduces the presence of aggregates and increases the formation of correctly assembled complexes. Single chain aggregates could be either intermediates in the process that leads to the formation of properly folded complexes or they could be the end product of an unproductive folding reaction that occurs in the absence of invariant chain. Pulse-chase experiments using very brief labelling times are in progress to distinguish between these possibilities. The incompletely folded or improperly folded MHC class II molecules are subsequently degraded in the ER.

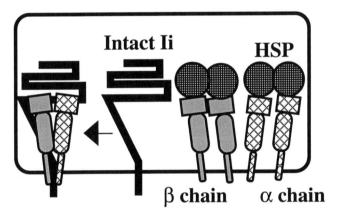

Figure 1: Invariant chain promotes assembly of class II heterodimers from free chains

Invariant Chain Inhibits Peptide Binding to MHC Class II Molecules

A second property of Ii is its ability to inhibit the tight binding of peptides to MHC class II molecules. This was first demonstrated using in vitro binding assays (13, 14). Following the discovery of a relationship between class II behavior in SDS-PAGE and peptide occupancy (15), similar findings were made for molecules newly synthesized in cells and en route to the endocytic pathway (16, 17). The fact that Ii inhibits the binding of peptides to MHC class II molecules led to the suggestion that another important role of Ii early after class II biosynthesis is to prevent the capture of ER peptides meant for presentation by MHC class I. In the cells of Ii-deficient mice, however, there is a failure of most class II α and β chains to stably assemble and a failure of those heterodimers that do form to undergo the structural transition to the SDS-stable state known to be associated with effective peptide binding. This suggests that inhibition of the interaction of newly synthesized class II molecules in the ER with the short peptides meant for class I molecule binding is not a primary role of Ii. Thus, the interesting conclusion drawn from studying the invariant chain-deficient animals is that the primary function of invariant chain in the ER is to promote proper folding of MHC class II molecules and that inhibition of peptide binding seems to be a secondary event to this assembly / folding property. We were especially intrigued by the notion that these functions of invariant chain might be linked to intrinsic structural and biochemical properties of MHC class II molecules that demand some type of interaction with the binding site to attain stability and transport-competence.

To test whether there is any direct relationship between the formation of transport-competent αβIi complexes and peptide-binding site occupancy of MHC class II molecules, cDNA constructs encoding mutant invariant chain proteins with truncations of the lumenal portion were made in our laboratory. This strategy was chosen because our previous data on invariant chain promotion of export of haplotype-mismatched class II dimers suggested a lumenal site for this function of invariant chain (6), and because the inhibition of peptide binding by already assembled class II dimers was clearly a function involving this portion of the class II and invariant chain proteins. These constructs were transfected into COS cells with and without cDNA vectors for class II α and β chains, and the localization and biochemical behavior of the corresponding expressed proteins were analyzed by confocal microscopy and immunoprecipitation. A short fragment of Ii (residues 1-107 of the mouse p31 form, called mIi1-107) was found to be necessary and sufficient for the association of invariant chain with MHC class II molecules, for the formation of a transport-competent αβIi complex, and for vesicular localization.

Work by others had demonstrated that removal of 15 or more residues from the cytoplasmic tail of Ii led to its localization on the cell membrane rather than intracellular vesicles, and that class II was stoichiometrically associated with Ii on the cell surface when coexpressed with such mutant forms of Ii (18). We took advantage of this to examine whether the portion of Ii extending to residue 107 also was also necessary and sufficient for inhibiting peptide binding to class II molecules. Remarkably, surface coexpression of class II with the 19-107 form of Ii led to complete inhibition of the capacity of these molecules to bind or to present peptides. Thus, both the assembly/transport and the peptide inhibition functions of invariant chain were found to be inseparable functions of the same minimal segment of Ii necessary for stable interaction with class II molecules.

Peptide binding to MHC class II molecules could be inhibited either by occlusion or inactivation of the class II-peptide binding site. "Occlusion" of MHC class II molecules could be achieved either by covering the groove with a protein cap, or by "occupancy" of the groove with a peptide-like region of Ii. "Inactivation" of class II-peptide binding sites could be accomplished if Ii acts by binding outside the groove to force or hold the binding site closed and inaccessible to free peptide. The experimental data obtained so far make it unlikely that mIi1-107 is mediating its function by covering the peptide binding site of MHC class II molecules like a cap. In fact, in contrast to intact Ii (19) this fragment does not inhibit the interaction of the superantigen TSST-1 or of helix-specific mAb with the external portion of the peptide binding domain. It is more difficult to distinguish between physical occupancy and extrinsic inactivation of class II-peptide binding domain. Of interest in this regard is the finding that the region of Ii giving rise to peptides that can be eluted from several human (20-22) and murine (23, 24) MHC class II molecules (CLIP) is present in mIi1-107. Synthetic peptides corresponding to these Ii sequences can inhibit binding of antigenic peptides to class II in vitro (20, 21, 24). As a working model, we suggest that this region of Ii may play the role of a "generic" class II-binding ligand while it is still a part of the intact invariant chain molecule. Further experiments are in progress to test this hypothesis.

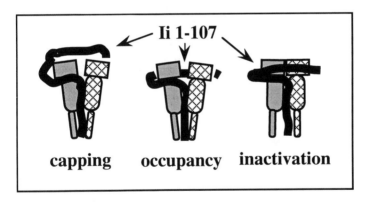

Figure 2: Models for the mechanism of invariant chain inhibition of peptide binding to MHC class II molecules

The intriguing outcome of these studies is that invariant chain mediates both a chaperon-like function and the inhibition of class II-peptide binding through the same discrete subdomain, indicating that these two properties of Ii are intimately connected. It is tempting to draw a parallel with the class I system: MHC class I molecules become transport-competent in the ER by loading peptide and associating with β2m. MHC class II molecules, instead, become transport-competent after "loading" an internal segment of Ii. This analogy suggests that the demand of peptide binding site occupancy in the same intracellular compartment (ER) is differentially achieved by the two classes of MHC molecules: MHC class I molecules occupy their grooves with ER peptides, while MHC class II proteins associate with Ii. By playing a direct role in the formation of proper class II heterodimers, Ii assures the efficient formation of class II binding sites only when the CLIP region is physically present to inactivate the newly formed site and prevent its interaction with potential target structures in the ER. This action of Ii may coincidently inhibit some binding to short cytosol derived peptides, as originally envisioned, but appears primarily designed to jointly meet the quality control standards for ER export by generating correctly folded class II heterodimers, and at the same time blocking interaction with the segments of other proteins in the ER that during folding resemble the desired denatured protein targets of the class II binding site present in the endocytic pathway.

FOLDING AND PEPTIDE LOADING OF MHC CLASS II MOLECULES IN THE ENDOCYTIC PATHWAY

Once having acquired a transport-competent conformation, the $\alpha\beta$Ii complexes leave the ER, passing through the Golgi complex to the trans- Golgi network, where the intracellular route followed by MHC class I and MHC class II molecules diverges (25). Class I molecules, already loaded with peptide, are rapidly transported to the cell surface through the secretory pathway. In contrast, biosynthetic studies indicate that the removal of invariant chain from class II and the arrival of such invariant chain-free MHC class II molecules to the cell surface is relatively slow. Class II molecules still associated with Ii are retained intracellularly in a post-Golgi acidic compartment, identified as part of the endocytic pathway, for two to three hours before reaching the cell surface (25). In the endocytic pathway, due to acidic pH and the presence of proteases, invariant chain is sequentially cleaved and dissociated from the $\alpha\beta$ complex, allowing the loading of peptide.

Sorting signals that mediate the localization of MHC class II-Ii complexes to the endocytic pathway have been identified in the cytoplasmic tail of Ii (26). In transfected fibroblasts lacking Ii, MHC class II molecules are mainly found in the ER, in the Golgi complex, and on the plasma membrane; Ii co-expression promotes their vesicular localization (26-28). Nevertheless, sorting events also seem to have a cell type-specific component. In the absence of Ii, MHC class II molecules in various transfected cell lines (mouse L cells, simian CV-1 cells, rat-2 epithelial cells) are found in vesicles, suggesting that MHC class II molecules themselves have an independent sorting signal. However, in these cell lines co-expression of Ii increases the numbers of MHC class II positive vesicles and favors MHC class II localization in more acidic compartments (29). An open question is whether the primary role of the Ii is to target MHC class II molecules to the endocytic pathway or to retain them there, thus leading to their accumulation and their consequent detection.

At the moment there is no consensus on the distribution of Ii and of MHC class II molecules in the different endocytic compartments, on the route(s) that they follow to enter and to egress the endocytic pathway, and on the site(s) of MHC class II peptide interaction. Immunoelectron microscopic analysis of EBV-transformed human lymphoblastoid cell lines show MHC class II molecules to accumulate in a specific pre-lysosomal compartment, named the MIIC compartment, where Ii has already been degraded (30). MHC class II molecules were also found in ER, Golgi complex, TGN and plasma membrane, but not in early endosomes. These data led to the suggestion that newly synthesized MHC class II molecules are directly transported from the trans-Golgi network to the MIIC compartment where they accumulate and are loaded with incoming peptides. This model is in contrast with a previous report on the co-localization of MHC class II molecules, Ii, antigen, and proteolytic enzymes in early endosomes (31).

In our laboratory, to better understand the influence of Ii on MHC class II localization, we analyzed by immunofluorescence the intracellular distribution of these molecules in COS cells transiently expressing Ii alone or together with class II molecules (28). In this system over-expression of Ii causes the appearance of large vesicular structures, referred to as "macrosomes", containing both MHC class II and uncleaved Ii. These vesicles are accessible to transferrin, and they possess other characteristics that led us to conclude they represented enlarged early endosomes. MHC class II and Ii are also detectable in a prelysosomal compartment, where Ii is C-terminally cleaved. These findings suggest that early endosomes are the first compartment of the endocytic pathway reached by $\alpha\beta I$ complexes. The complexes then move to more acidic and proteolytic compartments were Ii is cleaved and MHC class II loading should occur. The kinetics of accumulation of ovalbumin (used as an endocytic marker) in different intracellular compartments demonstrated that the presence of macrosomes delays the transport from early to late endosomes. However, the recycling of transferrin to the plasma membrane is not affected. This overexpression system may thus have uncovered an important capacity of the Ii to influence both the morphology and the function of the early endocytic compartment. Delaying anterograde traffic could favor mixing in early endosomes, enhancing the interaction between MHC class II molecules and incoming antigen.

The majority of the studies on the intracellular distribution of MHC class II have so far used a morphological approach leading to definition only of the steady state distribution of these molecules. This approach is limited by its inability to discriminate between newly synthesized MHC class II molecules transported to the endocytic compartments for loading and subsequent cell surface expression, and those present there but destined for degradation. Therefore, these studies do not provide direct information about the function and the fate of MHC class II molecules or of associated invariant chain.

In recent work from our laboratory, we attempted to address this question in normal murine B cell blasts by combining subcellular fractionation with biochemical analysis of the class II and Ii distribution in the various subcellular compartments. The different intracellular organelles were purified by density fractionation and MHC class II molecules were directly immunoprecipitated from these fractions after solubilization. The distribution of newly synthesized MHC class II molecules was followed over time by pulse-chase labeling analysis. In agreement with the results obtained in transfected COS cells, class II molecules colocalize with transferrin in a compartment with the characteristics of early endosomes (light density, transferrin-positive, β hexosaminidase-positive). Approximately one hour after synthesis, αβIi complexes access endocytic compartments with the characteristics of late endosomes (intermediate density, transferrin-negative, β hexosaminidase-positive) and of prelysosomes /lysosomes (high density, transferrin-negative, β hexosaminidase-positive, lamp 1-positive). Two to three hours after synthesis most of class II molecules, mainly lacking Ii, accumulate in the prelysosomal/lysosomal compartment.

To detect the site(s) of peptide loading we took advantage of the observation that many peptide-occupied MHC class II molecules adopt a compact dimer form that is stable in SDS-PAGE (sodium dodecyl sulphate polyacrylamide gel electrophoresis) under partial denaturing conditions (SDS and β-mercaptoethanol, without boiling) (15-17). By this criterion, at least in the murine B cell blasts examined, it was apparent that both the late endosomal and the prelysosomal /lysosomal compartments are involved in peptide loading of MHC class II molecules. Therefore, at least a subset lysosomes can be viewed as a functional compartment for peptide loading and not simply a site for terminal degradation of antigens and MHC class II molecules. This is supported by a recent study in which lysosomal fractions from monocytes fed with intact *Listeria* were demonstrated to contain MHC class II complexes stimulatory in a T cell presentation assay (32).

The ability of the MHC class II molecules to be loaded by peptides in different endocytic sub-compartments provides access of these molecules to the variety of peptides generated during the various stages of digestion of an antigenic protein. Such a system would lend economy and efficiency to antigen processing and presentation, whereas confining the loading event to a specific endosomal compartment would restrict the range of peptides available for binding. Based on these data, we suggest the following model: after passage through the trans-Golgi network, MHC class II molecules together with Ii first enter an early endocytic compartment, mix with incoming antigen, and are then further transported through the endocytic pathway along with the internalized antigens. As soon as Ii is detached, peptides available in that portion of the endocytic pathway can load, and such invariant chain -free, peptide-loaded class II molecules leave the endosomes for movement to the plasma membrane. The pathway that loaded MHC class II molecules follow to reach the plasma membrane from any of the endosomal loading compartments is still unclear.

To study the requirements for the formation of transport-competent MHC class II dimers in the endocytic pathway we tried to mimic in vitro the acidic environment found in endosomes (33). Lysates from pulse-labelled mouse spleen cells were exposed to acidic pH in the presence or absence of specific peptide antigen. After neutralization, class II molecules were immunoprecipitated and the acquisition of the stable compact dimer form was evaluated. Upon dissociation of Ii under acidic conditions, empty αβ complexes tend to associate with denatured cellular protein in aggregates. In the presence of specific peptides MHC class II molecules form stable compact dimers that do not associate with these aggregated proteins and hence are more likely to be in a state suitable for transport to the cell surface. These findings indicate that peptide loading is required for the acquisition of a stable configuration of MHC class II molecules in the endocytic pathway and suggest the existence of an editing mechanism that preferentially promotes cell surface expression of peptide-loaded MHC class II molecules that have dissociated from aggregates of denatured antigen. The parallel with class II folding reactions in the ER is evident: in both compartments (ER, endocytic pathway) retention is mediated by aggregation, while formation of transport-competent complexes depends on peptide binding site occupancy.

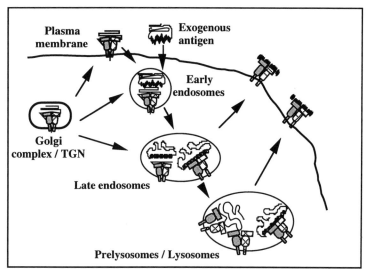

Figure 3: Model for the movement of class II and invariant chain through the endocytic pathway and for the acquisition of peptide

A prerequisite for class II peptide loading in the endocytic pathway is the dissociation of Ii. This dissociation is inhibited by leupeptin, a protease inhibitor that blocks the complete degradation of Ii (34, 35). The fragment of Ii produced in the presence of leupeptin (leupeptin induced fragment, LIP) remains associated with MHC class II molecules and does not allow peptide binding or cell surface expression of class II molecules (36). These findings suggest that Ii not only mediates sorting to the endocytic pathway, but also retention of class II at this site. An active or a passive mechanism could be responsible for MHC class II retention in the endocytic pathway: a) an active process could be mediated by the physical interaction of the cytoplasmic tail of invariant chain with a protein resident in the endocytic pathway; b) a passive process could reflect invariant chain masking of a sorting signal present in MHC class II molecules that mediates their localization on the cell surface. This sorting signal might only be active when MHC class II molecules are properly folded as loaded dimers.

Interestingly, in mouse, LIP is an N-terminal 9-10 KD fragment resembling the artificially truncated mIi1-107 invariant chain we studied in transfected COS cells. Thus, in normal antigen presenting cells a fragment of Ii similar to the minimal structure required for folding and assembly of class II in the ER is produced, and this fragment inhibits peptide binding to MHC class II molecules like the comparable mutant form analyzed in transfected cells. There is thus a reciprocal relationship between events in the ER and in endosomes: to exit the ER class II molecules need to interact with the 1-107 region of Ii; while to exit the endocytic pathway and to reach the plasma membrane MHC class II molecules need to dissociate from the analogous fragment, LIP.

CONCLUSION

MHC class II molecules evolved to bind antigenic fragments of proteins of pathogens resident in endosomes and to present them to T cells. To load antigenic peptides derived from the processing of endocytosed material, MHC class II molecules are transported from the biosynthetic compartment to the endocytic pathway. After loading they are then delivered to the cell surface. In the recent past, we have focused on determining whether there are any requirements for the formation of transport-competent class II dimers in the ER and in endosomes and whether there is any relationship between the generation of these two stable complexes. To answer this question we analyzed two different experimental models: transfected cells that provide a versatile and flexible system and allow a rapid evaluation of the data, and professional antigen presenting cells that represent physiological conditions.

The results obtained indicate that: a) in the ER Ii is necessary for the formation of transport-competent $\alpha\beta$ complexes, as shown by the ER-retention of MHC class II molecules as incompletely folded and assembled proteins in the cells of Ii-deficient animals; b) in an artificial endosomal environment established in vitro, peptides are critical for rescuing stable dimers from aggregation. When combined with the fact that Ii association inhibits class II peptide binding, these data suggested that there might be an association between peptide binding site occupancy and transport-competence of MHC class II molecules. To test this model we made deletions of the lumen of Ii, and analyzed their function. These experiments have revealed that a short internal segment of Ii mediates both its chaperon-like function and its ability to inhibit peptide binding to class II, showing the strict connection of these invariant chain tasks and strongly supporting our hypothesis.

The second issue that we examined concerns the distribution of MHC class II molecules along the endocytic pathway and the competence of distinct endocytic compartment for peptide loading. Based on data obtained in transfected fibroblasts and in splenic B cells we propose that early endosomes are the first endocytic compartment reached by $\alpha\beta$Ii complexes. In these vesicles the first encounter with antigens takes place. $\alpha\beta$Ii complexes and antigens then migrate together along the endocytic pathway. Depending on the complexity of the antigen and the affinity of Ii for MHC class II molecules, loaded dimers are formed in different compartments. This mechanism amplifies the possibilities for successful loading, providing an opportunity for different antigens and MHC class II molecules to find the suitable proteolytic and acidic environment for this process.

Several important questions remain to be addressed. In the early biosynthetic compartment, it is of particular interest to determine the sequence of events that leads to the formation of transport-competent $\alpha\beta$Ii complexes. A crucial issue to resolve concerns the effect of Ii trimerization on class II molecule fate. This peculiar structure could play a special role in the initial class II assembly process or it could be important in subsequential phases of the class II export pathway. To deepen our knowledge of the organization of the endocytic pathway of professional antigen presenting cells and of the characteristics of the compartments where the loading occurs, it is going to be essential to study the proteins that mediate intracellular transport. These proteins (rab, arf, SNAREs, etc.) are specific for certain subcellular compartments and the expression of dominant negative mutants can selectively abrogate the traffic between two successive compartments. Such studies will hopefully contribute to answering what signals mediate the localization of invariant chain and class II within the cell, how the cell arranges for effective mixing of antigen with class II molecules, and how loaded class II molecules reach the plasma membrane. Finally, additional attention needs to be paid to the details of how Ii is removed from the class II molecules, and how Ii-free class II binding sites find and stably bind to suitable determinants in antigens undergoing processing within the endocytic pathway.

ACKNOWLEDGEMENTS

Our thanks to E. Bikoff, E. Robertson, M. Marks, and J. Bonafacino for their collaboration in some of the work described here, and these individuals along with E. Long, P. Roche, and the members of the Lymphocyte Biology Section for many helpful discussions.

REFERENCES

1. **Germain, R. N.** 1986. The ins and outs of antigen processing and

presentation. *Nature 322:687.*

2. **Townsend, A., C. Öhlén, J. Bastin, H. G. Ljunggren, L. Foster and K. Kärre.** 1989. Association of class I major histocompatibility heavy and light chains induced by viral peptides. *Nature 340:443.*

3. **Degen, E., M. F. Cohendoyle and D. B. Williams.** 1992. Efficient dissociation of the p88 chaperone from major histocompatibility complex class-I molecules requires both β2-microglobulin and peptide. *J Exp Med 175:1653.*

4. **Roche, P. A., M. S. Marks and P. Cresswell.** 1991. Formation of a nine-subunit complex by HLA class II glycoproteins and the invariant chain. *Nature 354:392.*

5. **Claesson-Welch, L. and P. A. Peterson.** 1985. Implications of the invariant gamma-chain on the intracellular transport of class II histocompatibility antigens. *J Immunol 135:3551.*

6. **Layet, C. and R. N. Germain.** 1991. Invariant chain promotes egress of poorly expressed, haplotype-mismatched class II major histocompatibility complex AαAβ dimers from the endoplasmic reticulum/cis-Golgi compartment. *Proc Natl Acad Sci U S A 88:2346.*

7. **Anderson, M. S. and J. Miller.** 1992. Invariant chain can function as a chaperone protein for class II major histocompatibility complex molecules. *Proc Natl Acad Sci U S A 89:2282.*

8. **Bikoff, E. K., L.-Y. Huang, V. Episkopou, J. Van Meerwijk, R. N. Germain and E. J. Robertson.** 1993. Defective major histocompatibility complex class II assembly, transport, peptide acquisition, and CD4+ T cell selection in mice lacking invariant chain expression. *J Exp Med 177:1699.*

9. **Rose, J. K. and R. W. Doms.** 1988. Regulation of protein export from the endoplasmic reticulum. *Annu Rev Cell Biol 4:257.*

10. **Hurtley, S. M. and A. Helenius.** 1989. Protein oligomerization in the endoplasmic reticulum. *Annu Rev Cell Biol 5:277.*

11. **Bonnerot, C., M. S. Marks, P. Cosson, R. N. Germain and J. S. Bonifacino.** 1993. Association with BiP and aggregation of class II molecules synthesized in the absence of invariant chain. *submitted for publication*

12. **Schaiff, W. T., K. A. Hruska Jr., D. W. McCourt, M. Green and B. D. Schwartz.** 1992. HLA-DR associates with specific stress proteins and is retained in the endoplasmic reticulum in invariant chain negative

cells. *J Exp Med 176:657.*

13.　Roche, P. A. and P. Cresswell. 1990. Invariant chain association with HLA-DR molecules inhibits immunogenic peptide binding. *Nature 345:615.*

14.　Teyton, L., D. O'Sullivan, P. W. Dickson, V. Lotteau, A. Sette, P. Fink and P. A. Peterson. 1990. Invariant chain distinguishes between the exogenous and endogenous antigen presentation pathways. *Nature 348:39.*

15.　Sadegh-Nasseri, S. and R. N. Germain. 1991. A role for peptide in determining MHC class II structure. *Nature 353:167.*

16.　Germain, R. N. and L. R. Hendrix. 1991. MHC class II structure, occupancy and surface expression determined by post-endoplasmic reticulum antigen binding. *Nature 353:134.*

17.　Davidson, H. W., P. A. Reid, A. Lanzavecchia and C. Watts. 1991. Processed antigen binds to newly synthesized MHC class II molecules in antigen-specific B lymphocytes. *Cell 67:105.*

18.　Roche, P. A., C. L. Teletski, D. R. Karp, V. Pinet, O. Bakke and E. O. Long. 1992. Stable surface expression of invariant chain prevents peptide presentation by HLA-DR. *EMBO J. 11:2841.*

19.　Karp, D. R., C. L. Teletski, P. Scholl, R. Geha and E. O. Long. 1990. The α1 domain of the HLA-DR molecule is essential for high-affinity binding of the toxic shock syndrome toxin-1. *Nature 346:474.*

20.　Riberdy, J. M., J. R. Newcomb, M. J. Surman, J. A. Barbosa and P. Cresswell. 1992. HLA-DR molecules from an antigen-processing mutant cell line are associated with invariant chain peptides. *Nature 360:474.*

21.　Sette, A., S. Ceman, R. T. Kubo, K. Sakaguchi, E. Appella, D. F. Hunt, T. A. Davis, H. Michel, J. Shabanowitz, R. Rudersdorf, H. M. Grey and R. Demars. 1992. Invariant chain peptides in most HLA-DR molecules of an antigen-processing mutant. *Science 258:1801.*

22.　Chicz, R. M., R. G. Urban, J. C. Gorga, D. A. Vignali, W. S. Lane and J. L. Strominger. 1993. Specificity and promiscuity among naturally processed peptides bound to HLA-DR alleles. *J Exp Med 178:27.*

23.　Rudensky, A., P. Preston-Hurlburt, S. C. Hong, A. Barlow and C. A. Janeway, Jr. 1991. Sequence analysis of peptides bound to

MHC class II molecules. *Nature 353:622.*

24. **Hunt, D. F., R. A. Henderson, J. Shabanowitz, K. Sakaguchi, H. Michel, N. Sevilir, A. L. Cox, E. Appella and V. H. Engelhard.** 1992. Characterization of peptides bound to the class I MHC molecule HLA-A2.1 by mass spectrometry. *Science 255:1261.*

25. **Neefjes, J. J., V. Stollorz, P. J. Peters, H. J. Geuze and H. L. Ploegh.** 1990. The biosynthetic pathway of MHC class II but not class I molecules intersects the endocytic route. *Cell 61:171.*

26. **Lotteau, V., L. Teyton, A. Peleraux, T. Nilsson, L. Karlsson, S. L. Schmid, V. Quaranta and P. A. Peterson.** 1990. Intracellular transport of class II MHC molecules directed by invariant chain. *Nature 348:600.*

27. **Lamb, C. A., J. W. Yewdell, J. R. Bennink and P. Cresswell.** 1991. Invariant chain targets HLA class II molecules to acidic endosomes containing internalized influenza virus. *Proc Natl Acad Sci U S A 88:5998.*

28. **Romagnoli, P., C. Layet, J. Yewdell, O. Bakke and R. N. Germain.** 1993. Relationship between invariant chain expression and MHC Class II transport into early and late endocytic compartments. *J Exp Med 177:583.*

29. **Simonsen, A., F. Momburg, J. Drexler, G. J. Hämmerling and O. Bakke.** 1993. Intracellular distribution of the MHC class II molecules and the associated invariant chain (Ii) in different cell lines. *International Immunology 5:903.*

30. **Peters, P. J., J. J. Neefjes, V. Oorschot, H. L. Ploegh and H. J. Geuze.** 1991. Segregation of MHC class II molecules from MHC class I molecules in the Golgi complex for transport to lysosomal compartments. *Nature 349:669.*

31. **Guagliardi, L. E., B. Koppelman, J. S. Blum, M. S. Marks, P. Cresswell and F. M. Brodsky.** 1990. Co-localization of molecules involved in antigen processing and presentation in an early endocytic compartment. *Nature 343:133.*

32. **Harding, C. V. and H. J. Geuze.** 1993. Immunogenic peptides bind to class II MHC molecules in an early lysosomal compartment. *J Immunol 151:3988.*

33. **Germain, R. N. and A. G. Rinker Jr.** 1993. Peptide binding inhibits protein aggregation of invariant-chain free class II dimers and promotes

selective cell surface expression of occupied molecules. *Nature 363:725.*

34. Blum, J. S. and P. Cresswell. 1988. Role for intracellular proteases in the processing and transport of class II HLA antigens. *Proc Natl Acad Sci U S A 85:3975.*

35. Nguyen, Q. V., W. Knapp and R. E. Humphreys. 1989. Inhibition by leupeptin and antipain of the intracellular proteolysis of Ii. *Hum Immunol 24:153.*

36. Neefjes, J. J. and H. L. Ploegh. 1992. Inhibition of endosomal proteolytic activity by leupeptin blocks surface expression of MHC class II molecules and their conversion to SDS resistance alpha beta heterodimers in endosomes. *Embo J 11:411.*

9

INTRACELLULAR EVENTS REGULATING MHC CLASS II-RESTRICTED ANTIGEN PRESENTATION

Andrea J. Sant, Alexander Chervonsky, Marta Dykhuizen, John F. Katz, George E. Loss, Christopher Stebbins, and L.J.Tan

INTRODUCTION

Much has been learned in recent years regarding the intracellular events involved in class II restricted presentation of antigen (Ag). It is clear that many of the key events take place within acidic endocytic compartments of antigen presenting cells (APC). Exogenous Ags and some internally synthesized Ags enter these compartments and are degraded in a limited fashion into peptides by resident lysosomal proteases. MHC class II molecules have a specialized ability to access these compartments and to bind the derived peptides. The MHC class II:peptide complex is spared from degradation and transported to the cell surface where it can be recognized by CD4 positive T lymphocytes.

Despite tremendous efforts by many laboratories, there are a number of key questions regarding the intracellular events involved in MHC restricted antigen presentation that remain unanswered at the time. First, the intracellular trafficking pathway that leads to class II expression within endocytic compartments is not known with certainty. There is evidence for direct sorting from the trans golgi network into endosomal compartments (1) and also for a transient cell surface intermediate prior to class II expression within these compartments (2). Second, the structural elements within the class II multimeric complex that direct it into and ultimately allow egress from the endosomal compartments for final expression at the cell surface are poorly defined. Third, the identity and exact role of protein co-factors that participate in class II restricted antigen presentation are not understood. Finally, it is not known if there is a single, specialized compartment for Ag degradation and peptide binding to class II and if the same intracellular events and compartments are involved in endogenous and exogenous antigen presentation. In the following discussion, we will describe experimental approaches we have taken and results we have obtained regarding some of these issues.

MHC CLASS II LOCALIZATION IN THE ENDOCYTIC PATHWAY

MHC class II molecules have the unique ability to present peptides derived from exogenous antigen (Ag), an ability thought to reflect selective trafficking events that lead to localization of these molecules in the endocytic pathway of APC. Current data suggests that *en route* to the cell surface, the MHC class II molecules can divert from the constitutive transport pathway that guide most cell surface proteins to the plasma membrane (1). Unlike most cell surface proteins that are rapidly transported to the plasma membrane after terminal glycosylation, the MHC class II molecules are significantly delayed in their transport to the cell surface. During this delay, class II molecules are sensitive to internalized neuraminidase, indicating that they have been transported through the Golgi stacks and have access to endocytic compartments (3, 4). Thus, the post synthetic delay in the cell surface expression of class II molecules likely represents their diversion into the endocytic pathway.

Structural Control of Class II Sorting into the Endocytic Pathway

A major unanswered question at the present time is what structural features in the class II complex direct it into endocytic compartments where antigenic peptides are available. There are at least two possibilities such structures: sequences in the class II molecule itself (either the α or β chains) or sequences in the associated invariant chain glycoprotein (Ii). Studies employing transfection of Ii constructs in cells lacking class II have indicated that Ii localizes to endosomal compartments independently of class II (5, 6). Some experiments have failed to detect Class II molecules in the endocytic compartments when the molecule is synthesized in the absence of Ii (6), leading to the conclusion that it is Ii that bears the targeting sequence for localization of these molecules in the endocytic pathway. Some data however argue in favor of the idea that class II molecules contain some of the structural elements for targeting into the endocytic pathway. For example, it is known that Ii negative APC are able to present exogenous Ags to class II restricted T cells (7-10), suggesting that class II molecule gains access to the endocytic pathway independently of Ii. Also, several groups have demonstrated expression of class II in endosomal compartments in cells that lack constitutive Ii expression (11, 12), a phenotype that seems to be dependent on cell type. Collectively, these data argue that both class II and Ii contribute to the localization of the class II molecule within endosomal compartments and raise the question of whether Ii and class II have redundant signals for endocytic localization or whether they are able to facilitate expression of class II to these compartments by distinctive and complementary mechanisms.

Invariant Chain Retains Class II In The Endocytic Pathway. Recently, we have examined the possibility that the ability of Ii to enhance localization of class II in the endocytic pathway was due to an endosomal retention rather than an endosomal sorting function. Several groups (6, 13, 14), including our own have observed that endosomal localization of class II is enhanced by Ii, a biological activity apparently regulated by its cytosolic tail. Although these data have been interpreted as evidence that Ii contains a positive transport signal, directing the class II molecules to Ag-containing endosomes, an alternative possibility is that Ii serves to retain class II molecules within the endocytic compartments. This function of Ii is supported by the post-Golgi delay in class II transport to the cell surface observed in Ii-positive cells. According to this hypothesis, intact MHC/Ii complexes remain within a post-golgi compartment until encountering a proteolytic environment sufficient for Ii proteolysis and release from class II. We have examined this question using cells that express class II and differ only in their expression of Ii (15).

To test whether Ii bears an endosomal retention signal, we evaluated the consequences of persistent Ii association with class II. To determine whether failure to dissociate Ii from class II would lead to deposition of newly synthesized Ii/class II complexes at the cell surface or accumulation of these molecules within cells, fibroblasts transfected with genes encoding I-A^d and Ii were treated with lysosomotropic reagents and examined for changes in Ii surface staining. Lysosomotropic agents such as chloroquine and ammonium chloride have been shown to block the proteolytic dissociation of class II/Ii complexes (16). We found that long term treatment (48-72 hours) with these amines did not result in Ii expression at the cell surface, indicating that disruption of Ii proteolysis does allow accumulation of $\alpha\beta$/Ii complexes at the cell surface. Interestingly however, these studies did reveal that although cell surface Ii staining was unchanged by culture with lysosomotropic amines, class II surface expression was markedly decreased. This effect was seen in class II positive cells expressing genomic Ii or the p31 form of Ii. To determine whether the loss in surface class II induced by lysosomotropic reagents was dependent on Ii expression, we compared the effect of these reagents on cells that lacked Ii. In contrast to the effect on class II surface expression in Ii positive cells, such treatment had no effect on class II surface expression in cells that lacked Ii. While the inhibition of class II expression was Ii-dependent, it was possible that this dependence was not due to a retention event effected by Ii, but due to an Ii-mediated sorting of class II that was pH dependent. To circumvent the effects of lysosomotropic agents on events other than proteolysis, we examined the influence of the protease inhibitor leupeptin (LP) on class II transport. LP inhibits the proteolytic release of Ii from class II (17, 18) without interrupting pH-dependent cellular events (19). Ii-positive and negative transfectants were treated in parallel with LP and then examined for changes in surface MHC expression. LP treatment resulted in a striking decline in class II surface expression in the Ii-positive transfectant, arguing that the observed Ii-dependent decline in surface class II was due to the inability to dissociate Ii from class II rather than a non-specific effect of pH changes within the endosomal compartments. As had been observed for the lysosomotropic amines, no effect of LP was seen with Ii negative cells nor on class I expression, demonstrating the selectivity of these effects.

Kinetic studies indicated that the earliest time at which a decrease in class II expression was detectable was 36 hours, suggesting that the decline in cell surface class II does not result from interruption of a rapid MHC recycling pathway (20), but rather a time-dependent loss of pre-existing cell surface class II coupled with a failure to replace these molecules with newly synthesized class II molecules. Consistent with this hypothesis, we observed that coordinate with the loss in cell surface expression of class II was an increase in intracellular class II and Ii staining, which co-localized in peripheral and perinuclear vesicles. Interestingly, we also found that significant recovery of surface class II expression can be achieved within 2 hours of termination of treatment with lysosomotropic amines. These data suggest that Ii retains rather than irretrievably re-directs transport of the class II complex, supporting the conclusion that Ii dissociation is a prerequisite for class II surface expression and that Ii bears an endosomal retention signal.

Class II β Chains Express an Endosomal Localization Signal. The retention function of Ii described above is sufficient to account for all of the known ability of Ii to enhance endosomal localization of class II. This necessitates a re-investigation of the structural elements that control the initial sorting of class II to these compartments. The finding that APC lacking Ii are capable of exogenous Ag presentation is consistent with the idea that class II molecules contain signals that direct its entry into the endocytic pathway. We have investigated this possibility by a mutagenesis and gene transfer system (21,22).

To identify putative trafficking signals within the class II molecule, we have focused on a particular region of the class II β chain (a. a. 80-84) because a chemically induced mutant at position 82 in the β chain had been shown to be arrested late in biosynthesis (23). Strikingly, this residue falls within a highly conserved region of both human and mouse class II, which is not present in class I, a molecule thought not to access the endosomal compartments. These characteristics suggested to us that this segment might control its selective transport to the endosomal compartments of antigen presenting cells. To investigate this possibility, site directed mutagenesis was used to conservatively mutate this region of the β chain, to alter, rather than totally block, transport of the β dimer. The substitutions chosen were based on Dayhoff's analysis of naturally occurring sequence variants (24), which would be least likely to disrupt biological function. Described below are the phenotypes of three of these mutant class II molecules.

The relative ability of the mutant and wild type β chain to be co-expressed with Aα^d at the cell surface was first evaluated using transfection systems and flow cytometry. These studies have indicated that the mutation at position 82, previously shown to block transport in B cell lines (23) also appears to cause arrest in class II transport in several of the non-B cell lines that we have analyzed, resulting in low or undetectable levels of class II at the cell surface. In contrast to alteration at position 82, mutations at position 80, 81, and 83 do not appear to block transport, despite the fact that these are highly conserved residues in both mouse and human class II β chains.

Intracellular distribution of wild type and mutant class II molecules. The transfectants bearing wild type Aα^d and either wild type or mutant class II β chains (denoted by the position of the amino acid change in β) were studied by intracellular immunofluorescent staining to determine the consequences of these mutations on class II trafficking. Discussed below are the results obtained with the mouse L cell line Ltk-, but essentially the same subcellular localization patterns have been obtained with several other cell lines.

Intracytoplasmic staining of class II in cells that lack Ii reveals a significant amount of diffuse, reticular ER staining pattern, some golgi staining and some vesicular staining. This pattern was not altered by mutation at position 83. Interestingly, however, mutation at position 80, 81 and 82 each led to a characteristic and distinctive intracellular localization phenotype. Mutation at position 81 led to a diminution in vesicular staining and more pronounced ER/Golgi staining. Class II molecules bearing mutation at position 82 in the β chain are trapped intracellularly in a perinuclear compartment, a staining pattern resembling that obtained for resident Golgi proteins (25-27). Two color immunofluorescence using anti-class II mAb and the lectin wheat germ agglutinin confirmed this as Golgi localization. This type of arrest in intracellular transport is unusual. Our work and the work of others have indicated that the most frequent site of retention of glycoproteins is within the ER, where unassembled components of multimeric complexes, misfolded or mutant glycoproteins are typically retained (28-30). These data suggest that the blockade in transport is not due to misfolding of the class II α and β chains but rather to an alteration in a structure necessary for successful egress from the golgi or trans golgi network. Strikingly, alteration at position 80 leads to a dramatic shift of class II from early compartments to predominantly endosomal/lysosomal staining, as indicated by the high number of discreet peripheral vesicles detected with anti-class II mAbs. These vesicles appear to be late endosomes based on the findings that 80m class II molecules co-localize with internalized ligands following 60 minutes but not 15 minutes of uptake and that the 80m class II molecules co-localize with the lysosomal membrane protein LAMP-1. Collectively, the analysis of these class II mutants have demonstrated that it is possible to positively or negatively modulate expression of class II in endocytic compartments in the absence of Ii by introducing single amino acid changes in the 80-82 segment of the class II β chain.

Effect of Ii co-expression on class II transport of wild type and mutant class II molecules. To examine whether Ii co-expression with mutant or wild type class II would modify their subcellular distribution, stable transfectants expressing wild type class II, 81m and 82m were supertransfected with the gene encoding the p31 form of Ii. Co-expression of Ii with the 81 mutant caused a dramatic change in the distribution of class II within these cells. While the 81m typically is detected as a diffuse reticular and golgi-like staining pattern, co-transfection of Ii resulted in a striking change to a highly vesicular staining pattern. These results indicate that Ii corrects or compensates for the defect in 81m. At the present time, it is unknown whether 81m traffics to the cell surface via the default exocytic pathway like MHC class I or whether it ordinarily traverses the endocytic pathway so rapidly that it is not detectable in these compartments at steady state by intracellular staining. In the former case, the presence of Ii may redirect transport of 81m into the endocytic pathway, while in the latter case, it would retain it in these compartments so that its steady state expression is increased. Of interest in this regard is that we have found that the cytoplasmic tail of Ii is not necessary for it to localize 81m in endosomal compartments. Because this truncated Ii has the chaperone function of wild type class II, facilitating its folding and egress from the ER, but lacks endosomal retention (31,32), this result suggests that one of the ways that Ii can enhance class II expression in the endocytic pathway is by facilitating its folding early in biosynthesis so that more class II is exported and is available for sorting in later Golgi compartments.

In contrast to the results obtained with the 81m, co-expression of Ii with the 82m did not cause any detectable change in its subcellular distribution. Cells that co-expressed class II and Ii now accumulate both molecules in the Golgi. These results indicate that Ii expression *per se* is insufficient to direct class II expression in the endocytic pathway, and that the correct sequences in β is necessary to allow Ii to enhance its localization in endocytic compartments.

Invariant chain is Expressed In The Endocytic Pathway Independently of MHC Class II. The preceding data on the sorting function of class II coupled with folding/retention function of Ii both reconciles much of the published data class II subcellular localization and supports the view that both molecules contribute in different ways to endosomal localization. However, one set of observations is inconsistent with the idea that Ii functions mainly by first enhancing folding of class II in early compartments and then retaining class II in the endosomes after sorting. Previous studies have shown that when Ii is transiently transfected into class II negative cells, it can be detected in endocytic compartments (5, 6). These results provide support for the idea that Ii expresses a dominant targeting signal for endosomal localization. To determine whether this expression could be visualized in stable transfectants expressing physiologic levels of Ii, we derived Ii p31 transfectants of Ltk- cells. As had been observed with human Ii in transient assays, murine Ii displays both ER staining and some vesicular staining. Co-localization of internalized ligands confirms these vesicular compartments as being related to endosomes or lysosomes, indicating that, murine Ii does localize within endosomal compartments in the absence of class II. Any model to describe the control of class II trafficking must thus account for the fact that both class II and Ii can be expressed in the endocytic pathway independently of each other.

In the absence of class II, Ii is very inefficiently transported out of the ER, as evidenced by a persistence in sensitivity to endoglysidase-H. This suggested to us that when Ii is expressed without class II, it may gain access to the lysosomal compartments by a mechanism that by-passes the Golgi. One such mechanism is autophagy, in which ER membranes sequester cytosolic material into vacuoles that ultimately fuse with lysosomal compartments (reviewed in

33,34). We have performed several types of experiments that argue in favor of the possibility that Ii can gain access to lysosomes by autophagy (35). First, degradation of immature Ii is sensitive to the lysosomotropic amine ammonium chloride, suggesting that the immature Ii is ultimately degraded by lysosomal proteases. Second, expression of Ii in endosomal vesicles is sensitive to 3-methyl-adenine, a hallmark of the process of autophagy. The endosomal expression of class II or Ii synthesized in the presence of class II is not blocked by 3-methyladenine. These findings suggest to us that the route by which Ii localizes to endosomes in the absence of class II may be distinct from that which occurs when the two molecules are synthesized together.

Based collectively on the data presented above, we hypothesize that the MHC class II molecule itself bears the targeting signal required for segregation of class II in the trans golgi network (TGN) and sorting to the endocytic pathway. This signal is controlled by the 80-82 region of the β chain. This structure in class II could directly control recognition by an as yet unidentified transport protein, or it may control multimerization of class II, and event that may be requisite for sorting from the TGN. Mutation at position 81 has resulted in a loss or diminution in this recognition, and so the class II molecule traffics to the cell surface by the constitutive exocytic pathway or rapidly dissociates in the endocytic pathway and is quickly expressed at the cell surface. In contrast, the mutation at position 80 results in a higher affinity interaction either with itself, in the self aggregation model, or a heterologous transporter protein. This leads to more efficient sorting into the endocytic pathway, or more persistance this compartment, which then leads to readily detectable co-localization with late endosomal markers. The mutation at position 82 could lead to formation of multimers that are segregated from the default pathway yet are not able to sort into endosomes, a phenotype that leads to accumulation of the 82 mutant in the Golgi complex.

With this model, we can conceive of at least three possible roles for Ii in class II sorting. The first is to facilitate class II folding in the ER so that the transport signal on class II is expressed correctly on a large fraction of the newly synthesized molecules. The "chaperone" function of Ii in the ER is well documented (36-38). Secondly, Ii may serve to retain class II in the endocytic pathway. Our data involving the use of lysosomotropic amines and protease inhibitors to block Ii dissociation support this as at least one dominant activity of this glycoprotein. Finally, Ii may direct class II into the endocytic pathway by a mechanism that is distinct from that controlled by the β chain residues that we have mapped, possibly by allowing for translocation to the plasma membrane and then rapid internalization (2). Testing of this model of class II transport and the potential functions of Ii in the intracellular sorting of the class II molecule provides a framework for future experiments.

ENDOGENOUS ANTIGEN PRESENTATION BY MHC CLASS II MOLECULES

A second major interest in this laboratory is to delineate the events regulating endogenous antigen presentation by MHC class II molecules. Although class II is specialized to sample extracellular Ags which enter the cell by endocytosis, it has become clear that class II can also efficiently present peptides derived from proteins synthesized within the APC. Of particular interest in this regard is the analyses of peptides eluted from class II MHC molecules that has revealed a preponderance of peptides derived from internally synthesized proteins (39-41). This observation suggests that in the absence of receptor-mediated uptake internally synthesized proteins may be the principal source of class II peptides. Although the extent to which class II displays peptides derived from internally synthesized proteins is becoming increasingly apparent, the mechanisms governing their presentation are poorly understood at the present time.

The Model Endogenous Antigen H-2L^d Reveals a Novel Presentation Pathway

To examine the regulation of class II presentation of an intracellular Ag, we have developed an experimental system in which a peptide processed from an MHC class I molecule (comprised of the segment from amino acids 61-85 of H-2L^d) is displayed by the murine class II molecule I-A^d. The L^d peptide/I-A^d complex is recognized by a T cell hybridoma derived from the L^d-negative, dm2 mouse. A hybridoma ("GTH") was derived to lessen the effect of accessory molecules on T cell activation, thus allowing a more direct method to measure surface MHC/peptide complexes. Because the target Ag (MHC class I) is constitutively expressed by a variety of nucleated cells, we are able to study the processing requirements of a normally expressed protein in cells of different lineages. Furthermore, T cells recognizing this epitope are stimulated by unmanipulated BALB/c splenocytes indicating that this peptide/class II complex is expressed *in vivo*. This system affords us the opportunity to examine presentation of a typical ubiquitous Ag which, under physiologic levels of expression, gains access to class II. Using this model we have made a number of unexpected observations (42), which we present below.

Importance Of Subcellular Localization of the Target Antigen For Class II Restricted Presentation. We anticipated that class II molecules might preferentially display proteins that are expressed in locations that give them ready access to the classical endocytic compartments. Such proteins would include rapidly turned-over cell surface proteins that are degraded by lysosomal proteases and also resident endosomal/lysosomal proteins. However, several types of experiments indicate that cell surface expression of H-2L^d was not required for presentation by I-A^d. First, a secreted form of L^d was tested by utilizing a chimeric L^dQ10 hybrid construct (43). This construct encodes a protein in which the α1 and α2 domains are derived from H-$2L^d$ (thus providing the 61-85 peptide epitope) while the α3 domain is contributed by the normally secreted class I molecule H-2Q10. When transfected into APC bearing the class II restriction element, this form of L^d was able to sensitize cells for recognition by the GTH hybridoma. This presentation event did not appear to involve secretion and re-uptake, in that cells expressing the class II restriction element could not be sensitized for recognition by co-culture with supernatant from cells expressing L^dQ10 or by co-culture with L^dQ10 producing cells. Moreover, even when purified soluble L^dQ10 was used for sensitization at levels 1000 fold greater than produced by the transfectants, I-A^d positive cells did not generate the appropriate T cell epitope of L^d/A^d. However, the soluble form of L^d was shown to be capable of sensitizing APC for recognition by T cells specific for other epitopes. The failure to generate the epitope seen by GTH by soluble L^d indicates that not only is secretion and re-uptake of L^d not required for presentation, but that the soluble form of L^dQ10 cannot be readily used for presentation. These data suggest that L^dQ10 gains access to class II *en route* to secretion, during its transport through normal biosynthetic compartments.

Another line of data that argues against the importance of a cell surface intermediate for endogenous antigen presentation by class II involves the utilization of a cell line deficient in TAP gene expression (44), EE2H3. When transfected with I-A^d and H-$2L^d$, there was little or no class I expression detectable at the cell surface. Despite the lack of cell surface expression of H-$2L^d$, the cell efficiently generated the epitope recognized by the T cell hybridoma. Finally, attempts to modulate L^d and $L^d/Q10$ presentation using exogenously added anti-L^d mAb were unsuccessful. Altogether, these data indicate that a cell-surface intermediate is not required for generating this endogenous antigenic determinant and that the compartment where processing of L^d occurs is not readily accessed by mAb added to the extracellular medium. Thus, there is little evidence for a cell surface or secreted intermediate in L^d processing for class II restricted presentation and it appears that that this internally synthesized Ag may be processed early in its biosynthesis. This processing pathway is therefore distinct from the previously described classical exogenous class II presentation pathway.

Endogenous Antigen Presentation Segregates From Exogenous Antigen Presentation in Some Cell Lines. To further examine whether class II presentation of L^d occurs by pathway distinct from the classical endosomal pathway of antigen presentation, we examined cell lines defective at exogenous antigen presentation to present endogenous L^d. The T cell lymphoma, EL-4, when transfected with A^d, fails to process Ag for recognition by Ag specific T cells (42). It is, however, competent to stimulate these T cells when antigen processing was by-passed by providing peptide, arguing that the defect lies with processing, rather than with accessory molecules needed for T cell interactions. When this cell was transfected with the gene encoding H-2L^d, it was found to present endogenous L^d as effectively as control L-cells. . Through cell surface staining, we established that EL-4 cells expressed levels of I-A^d and L^d comparable to control L cells. A series of mAb blocking experiments utilizing both exogenously provided L^d peptide or the internally derived Ag allowed us to conclude that the EL-4 cells process and presents endogenous L^d with an efficiency similar to that of L-cells. These experiments therefore strongly suggest that different class II antigen presentation pathways exist, and that factors critical for the presentation of exogenous Ag by EL-4 cells are not critical for efficient presentation of endogenous L^d.

Although the data obtained using EL-4 suggests that different cofactors may be required for exogenous and endogenous antigen presentation, it was only after examining the presentation characteristics of βTC3-AD, a pancreatic islet cell line, that we established that the inverse is also true. βTC3-AD displays a class II presentation phenotype which is the reciprocal of that displayed by EL-4. This islet cell line can present exogenous Ag to I-A^d-restricted T cell hybridomas, but is incapable of presenting endogenous L^d to GTH. Using exogenously added synthetic L^d peptide to look for shifts in the peptide dose-response curve induced by L^d synthesis within the islet cell, we detected no endogenously derived L^d peptide/MHC complexes on the cell surface. These data establish that it is insufficient for the APC to merely express the target Ag and restriction element to achieve effective presentation of endogenous L^d. In conjunction with the results obtained using EL-4, it is likely that different cofactors regulate exogenous and endogenous class II antigen presentation and that these cofactors may be regulated in a tissue-specific way.

Presentation of Ld Does Not Occur by the Class I Presentation Pathway. The preceding data support our hypothesis that the endogenously derived L^d peptide recognized by GTH is not generated by the exogenous antigen presentation pathway. We next sought to determine whether or not this epitope is generated by the other known antigen presentation pathway, the endogenous class I pathway. If true, three experimental results would be expected: (a) a cytosolic form of L^d should stimulate GTH; (b) a cell line deficient in class I antigen presentation should not present the L^d target peptide and (c) expression of Ii should block binding of target L^d peptide to I-A^d and, as a result, inhibit stimulation of the L^d/A^d specific T cell. The results obtained fulfilled none of these predictions. First, we found that a cytosolic form of L^d (lacking the leader sequence) was not presented to GTH. Attempts to detect low levels of surface MHC/peptide complexes by evaluating peptide dose response curves revealed no detectable endogenously generated complexes recognized by GTH. Second, as discussed in the previous section, when EE2H3 was tested, a cell line that is deficient in TAP-1 gene expression and thus its class I presentation pathway (44), we were able to demonstrate effective class II restricted presentation of L^d. This suggests that a functional TAP-1 protein and an intact class I presentation pathway are not required for class II restricted presentation of the L^d peptide. Finally, we examined the ability of Ii to block antigen presentation of the L^d peptide recognized by GTH. Current evidence suggests that one function of Ii is to block peptide binding to class II prior to Ii dissociation in the endocytic pathway (45,46). If the endogenous L^d peptide binds to MHC class II at a point proximal to the endocytic compartment, we predicted that cells which express saturating levels of Ii should display diminished presentation of the L^d peptide. Using both genomic and p31 Ii transfectants and several different cell lines expressing A^d and L^d, we have been unable to demonstrate an inhibitory effect of Ii on the ability of L^d-positive APC to stimulate GTH. This suggests that either the L^d peptide fragment effectively competes with Ii for class II binding sites or the L^d peptide is generated early in biosynthesis but associates with class II only after Ii dissociation. This latter model requires that the free peptide gain access to the typical endosomal compartments at levels sufficient to bind to the now available peptide binding site of class II. To discriminate among these possibilities, experiments are currently in progress defining the importance of the endocytic pathway in L^d peptide presentation and the role of early biosynthetic compartments in formation of this and other peptide:class II peptide complexes.

Collectively, our experiments utilizing H-$2L^d$ as a model internal Ag presented by MHC class II have revealed a novel presentation pathway that is distinct from the classical class II and classical class I pathways. Much of our data argues that one key feature that distinguishes the L^d pathway from these other two pathways is the site of Ag proteolysis. A cytosolic proteolytic complex appears not to be utilized based on the failure of the soluble, cytosolic form of L^d to sensitize for recognition. Similarly, an endosomal proteolysis event appears to be unlikely based on the failure of an extracellular, internalized form of the Ag to sensitize for recognition.

Our data is most consistent with the hypothesis that early biosynthetic compartments are the site of L^d degradation, consistent with its relatively poor export from, and consequently relatively high expression within these compartments (47). If ER or Golgi proteases are indeed responsible for L^d degradation, then it is possible that A^d:L^d complex formation takes place within these compartments. Experiments to test the importance of acidic compartments for peptide generation or binding, such as treatment with lysosomotropic amines, have not yet yielded unequivocal results. Currently we are investigating the possibility that the L^d peptide is generated and binds to class II early in biosynthesis by testing the ability of class II isolated from the early compartments of biosynthesis to stimulate the GTH hybridoma.

Endogenous Antigen Presentation and Allorecognition

Our experiments studying exogenous and endogenous Ag presentation by MHC class II using model Ags suggest that the intracellular events regulating peptide:MHC class II expression does not depend exclusively on cellular machinery that cells require to survive, such as endocytosis and acid dependent proteases. Rather, it appears that some of the components that mediate the function of class II are differentially expressed in different APC and may be selectively important for exogenous or endogenous antigen presentation.

The finding that MHC class II restricted antigen presentation depends on proteins of limited tissue expression raises the possibility that highly differentiated cells may express unique sets of self peptides in association with the MHC class II molecule, an idea that now has significant experimental support (48-51). Under the premise that some recognition of class II alloantigens is peptide dependent, we predicted that alloreactive T cells raised against a given cell type would yield T cells specific for its self peptides. A panel of self peptide-specific T cells could thus be used as probes for MHC-associated peptides on cells of different lineages. An inability of a T cell to recognize a given lineage APC could reflect the APC's lack of a specific self peptide. To test this possibility, we generated a panel of 25 I-A^d restricted, alloreactive T cell hybridomas. Upon screening the these T cells on a large (10 member) panel of established cell lines representing different lineages, (ie embryonic stem cells, B and T cells, thymic epithelial cells, pancreatic islet cells) we found that the majority (85%) of the T cells displayed selective reactivity with subsets of the APC (52). Interestingly, their reactivity can be grouped into a hierarchal pattern in which some APC stimulate the majority of the panel, while others stimulate only a few. In addition, a number of cases of reciprocal selective reactivity were observed, in which different T cells reacted reciprocally on a set of APC.

A hierarchical pattern of reactivity could reflect the allorecognition of different cell type-specific self peptides or could reflect a hierarchy of other differences (i.e. adhesion and accessory molecules, signaling machinery, etc.) among allogeneic targets and individual hybridomas. To assess these latter differences, we have measured the alloreactive T cell hybridomas' T cell receptor "affinities" for specific class II alloantigen, ability to produce lymphokines, and the expression of known adhesion/accessory molecules important for T cell recognition. We also measured the expression of T cell recognition molecules on the target cells, and the ability of the allogeneic target cells to stimulate Ag specific T cell hybridomas. Although these experiments revealed differences among APC and hybridomas, the differences did not resemble the hierarchal reactivity pattern.

These results suggest first that a significant fraction of allorecognition of MHC class II may be peptide specific, and second, that different APC types display distinct sets of peptides. We are currently trying to directly test these hypotheses by peptide sensitization experiments. If these studies demonstrate that the failure to stimulate alloreactive T cells is accounted for by differences in the peptides presented by the different APC, it will suggest an extensive degree of microheterogeneity in the MHC class II molecules expressed by different cell types. Clearly, if these differences in peptide repertoire extend to naturally occurring APC *in vivo*, then they could significantly influence T cell education, tolerance, transplantation responses and the ability of different APC to cooperate in the response to pathogenic organisms.

FUTURE DIRECTIONS

Our current experiments on endogenous antigen presentation by class II are focussed on several issues. We are exploring the possibility that the endocytic pathway may not be uniformly required for endogenous antigen presentation as it is for exogenous antigen presentation. Molecular techniques are being used to sequester the internal antigen and the class II restriction element to early compartments of biosynthesis so that we can evaluate whether self peptide: class II complexes form in these early biosynthetic sites. In addition, we want to identify the molecular mechanisms underlying cell dependent antigen presentation and the proteins that regulate these events.

Biochemical and cell biology techniques are being used to analyze the dynamics of class II trafficking and the sequence of events that takes place when class II is sorted into, retained within and ultimately released from endosomal compartments. We are developing methods to selectivity isolate endosomal vesicles, so that the biosynthetic source of class II and Ii in these compartments can be determined and the proteins responsible for their localization here can be identified. In addition, we are developing assays that will allow us to examine the fate of empty vs peptide-bound class II. Ultimately, these experiments should provide considerable insight into the mechanisms that regulate the ability of class II to access both foreign and self antigens and display them for recognition by T lymphocytes.

REFERENCES

1. **Peters, P., J. Neefjes, V. Oorschot, H. Ploegh and H. Geuze.** 1991. Segregation of MHC class II molecules from MHC class I molecules in the Golgi complex for transport to lysosomal compartments. *Nature. 349*: 669.

2. **Roche, P., C. Teletski, E. Stang, O. Bakke and E. Long.** 1993. Cell surface HLA-DR-invariant chain complexes are targeted to endosomes by rapid internalization. *Proc. Natl. Acad. Sci. (USA). 90*:8581.

3. **Cresswell, P.** 1985. Intracellular class II HLA antigens are accessible to transferrin-neuraminidase conjugates internalized by receptor mediated endocytosis. *Proc. Natl. Acad. Sci. (USA). 82*:8188.

4. **Neefjes, J., V. Stollorz, P. Peters, H. Geuze and H. Ploegh.** 1990. The biosynthetic pathway of MHC class II but not class I molecules intersects the endocytic route. *Cell. 61*: 171.

5. **Bakke, O. and B. Dobberstein.** 1990. MHC class II-associated invariant chain contains a sorting signal for endosomal compartments. *Cell.*

    63: 707.

6. **Lotteau, V., L. Teyton, A. Peleraux, T. Nilsson, Karlsson, S. Schmid, V. Quaranta and P. Peterson.** 1990. Intracellular transport of class II MHC molecules directed by invariant chain. *Nature. 348*:600.

7. **Sekaly, R., S. Jacobson, J. Richert, C. Tonnelle, H. McFarland and E. Long.** 1988. Antigen presentation to HLA Class II-restricted measles virus-specific T cell clones can occur in the absence of the invariant chain. *Proc. Natl. Acad. Sci. (USA). 85*:1209.

8. **Hammerling, G. and J. Moreno.** 1990. The function of the invariant chain in antigen presentation by MHC class II molecules. *Immunol. Today. 11*:337.

9. **Nadimi, C., J. Moreno, F. Momburg, A. Heuser, S. Fuchs, L. Adorini and G. Hammerling.** 1991. Antigen presentation of hen egg-white lysozyme but not of ribonuclease A is augmented by the major histocompatibility complex class II-associated invariant chain. *Eur. J. Immunol. 21*:1255.

10. **Peterson, M. and J. Miller.** 1992. Antigen presentation enhanced by the alternatively spliced invariant chain gene product p41. *Nature. 357*:596.

11. **Simonsen, A., F. Momburg, J. Drexler, G. Hammerling and O. Bakke.** 1993. Intracellular distribution of the MHC class II molecules and the associated invariant chain in different cell lines. *Int. Immunol. 5*: 903.

12. **Salamero, J., M. Humbert, P. Cosson and J. Davoust.** 1990. Mouse B lymphocyte specific endocytosis and recycling of MHC class II molecules. *EMBO J. 9*:3489.

13. **Lamb, C., J. Yewdell, J. Bennink and P. Cresswell.** 1991. Invariant chain targets HLA class II molecules to acidic endosomes containing internalized influenza virus. *Proc. Natl. Acad. Sci. (USA). 88*:5998.

14. **Romagnoli, P., C. Layet, J. Yewdell, O. Bakke and R. N. Germain.** 1993. Relationship between invariant chain expression and major histocompatibility complex class II transport in to early and late endocytic compartments. *J. Exp. Med. 177*:583.

15. **Loss, G. E. and A. J. Sant.** 1993. Invariant chain retains MHC Class II molecules in the endocytic pathway. *J. Immunol. 150*:3187.

16. **Nowell, J. and V. Quaranta.** 1985. Chloroquine affects biosynthesis of Ia molecules by inhibiting dissociation of invariant g chains from ab dimers in B cells. *J. Exp. Med. 162*:1371.

17. **Blum, J. and P. Cresswell.** 1988. Role for intracellular proteases in the processing and transport of class II HLA antigens. *Proc. Natl. Acad, Sci. (USA) 85*:3975.

18. **Nguyen, Q. V., W. Knapp and R. E. Humphreys.** 1989. Inhibition by leupeptin and antipain of the intracellular proteolysis of Ii. *Human Immun. 24*: 153.

19. **Seglen, P. O.** 1983. Inhibitors of lysosomal function. *Methods Enzymol. 96*:737.

20. **Reid, P. A. and C. Watts.** 1990. Cycling of cell surface MHC glycoproteins cycle through primaquine sensitive intracellular compartments. *Nature. 346*:655.

21. **Chervonsky, A., L. Gordon, and A. Sant.** 1993. A segment of the MHC Class II β chain modulates expression of MHC Class II molecules in the endocytic pathway. *Manuscript submitted for publication.*

22. **Chervonsky, A., L. J. Tan, and A. J. Sant.** A lumenal sequence in the Class II MHC β chain controls targeting of Class II molecules into endocytic compartments of antigen presenting cells. *Manuscript submitted for publication.*

23. **Griffith, I. J., N. Nabavi, Z. Ghogawala, C. G. Chase, M. Rodriguez, D. J. McKean and L. H. Glimcher.** 1988. Structural mutation affecting intracellular transport and cell surface expression of murine class II molecules. *J. Exp. Med. 167*:541.

24. **Dayhoff, M. O., R. M. Schwartz and B. C. Orcutt.** "A model of evolutionary change in proteins." Atlas of Protein Sequence and Structure. 1978.

25. **Reaves, B., O. Wilde and G. Banting.** 1992. Identification, molecular characterization and immunolocalization of an isoform of the *trans*-Golgi-network (TGN)-specific integral membrane protein TGN38. *Biochem. J. 283*:313.

26. **Colley, K. J., E. U. Lee and J. C. Paulson.** 1992. The signal anchor and stem regions of the bβ-galactosidase a2,6-sialyltransferase may each act to localize the enzyme to the Golgi apparatus. *J. Biol. Chem. 267*: 7784.

27. **Waters, M. G., D. O. Clary and J. E. Rothman.** 1992. A novel 115-kD peripheral membrane protein is required for intercisternal transport in the Golgi stack. *J. Cell Biol. 118*:1015.

28. **Gething, M. J., K. McCammon and J. Sambrook.** 1986. Expression of wild-type and mutant forms of Influenza Hemagglutinin: the role of folding in intracellular transport. *Cell. 46:*939.

29. **Pelham, H. R. B.** 1989. Control of protein exit from the endoplasmic reticulum. *Annu. Rev. Cell Biol. 5*:1.

30. **Sant, A., L. Hendrix, J. Coligan, W. Maloy and R. Germain.** 1991. Defective Intracellular Transport as a Common Mechanism Limiting Expression of Inappropriately Paired Class II Major Histocompatibility Complex α/β Chains. *J. Exp. Med. 174*:799.

31. **Anderson M., K. Swier, L. Arneson and J. Miller.** 1993. Enhanced Antigen Presentation in the Absence of the Invariant Chain Endosomal Localization Signal. *J. Exp. Med.*, In Press.

32. **Tan, L. J., J. Miller, A. Chervonsky and A.J. Sant.** 1993. Invariant chain can facilitate localization of MHC Class II molecules in endocytic compartments in the absence of its cytoplasmic tail. *Manuscript in preparation.*

33. **Seglen, P.O. and P.B. Gordon.** 1982. 3-Methyladenine: a specific inhibitor of autophagic/lysosomal protein degradation in isolated rat hepatocytes. *Proc. Natl. Acad. Sci. (USA). 79*:1889.

34. **Seglen, P. O. and P. Bohley.** 1992. Autophagy and other vacuolar protein degradation mechanisms. *Experientia 48*:158.

35. **Chervonsky, A. and A. Sant.** Invariant chain can localize to endocytic compartments by an autophagic mechanism. *Manuscript in preparation.*

36. **Layet, C. and R. Germain.** 1991. Invariant chain promotes egress of poorly expressed, haplotype-mismatched class II major histocompatibility complex AαAβ dimers from the endoplasmic reticulum/ cis-Golgi compartment. *Proc. Natl. Acad. Sci. (USA) 88*:2346.

37. **Schaiff, T., K. Hruska, C. Bono, S. Shuman and B. Schwartz.** 1991. Invariant Chain Influences Post-Translational Processing of HLA-DR Molecules. *J. Immunol. 147*:603.

38. **Anderson, M. and J. Miller.** 1992. Invariant chain can function as a chaperone protein for class II major histocompatibility complex molecules. *Proc. Natl. Acad. Sci. (USA). 89*:2282.

39. **Rudensky, A., P. Preston-Hurlburt, S.-C. Hong, A. Barlow and C. A. Janeway.** 1991. Sequence analysis of peptides bound to

MHC class II molecules. *Nature. 353*:622.

40. **Hunt, D., H. Michel, T. Dickinson, J. Shabanowitz, A. Cox, K. Sakaguchi, E. Appella, H. Grey and A. Sette.** 1992. Peptides Presented to the Immune System by the Murine Class II Major Histocompatibility Complex Molecule I-A[d]. *Science. 256*:1817.

41. **Chicz, R., R. Urban, W. Lane, J. Gorga, L. Stern, D. Vignali and J. Strominger.** 1992. Predominant naturally processed peptides bound to HLA-DR1 are derived from MHC-related molecules and are heterogeneous in size. *Nature. 358*:764.

42. **Loss, G. E., C. E. Elias, P. E. Fields, R. K. Ribaudo, M. McKisic and A. J. Sant.** 1993. MHC Class II-restricted presentation of an internally synthesized antigen displays cell-type variability and segregates from the exogenous Class II and endogenous Class I antigen presentation pathways. *J. Exp. Med. 178*:73.

43. **Corr, M., L. Boyd and D. Margulies.** 1992. Isolation and characterization of self peptides eluted from engineered soluble major histocompatibility antigens H-2Ld and H-2Dd. *J. Cell Biol. 101*:725.

44. **Bikoff, E. K., L. Jaffe, R. K. Ribaudo, G. R. Otten, R. N. Germain and E. J. Roberts.** 1991. MHC Class I expression in embryo-derived cell line inducible with peptide or interferon. *Nature. 354*:506.

45. **Roche, P. and P. Cresswell.** 1990. Invariant chain association with HLA-DR molecules inhibits immunogenic peptide binding. *Nature. 345*:615.

46. **Teyton, L., D. O'Sullivan, P. Dickson, V. Lotteau, A. Sette, P. Fink and P. Peterson.** 1990. Invariant chain distinguishes between the exogenous and endogenous antigen presentation pathways. *Nature. 348*: 39.

47. **Smith, J. D., W.-R. Lie, J. Gorka, C. S. Kindle, N. B. Meyers and T. H. Hansen.** 1992. Disparate interaction of peptide ligand with nascent versus mature class I Major Histocompatibility Complex molecules: Comparisons of peptide binding to alternative forms of L[d] in cell lysates and the cell surface. *J. Exp.Med. 175*:191.

48. **Marrack, P. and J. Kappler.** 1988. T cells can distinguish between allogeneic MHC products on different cell types. Nature. 332:840.

49. **Hogan, K., C. Clayberger, E. Berhard, S. Walk, J. Ridge, P. Parham, A. Krensky and V. Engelhard.** 1989. A panel of unique HLA-A2 mutant molecules define epitopes recognized by HLA-A2-specific antibodies and cytotoxic T lymphocytes. *J. Immunol. 142*:2097.

50. **Heath, W., M. Hurd, F. Carbone and L. Sherman.** 1989. Peptide dependent recognition of H-2K^b by alloreactive cytotoxic T lymphocytes. *Nature. 341*:749.

51. **Bonomo, A. and P. Matzinger.** 1992. Thymus epithelium induces tissue-specific tolerance. *J. Exp. Med. 177*:1153.

52. **Katz, J. F. and A. J. Sant.** 1993. T cell receptor recognition of MHC Class II alloantigens is highly cell type dependent. *J. Immunol.* In press

10

REGULATION OF ANTIGEN PROCESSING AND PRESENTATION IN B LYMPHOCYTES

Susan K. Pierce, Anne E. Faassen, Yin Qiu, Xiao Xing Xu,
Peter Schafer, and David Dalke

Helper T cell-dependent B cell activation requires that B cells process and present antigen. It is now firmly established that B cells are able to internalize antigen, degrade it and assemble complexes containing processed antigen bound to MHC class II molecules for cell surface display (1). Although B cells as antigen presenting cells (APC) are able to assemble and display processed antigen-class II complexes, B cells may not express all costimulatory molecules necessary for activation of naive resting T cells and thus, may be restricted *in vivo* to interacting with T cells which have been activated by other more potent APC such as macrophages or dendritic cells (2). Indeed, it is speculated that B cells, failing to express costimulatory molecules, may function *in vivo* to tolerize naive T cells. However, regardless of the possible ramifications of the interactions of B cells with naive T cells, one cannot lose sight of the fact that B cells are able to process antigen and that this function is essential for B cell interactions with helper T cells in humoral immune responses.

Unlike other class II-expressing APC, B cells express specific receptors for antigen, and processing is initiated when antigen is bound to the surface Ig. The bound antigen is subsequently internalized for processing and presentation. Antigen-specific processing is critical for the ultimate stimulation of B cells to proliferate and to differentiate into antibody-secreting cells in response to T cell-dependent antigens. Indeed, Lanzavecchia (3) first demonstrated that antigen-specific B cells are far more efficient APC requiring 1/1000th the concentration of antigen to activate a helper T cell as compared to nonspecific B cells which rely on fluid phase pinocytosis to internalize antigen. The antigen-specific nature of B cell processing may well account for the antigen-specific activation of B cells in collaboration with helper T cells. Thus, under conditions of limiting concentrations of antigen, as is likely to occur *in vivo*, only antigen-specific B cells capture sufficient antigen for processing and presentation to helper T cells. In this way helper T cells are restricted to interacting only with antigen-specific B cells. The ability of B cells to activate T cells and in turn be activated is dependent on the B cell's ability to produce a sufficient number of processed antigen-class II complexes on the cell surface for a sufficient period of time allowing an antigen-specific T cell to find and interact with the B cell. Thus, minimally, both the number of complexes expressed and the duration of their expression are likely to be key factors in determining the outcome of B cell-T cell interactions and thus the outcome of B cell exposure to antigen. Given the central role for processed antigen-class II complexes in B cell responses, one might predict that B cells carefully control the production of processed antigen-class II complexes during immune responses. It may be accurate to view the APC function of B cells as a differentiated function and one over which the B cell exerts considerable control. This view would predict that cellular and molecular mechanisms which underlie processing in B cells may have uniquely evolved in B cells and that the B cell employs novel strategies to regulate the expression of processed antigen-class II complexes. Here we discuss a number of phenomena in B cell antigen processing which suggest that B cells do indeed regulate the processing and presentation of antigen. These observations indicate the direction of future studies to define the molecular basis of such regulation.

The Phenomenon of Ig-Mediated Antigen Processing in B Lymphocytes

All mature B cells appear to have the ability to process and present antigen (1). Initially believed to be a function only of activated B cells, it is now clear that resting cells act as APC (4,5). Also shown to function as APC are B cells of the CD-5 lineage (1) and naive as well as memory B cells, defined by the JIID marker (6). All B cells are also able to facilitate processing by binding antigen through the cell surface Ig. In this regard both the IgM and IgD cell surface receptors have been shown to be equally as effective (1). We (7) and Davidson, *et al.*

(8) followed the kinetics of the assembly of processed antigen-class II complexes directly using a biochemical assay for the binding of a fragment of a radiolabeled antigen with the class II molecules. In normal mouse splenic B cells we found that within 30-60 min of the internalization of antigen, processed antigen-class II complexes are detected and that maximal numbers of complexes are formed by 2 hr. The assembly of processed antigen-class II complexes does not require new protein synthesis and thus, presumably involves class II molecules existing at the time of antigen binding. Davidson, *et al.* (8) showed that class II already expressed on the cell surface was not involved.

A somewhat unexpected finding from the studies of the time course of antigen processing is that the processed antigen-class II complexes are disassembled or degraded relatively rapidly, within 4-6 hours, after formation (7). The rate of disassembly of the complexes is far more rapid than the turn over rate for the bulk of class II in B cells. Thus, the class II molecules which have bound peptides derived from antigen entering the B cell through surface Ig appear to have a different half life than the bulk of class II. Using a very similar biochemical assay for the assembly of processed antigen-class II complexes, Davidson, *et al.* observed that in EBV-transformed B lymphoblastoid cell lines, processed antigen-class II complexes are extremely long lived. The rapid turn over of complexes in normal splenic B cells and the longevity of the complexes in transformed cells suggest that the processed antigen-class II complexes are not inherently unstable and must require active degradation or disassembly mechanisms which are absent in transformed cells. Our studies so far indicate that in order to degrade or disassemble the complexes, splenic B cells must internalize the class II into an intracellular acidic compartment (7).

If one views the B cell APC phenomenon from the T cell's point of view, rather than a biochemical view, a similar picture emerges. When normal splenic B cells are incubated with antigen that binds to surface Ig in the form of a model antigen, cytochrome *c*, covalently coupled to antibodies specific for Fab (*c*-anti-Ig), and excess antigen is washed away, antigen presentation to T cells is evident within 1-2 hr. Presenting ability reaches a peak by 4-6 hrs and is gradually lost by 12 h (7). Thus, the kinetics of antigen presentation to T cells mirror that of the assembly process measured biochemically. An interesting picture emerges when B cells are continuously incubated with *c*-anti-Ig conjugates (Fig. 1). In this case, both the acquisition as well as the loss of complexes mirrors that for a single round of antigen processing. This is despite the fact that antigen is not depleted from the culture over the time course of the experiment and the amount of antigen added in culture is small (20 nM), far below what is needed to saturate the surface Ig. This is not the observation when B cells are continuously cultured in the presence of antigen which is not conjugated to anti-Ig and which, consequently, enters the B cell through fluid phase pinocytosis (Fig. 1). In this case presenting ability is acquired more or

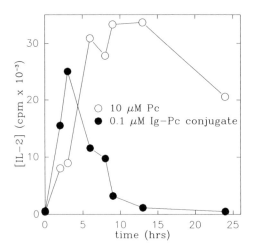

Figure 1. Fluid phase and receptor mediated antigen processing in B cells are not identical. B cells were incubated with the antigen cytochrome c alone (o) or cytochrome c covalently coupled to antibodies specific for Ig (•). At various times afterward the cells are washed, fixed with glutaraldehyde to block further processing and tested for their ability to activate an I-E^k-restricted, cytochrome c-specific T cell hybrid to secrete IL-2.

less in the same time frame but decreases only very gradually over time. The continued ability of B cells to present pinocytosed antigen appears to be due to the continued assembly of new processed antigen-class II complexes rather than an increased longevity of complexes once formed. Indeed, if B cells are washed free of antigen after becoming maximally stimulatory to T cells and new assembly is blocked by blocking acidic vesicle function, presenting ability rapidly decreases and is no longer measurable 1-2 hours later (data not shown). Thus, the half life of processed antigen-class II complexes is short. We conclude that the binding of antigen to the surface Ig both facilitates the processing of the bound antigen and precludes the subsequent processing of antigen. Thus, B cells apparently limit the display of processed antigen-class II complexes by an Ig-mediated mechanism.

What is the role of surface Ig in B cell antigen processing and how might antigen binding to surface Ig influence processing? The most dramatic effect of Ig-mediated antigen processing is a 1,000- to 10,000-fold reduction in the quantity of antigen required to produce the number of processed antigen-class II complexes necessary for T cell activation (9). This effect is observed with either bivalent or monovalent antigen binding to the surface Ig and is not unique to Ig in that antigen artificially bound to a number of different cell surface proteins, including the class II and class I molecules, has a similar effect (9,10). This enhancement of processing is presumably due to the mechanical concentration and transport of antigen to processing compartments. However, not all cell surface receptors have access to processing compartments suggesting that the molecules which can facilitate processing are a special class (11). Recent studies show that the Ig-associated $\alpha\beta$ sheath is required for Ig-mediated antigen processing (12) and suggest that the sheath acts to stimulate internalization of the antigen. Other receptors may have similar requirements for facilitating processing. However, antigen binding to surface Ig also has an effect on processing independent of the internalization of bound antigen. Indeed, treatment of B cells with Fab-specific antibodies increases APC function 10 to 50 fold (13). Given that Ig is a signal transducing receptor it is reasonable to conclude that Ig has the potential to signal for enhanced processing. Whether such signalling might account for the phenomenon described above awaits a better delineation of the discrete steps in the pathway.

Another interesting aspect of the presentation of antigen in B cells brought out by the use of a biochemical assay to detect processed antigen-class II complexes is that only approximately 0.2% of the class II molecules in B cells assemble with antigen (14). This small number of complexes is sufficient for T cell activation (15) and appears to be the maximal number of complexes B cells can produce. It is somewhat curious that B cells fill such a small percent of their class II with antigen. Presumably the antigen bound to the surface Ig is the only relevant antigen to a B cell and yet despite this very few complexes are filled with the peptides derived from this antigen. This observation raises several questions concerning the function of the remaining, majority of class II and the factors which limit the number of complexes which are filled. The class II in B cells which does not contain internalized antigen bound to Ig is presumably filled with a variety of peptides derived from the endocytic pathway or from exogenous proteins as has been shown for B cell tumors (16,17), although this has not yet been directly demonstrated for normal B cells. Moreover, it is not yet known if the assembly of such complexes occurs in the same subcellular compartments where Ig-bound antigen-class II complexes are assembled. It is possible that the small number of class II molecules which bind to processed antigen in B cells represent a subpopulation which traffics in the B cell and is expressed on the plasma membrane differently than the bulk of class II. Future studies coupling assays to detect processed antigen-class II complexes with subcellular fractionation may shed light on these issues.

The Subcellular Compartments in which Processed Antigen-Class II Complexes Are Assembled in B Cells

An understanding of the microenvironment in which processing occurs is likely to be important to an understanding of the regulation of the assembly of processed antigen-class II complexes in B cells. The conditions within subcellular compartments may differ dramatically and influence the assembly process. The cell's ability to alter the processing microenvironments may represent important control mechanisms in processing. Both the early and late endocytic compartments have been implicated in antigen processing (18). To identify the subcellular compartment in which processing occurred we used a combination of subcellular fractionation and assays for functional processed antigen-class II complexes (Qiu, Y., *et al.* "Separation of subcellular compartments containing distinct functional forms of MHC class II" submitted for publication). B cells were allowed to process the antigen *c*-anti-Ig for varying lengths of time (15 min, 2 and 4 hr). Processing was stopped, the cells were fractionated and the subcellular compartments separated by percoll density gradient centrifugation. The subcellular compartments were identified by a variety of enzymatic, functional and biochemical markers (Table I). The gradient allowed a clear separation of endosomes, ER, PM and lysosomes. The fractions

Table I. In B Cells, Separate Subcellular Compartments Contain Distinct Forms of Class II

Subcellular Fraction	Processed Antigen-Class II Complexes			Class II Receptive to Peptide Binding			Class II Conformation		Ii
	Time (h):			Time (h):					
	.25	2	4	.25	2	4	Floppy	Compact	
Lysosomes	+	+	−	−	−	−	−	+	−
X	−	−	−	+	+	−	+	−	+
PM/ER	−	−	+	+	+	+	+	+	+
EE/LE	−	+	−	−	−	−	+	+	+

Subcellular fractions were separated by percoll density centrifugation. The fractions are listed in order of decreasing density. The lysosomes are the densest, the endosomes the lightest. Lysosomes are positive for the lysosome marker LAMP-1, as well as β-hexosaminadase and contain horse radish peroxidase 25 min after pulsing. The compartment which we have termed X does not contain any of the markers tested including those for EE, LE, PM, ER, Golgi, lysosomes or mitochondria. The plasma membrane and endoplasmic reticulum (PM/ER) are identified as Na+/K+ ATPase positive and choline phosphotransferase positive. Early and late endosomes (EE/LE) are positive for β-hexosaminadase, and for the early and late endosome markers, rab 5 and rab 7, and contain horse radish peroxidase after a 4 min incubation.

were assayed *in vitro* for their ability to activate a T cell hybrid to secrete IL-2 as an indication of the presence of processed antigen-class II complexes. The findings are summarized in Table I. Functional complexes were first detected in the lysosome fraction, then subsequently in endosomes and finally in the plasma membrane. The order of appearance of the complexes suggests that complexes are first formed in a dense compartment which cosediments with lysosomes, then travel through endosomes to the plasma membrane. This dense loading compartment in B cells is similar to the MIIC compartment described by electron microscopy by Peters, *et al.* (19) and the loading compartment recently described by Harding, *et al.* (20) in macrophages which is the topic of an article in this volume as well.

An interesting feature of the assembly of the processed antigen-class II complexes in B cells is that the complexes are not continuously formed in the dense compartment. Thus, by 4 hours all complexes have left this compartment and are expressed on the B cell surface and no further assembly occurs. This is the case despite the fact that antigen is continuously present during the four-hour incubation. Moreover, the concentration of antigen present in culture (2 nM) is considerably less than would be required to saturate the surface Ig. The transient assembly of processed antigen-class II complexes in the dense compartment mirrors the transient nature of the presentation of antigen initially bound to the surface of B cells described above. Thus, it appears that the binding of antigen to B cells both enhances the processing of that antigen and precludes subsequent antigen processing. The results of subcellular fractionation show that the assembly occurs in a discrete subcellular compartment distinct from early and late endosomes. Future studies will focus on the effect of the binding of antigen to surface Ig on the appearance and longevity of this assembly compartment.

In addition to identifying the compartments in which functional, processed antigen-class II complexes are formed, it was of interest to determine in which compartments class II receptive to peptide binding resided. Subcellular fractions were analyzed for the presence of class II capable of presenting a peptide antigen which does not require processing to T cells *in vitro*. The results summarized in Table I show the somewhat surprising result that the majority of class II which can present peptide resides in a compartment which is different than the compartment in which processed antigen-class II complexes are formed. In addition to the PM/ER fractions, which contain the majority of the class II in the cell, and also contain some class II which can present peptide, the majority of class II capable of presenting peptide is found in a compartment which we refer to as X. This compartment does not have markers of the EE, LE, PM, ER, Golgi or lysosomes and apparently contains class II receptive to peptide binding but which does not have access to processed antigen in the B cell. Unlike the class II in the dense loading compartment, a portion of class II in the X compartment is associated with Ii (Table I). The class II in these compartments was further analyzed by SDS-PAGE under conditions where the α and β chains remain associated. Of some interest was the finding that the class II in the X compartment was in the so-called "floppy" form (21) while the class II in the dense compartment was in the "compact" form (Table I). Recent studies have correlated the formation of the compact form with peptide binding to class II (22). The observation that the floppy and compact forms reside in separable subcellular compartments indicates that in B cells the assembly process is staged, in physically separated steps. The class II which is still associated with Ii, and of which at least a portion binds peptide is contained in a compartment separate from the compartment in which the class II binds processed antigen. The X compartment only transiently contains class II capable of peptide binding (Table I) again reflecting the transient nature of Ig-mediated antigen processing. It is interesting to speculate that during biosynthesis the class II-Ii complexes enter the X compartment where they reside until Ii chain releases class II to enter the processed antigen-loading compartment. It will be of considerable interest to determine how the binding of antigen to the surface Ig influences the appearance and disappearance of this compartment. Future studies will also focus on defining the content of both the X and the loading compartments with regard to other molecules important in the processing and presentation of antigen to T cells.

The Regulation of B cell APC Function During Development

Although all mature B cells appear to function as APC, we have recently shown that the APC function is acquired late in neonatal development (6). Thus, B cells obtained from neonatal mice are not equivalent in their APC function to B cells from adult mice until mice are approximately 2 weeks of age. This is despite the fact that the expression of class II and Ig as well as the adhesion molecules ICAM-1 and LFA-1 are adult like in B cells from one week old mice. Moreover, neonatal B cells' presentation of peptide fragments of antigen which do not require processing is equivalent to that of B cells obtained from adult mice. In addition, B cells from neonatal mice are able to bind antigen through surface Ig, internalize and degrade it long before they are competent APC. Thus, during development the ability to assemble or display processed antigen-class II complexes is delayed in B cells. This may have important repercussions in the developing immune system, limiting the interactions of helper T and B cells and the potential for expansion of B cells prior to the complete establishment of tolerance. At present we do not know what steps in antigen processing are blocked in B cells from neonatal mice. One potentially interesting recent observation in our laboratory is that the expression of the chondroitin sulfated form of Ii (Cs-Ii) is also delayed in B cells during neonatal development and the time course of the synthesis and appearance of Cs-Ii closely mirrors that of the APC function (data not shown). It has been recently shown that the Cs-Ii dramatically enhances APC function (23). Cs-Ii was shown to be a ligand for T cell CD44 molecules and thus may play a role in APC and T cell adhesion (23). It is interesting to speculate that the delay in the production of the Cs-Ii may contribute to the poor APC ability of neonatal B cells.

The Regulation of Antigen Processing by Stress and by Virus Infection

In addition to the inherent control of antigen processing under normal conditions, B cells may regulate antigen processing in response to environmental signals. We have investigated the impact of two factors on the B cells' APC function. These are stress, in the form of heat shock, and virus infection.

In terms of the effect of stress on B cell antigen processing, it is well established that all cells show a highly conserved cellular response to environmental stress, termed the heat shock response (24). The effect of heat shock on the B cell APC function is likely to be relevant to processing *in vivo* which may often occur under conditions in which the heat shock response is induced, such as during viral or bacterial infections, inflammation and fever. We recently investigated the effect of heat shock induced by elevated temperatures on the APC function of B cells (25). We found that heat shock enhanced the processing and presentation of antigen by B cells. B cells undergoing a stress response required considerably less time to process and present antigen and achieved higher levels of T cell activation as compared to untreated B cells. The enhancement of processing required the presence of antigen during the heat shock response in that heat shock had no effect on B cells which had already processed antigen nor did heat shock prior to antigen processing affect subsequent processing. The acceleration of antigen processing during heat shock was reflected in an acceleration of the formation of compact $\alpha\beta$ dimers in B cells. As discussed above, the compact forms are presumably class II molecules which have bound peptide. Moreover, the class II purified from heat shocked cells was better able to present peptide *in vitro* as compared to class II purified from untreated cells. This latter finding suggested that during the stress response more class II may bind to peptides which are readily exchanged with peptide *in vitro* or more class II may be left empty as compared to class II in untreated cells. Although, the precise steps in the processing pathway affected during the stress response have not been delineated, these findings indicated that the processing of antigen changes in B cells in response to changes in the environment.

We have also investigated the effect of virus infection on the ability of B cells to process and present antigen. We found the infection of B cells by viruses of two different families, namely influenza A and vaccinia, completely blocked processing and presentation of exogenous antigen (26). The block was within the processing pathway as virus infection did not affect the presentation of a peptide which did not require processing. The block required live, infectious virus and is likely to affect several steps along the processing pathway. This observation is important in indicating that the ability to process and present an antigen can be severely limited by the antigen itself.

Future Directions

There is accumulating evidence indicating that the presentation of antigen is a regulated process in B cells and that the surface Ig is likely to play an important role in such regulation. The ability to isolate the discrete subcellular compartments involved in the assembly of processed antigen-class II complexes offers an opportunity to pin point the steps in processing which are regulated in B cells. Further biochemical characterization of these compartments will hopefully lead to a description of the cellular and molecular machinery employed by B cells in antigen processing. Thus, future studies will attempt to further delineate the discrete steps in the processing pathway and identify the compartments in which these occur. We hope to: 1) define the effect of Ig-mediated signalling on the formation and longevity of processing compartments; 2) determine if other cell surface receptors regulate processing in B cells; and 3) further evaluate the effects of both developmental and environmental factors on B cell antigen processing.

REFERENCES

1. **Pierce, S.K., J. F. Morris, M. J. Grusby, P. Kaumaya, A. van Buskirk, M. Srinivasan, B. Crump, and L. A. Smolenski.** 1988. Antigen-presenting function of B lymphocytes. *Immunol. Rev. 106*:149.

2. **Lassila, O., O. Vainio, and P. Matzinger.** 1988. Can B cells turn on virgin T cells? *Nature 334*:253.

3. **Lanzavecchia, A.** 1985. Antigen-specific interactions between T and B cells. *Nature 314*:537.

4. **Jelachich, M.L., E. K. Lakey, L. A. Casten, and S. K. Pierce.** 1986. Antigen presentation is a function of all B cell subpopulations separated on the basis of size. *Eur. J. Immunol. 16*:411.

5. **Gosselin, E.J., H. P. Tony, and D. C. Parker.** 1988. Characterization of antigen processing and presentation by resting B lymphocytes. *J. Immunol. 140*:1408.

6. **Morris, J.F., J. T. Hoyer, and S. K. Pierce.** 1992. Antigen presentation for T cell interleukin-2 secretion is a late acquisition of neonatal B cells. *Eur. J. Immunol. 22*:2923.

7. **Marsh, E., D. P. Dalke, and S. K. Pierce.** 1992. Biochemical evidence for the rapid assembly and disassembly of processed antigen-MHC Class II complexes in acidic vesicles of B cells. *J. Exp. Med. 175*:425.

8. **Davidson, H.W., P. A. Reid, A. Lanzavecchia, and C. Watts.** 1991. Processed antigen binds to newly synthesized MHC class II molecules in antigen-specific B lymphocytes. *Cell 67*:105.

9. **Lanzavecchia, A.** 1990. Receptor-mediated antigen uptake and its effect on antigen presentation to class II-restricted T lymphocytes. *Ann. Rev. Immunol. 8*:773.

10. **Casten, L.A. and S. K. Pierce.** 1988. Receptor-mediated B cell antigen processing: increased antigenicity of a globular protein

covalently coupled to antibodies specific for B cell surface structures. *J. Immunol. 140*:404.

11. **Niebling, W.L. and S. K. Pierce.** 1993. Antigen entry into early endosomes is insufficient for MHC class II processing. *J. Immunol. 150*:2687.

12. **Patel, K.J. and M. S. Neuberger.** 1993. Antigen presentation by the B cell antigen receptor is driven by the α/β sheath and occurs independently of its cytoplasmic tyrosines. *Cell 74*:939.

13. **Casten, L.A., E. K. Lakey, M. L. Jelachich, E. Margoliash, and S. K. Pierce.** 1985. Anti-immunoglobulin augments the B-cell antigen-presentation function independently of internalization of receptor-antigen complex. *Proc. Natl. Acad. Sci. USA. 82*:5890.

14. **Srinivasan, M., E. W. Marsh, and S. K. Pierce.** 1991. Characterization of naturally processed antigen bound to MHC-class II. *Proc. Natl. Acad. Sci. USA 88*:7928.

15. **Srinivasan, M. and S. K. Pierce.** 1990. Isolation of a functional antigen-Ia complex. *Proc. Natl. Acad. Sci. USA. 87*:919.

16. **Hunt, D.F., H. Michel, T. A. Dickinson, J. Shabanowitz, A. L. Cox, K. Sakaguchi, E. Appella, H. M. Grey, and A. Sette.** 1992. Peptides presented to the immune system by the murine class II major histocompatibility complex molecule I-A^d. *Science 256*:1817.

17. **Chicz, R.M., R. G. Urban, W. S. Lane, J. C. Gorga, L. J. Stern, D. A. A. Vignali, and J. L. Strominger.** 1992. Predominant naturally processed peptides bound to HLA-DR1 are derived from MHC-related molecules and are heterogeneous in size. *Nature 358*:764.

18. **Neefjes, J.J. and H. L. Ploegh.** 1992. Intracellular transport of MHC class II molecules. *Immunol. Today 13*:179.

19. **Peters, P.J., J. J. Neefjes, U. Oorschot, H. Ploegh, and H. J. Geuze.** 1991. Segregation of MHC class II molecules from MHC class I molecules in the Golgi complex for transport to lysosomal

compartments. *Nature 349*:669.

20. **Harding, C.V. and H. J. Geuze.** 1993. Immunogenic peptides bind to class II MHC molecules in an early lysosomal compartment. *J. Immunol. 151*:3988.

21. **Dornmair, K., B. Rothenhausler, and H. M. McConnell.** 1989. Structural intermediates in the reactions of antigenic peptides with MHC molecules. *Cold Spring Harbor Sym. Quant. Biol. 54*:409.

22. **Sadegh-Nasseri, S. and R. N. Germain.** 1991. A role for peptide in determining MHC-class II structure. *Nature 353*:167.

23. **Naujokas, M.F., M. Morin, M. S. Anderson, M. Peterson, and J. Miller.** 1993. The chondroitin sulfate form of invariant chain can enhance stimulation of T cell responses through interaction with CD44. *Cell 74*:257.

24. **Hanover, J.A. and R. B. Dickson.** 1985. Transferrin: receptor-mediated endocytosis and iron delivery. In Receptor-mediated Endocytosis. I. Pastan and M. Willingham, editors. *Plenum Press. N. Y.* 341.

25. **Cristau, B., P. H. Schafer, and S. K. Pierce.** 1993. Heat shock enhances antigen processing and accelerates the formation of compact class II αß dimers. *J. Immunol. (in press)*:

26. **Domanico, S.Z. and S. K. Pierce.** 1992. Virus infection blocks the processing and presentation of exogenous antigen with the major histocompatibility complex class II molecules. *Eur. J. Immunol. 22*:2055.

11

REGULATION OF ANTIGEN/CLASS II MHC COMPLEX FORMATION

Peter E. Jensen, Melanie A. Sherman, Herbert A. Runnels, and S. Mark Tompkins

INTRODUCTION

CD4-positive T lymphocytes play a central role in the regulation of immune responses. These cells recognize peptides stably associated with class II major histocompatibility (MHC) glycoproteins expressed on the surface of specialized antigen presenting cells (APC), such as macrophages, B lymphocytes, and dendritic cells. As a general rule, protein antigens do not directly bind to cell surface MHC molecules. Instead, proteins are taken into the endocytic pathway of APC where they unfold and are cleaved into short peptides that bind to newly synthesized or recycling class II proteins. The final complexes are transported to the cell surface where they reside for a considerable period of time. A major interest of our laboratory has been to investigate factors that regulate the formation of peptide/class II MHC complexes under physiological conditions.

Many studies have demonstrated that antigen processing requires metabolic activity that can be inhibited by fixing APC with aldehydes or by the neutralization of internal acidic compartments such as endosomes and lysosomes. Fixation inhibits all metabolic activity, including antigen uptake. Agents that raise the pH of acidic vesicles inhibit the activity of acid proteases that degrade proteins antigens. The antigen processing requirement can be bypassed by using short peptides 10-20 amino acids in length which can bind directly to class II molecules on the surface of APC. Short peptides can bind to purified class II proteins in solution or in artificial membranes (1,2). The specificity of binding, as determined from functional studies, is preserved in experiments with purified class II. However, the binding kinetics are unusually slow, requiring days to reach equilibrium (3). Similar results are obtained in functional assays with fixed APC. Long incubation periods are generally required to generate a number of peptide/MHC complexes on the surface of fixed APC that is sufficient to activate T cells. By contrast, peptide/MHC complex formation appears to be relatively rapid in viable APC exposed to native protein antigen or peptide (4). Antigen handling by viable APC only requires hours rather than days to trigger maximal T cell stimulation. These general observations raise the possibility that conditions or co-factors present in viable APC serve to facilitate physiological peptide loading of class II molecules.

ENHANCED PEPTIDE BINDING AT LOW pH

Membrane-bound vesicles in the endocytic pathway are clearly involved in antigen processing. In these compartments antigens are degraded to form peptides. Newly synthesized class II molecules are known to traffic through endosomes on their way to the cell surface. Release of invariant chain (Ii) and peptide loading are thought to occur in these compartments. One common component of endosomes and lysosomes is the presence of H^+-ATPases responsible for maintaining an internal acidic environment (5). There is a gradient of hydrogen ion concentrations such that pH becomes progressively lower in later compartments in the endocytic pathway. Early endosomes are mildly acidic (pH 6-6.5) and lysosomes are the most acidic organelles with a pH of 4.5-5.0. The proteolytic enzymes responsible for degradation of antigen and release of Ii have optimal activity at low pH. The acidic environment also plays a direct role in facilitating the binding of peptides to class II molecules. This was first observed in functional experiments with both fixed APC and APC membrane fragments. We found that the capacity of the membranes to stimulate T cells was enhanced if the membranes were exposed to peptide antigen at low pH. This observation prompted us to carry out a careful investigation of the effect of pH on the functional association of peptide antigen with APC membrane (6). Aldehyde-fixed B lymphoblastoid cells were exposed to peptides under various conditions and the extent of cell-surface peptide/class II complex formation was assessed using T cell hybridomas. The rate and maximal extent of peptide binding was strikingly enhanced at pH 5 as compared to neutral pH. Pretreatment of the cells at low pH in the absence of peptide had no effect on the subsequent capacity to form peptide/MHC complexes. Similar results were obtained with several peptides and with IA^d and IE^d-restricted T cells. These findings led to the conclusion that pH had a direct effect on the interaction of peptide with either the class II protein or with other components of the APC membrane involved in facilitating peptide/MHC complex formation. The rate of peptide binding in these experiments was quite rapid, with apparent saturation reached within a few hours. This was in good agreement with the rate of antigen processing observed in functional experiments with live APC (4,7). It was much more rapid than the rate of peptide binding to purified class II proteins measured in several laboratories. Our results raised several questions: 1) Does pH directly regulate the peptide binding activity of class II molecules or are other co-factors involved? 2) Could saturation artifacts account for the rapid peptide loading that was observed in our functional assays? 3) Can pH fully account for the difference between the observed kinetics of antigen processing in viable APC and the much slower rate of complex formation that was observed with purified class II incubated with peptide at neutral pH?

Experiments with purified class II proteins were used to address these questions (8). Purified IE^d and IA^d were reconstituted into liposomes containing defined lipids by the detergent dialysis technique. Silica particles (10 μm) were coated with the liposomes and class II density in the supported membranes was determined by flow cytometry. The coated particles, which can be handled like cells, were incubated with peptide in buffers at various pH, washed, and cultured with T cell hybridomas. We found that formation of functional peptide complexes with purified class II in artificial membranes was enhanced at pH 5. Thus, pH had a direct effect on the interaction of peptide with the class II molecule. The kinetics were relatively slow compared to the results that we had obtained with fixed APC. The possibility that T cell responses, but not peptide/MHC binding, was rapidly saturated in experiments with fixed cells was further addressed in titration experiments. Fixed APC were exposed to peptide for 4 or 18 h at pH 5. T cell assays were performed with varying numbers of APC and surface class II density. The results suggested that the quantity of peptide/MHC complexes generated was the same after 4 h or 18 h of incubation. Since vesicular acidification in APC can only partially account for the rapid kinetics of antigen processing, the results suggest that other factors or conditions may be present in the normal membrane to facilitate peptide binding.

We developed a new method for measuring peptide/MHC binding to further study the mechanism responsible for enhanced binding of peptide to purified class II protein at low pH. Biotin-labeled peptides were incubated with purified class II under various conditions. The samples were neutralized and complexes were captured by incubation in microtiter plates coated with appropriate anti-class II mAb. Bound biotin-peptide was detected by incubation with excess avidin-alkaline phosphatase followed by substrate. Bell-shaped profiles were obtained in experiments measuring the pH-dependence of binding. These were very similar to the profiles we had obtained in functional assays. IE^d bound peptide optimally at pH 4.5 whereas binding to IA^d was optimal at pH 5.5. The results of experiments with several peptides, including one that bound both class II proteins, suggested that different class II alleles or isotypes may have different, characteristic pH optima for peptide binding. Both the rate of peptide association and the maximal extent of binding were increased at low pH. The effect of pH was completely reversible. No increase in the quantity of peptide bound at pH 7 was observed after lengthy pretreatment at pH 5 in the absence of peptide. Ionic interactions between peptide and MHC were not required for the enhancement in peptide binding observed at low pH. Our results strongly suggested that increased peptide binding activity is a consequence of a direct effect of low pH on the structure of the class II molecule.

We proposed a model in which the ionization of critical groups in class II molecules regulate the conformation or flexibility of the protein (8). In our model, the protein is assumed to have increased flexibility at low pH which facilitates a peptide-dependent conformational transition to form stable peptide/class II complexes. Sadegh-Nasseri and Germain (9) proposed a similar

160

Peter E. Jensen *et al.*

model based on the observation that long-lived peptide/class II complexes could be rapidly generated by initial incubation at pH 4.5 followed by a rapid stabilization event that occurs during pH neutralization. They suggested that at low pH peptides rapidly associate and dissociate and when pH is neutralized the bound peptides become trapped. Thus, neutralization of pH as class II molecules are transported from a more acidic to less acidic compartments may play a role in the formation of stable peptide/MHC complexes. However, our own results suggested that neutralization was not required for the efficient formation of stable complexes at mildly acidic pH in the range of pH 5.0 to 6.5. We had observed that very low pH (<4.5) is required for peptide dissociation whereas enhanced binding is observed at mildly acidic pH (10,11). Thus, there was a discordance between the conditions required for enhanced binding and those required for enhanced dissociation. We devised an assay system to directly determine whether pH neutralization is required for the formation of stable peptide complexes. All steps in the binding assay were carried out at low pH, neutral pH, or with pH shifts during incubation with peptide. The results demonstrated that stable complexes were readily formed under conditions where acidic pH is never neutralized and that neutralization or pH shifts did not increase binding (11). We therefore proposed that although low pH increases the flexibility of the class II protein, this flexibility does not result in enhanced peptide dissociation because interactions involving peptide stabilize the complex. This is reasonable because we know that peptide is intimately involved in the conformation of the protein after complex formation based on early studies demonstrating the remarkable stability of the complexes (10,12) and data from the crystal structure (13). It should be noted that some peptides/class II complexes do have an increased rate of dissociation at mildly acidic pH (14,15). These peptides may represent non-physiological or non-dominant determinants that do not form sufficient interactions to stabilize the complex under conditions of low pH where the class II protein is assumed to have increased flexibility. Alternatively, a change in the charge of specific residues in peptide or the peptide-binding site at low pH could lead to repulsive forces that destabilize the complex. If we assume that physiological peptide loading occurs in acidic compartments in vivo, then antigenic determinants with high dissociation rates at low pH would be at a disadvantage among peptides competing for class II binding.

Direct evidence for a change in the conformation of class II molecules at mildly acidic pH has come from an analysis of the sensitivity of purified class II proteins to SDS-induced denaturation at 24°C (Runnels, et al., manuscript in preparation). Class II proteins are generally stable in SDS if the samples are not heated. The proportion of class II heterodimers that dissociate in SDS into monomers varies between different isotypes and alleles. For example, IA[d] molecules are quite unstable in SDS while IE[d] is very stable. We have observed a gross correlation between stability and pH optimum for peptide binding. Class II proteins with high pH optimum tend to be less stable than those with optimum at low pH. We found that stability in SDS was very dependent upon pH. Class II proteins are more susceptible to SDS-induced denaturation at mildly acidic pH. Enhanced subunit dissociation is observed in the same pH range that is associated with enhanced peptide binding activity. Acidic pH alone is not responsible for subunit dissociation because exposure to low pH, followed by neutralization before exposure to SDS, does not result in increased dissociation. We believe that the increased sensitivity to SDS reflects an altered conformation or conformational flexibility. It should be noted that recent studies have suggested that "empty" class II molecules are uniformly unstable in SDS and that stability in SDS is acquired after peptide loading. Therefore our data reflect a structural transition in class II/peptide complexes rather than empty molecules. However, Reay, et al. (15) have demonstrated that the binding of a high affinity peptide to recombinant IE^k molecules is markedly facilitated at pH 5 as compared to neutral pH. These recombinant molecules appear to be empty (15). Thus, we presume that pH can regulate the structure of empty as well as peptide-loaded class II molecules.

Membrane Interactions Can Influence Peptide Binding

Recent reports using several purified DR proteins raise the possibility that the peptide-binding activity of human class II molecules may be optimal at neutral rather than acidic pH (16-18). These results prompt us to question the hypothesis that vesicular acidification is generally important in regulating peptide loading in human APC under physiological conditions (6). We have studied the peptide binding behavior of DR1 in detail using the T cell-defined influenza epitope, MAT(17-31). Variation in hydrogen ion concentration within the physiological range had very little effect on peptide binding to purified DR1. Experiments were done to determine if DR1 would behave similarly in the natural environment of the cellular membrane. We have developed a fluorescence immunoassay for quantification of peptide binding to class II in various membrane environments (19). Purified class II, fixed or viable APC, membrane fragments, or liposomes are incubated with biotin-labeled peptide. The membranes are solubilized in detergent and specific class II proteins are captured by incubation on microtiter plates coated with appropriate mAb. Biotin-peptide/class II complexes are quantified by incubation with europium-labeled streptavidin and time-resolved fluorescence of chelated europium is measured.

Our results demonstrated that by contrast to soluble DR1, peptide binding by cell-surface DR1 was strikingly dependent on pH (20). Optimal binding was observed at pH 4.0 and little or no binding was detected at pH 6.5 and 7.0. The pH-dependence of peptide binding was largely recovered after reconstitution of purified DR1 into mouse B cell membranes by detergent dialysis. Reconstitution required the transmembrane domain and was not observed with truncated forms of DR1 lacking the transmembrane domain. The pH-dependence of peptide binding was partially recovered by reconstitution in artificial membranes containing phosphatidylcholine and cholesterol. This suggests that interactions with the lipid bilayer may reduce the flexibility of DR1 at neutral pH. Formation of the appropriate conformation for efficient peptide binding in the membrane environment appears to require acidic pH. These results strengthen the hypothesis that pH plays a general role in the physiological regulation of peptide loading. The fact that a more complete recovery of pH-dependent peptide binding was observed after reconstitution into B cell membranes suggests that other co-factors or membrane components may also regulate peptide binding by DR1 (21,22). We have obtained some preliminary evidence that the rate of peptide binding at pH 5 can be influenced by the local environment of class II proteins. Cell surface IA^k appears to bind peptide more rapidly than purified IA^k in detergent solutions. We hypothesize that specific membrane components may interact with the class II molecule or with peptide to facilitate peptide binding. The existence of this activity and the identity of the putative co-factor remain to be firmly established. Our preliminary experiments have suggested that rapid peptide binding can be partially recovered after reconstitution of IA^k into liposomes containing B cell membrane components but not defined lipid, alone. The detergents used to solubilize membranes and create liposomes could irreversibly disrupt interactions between class II and other membrane components. Co-factors may be distributed in selected compartments and they would be diluted during reconstitution of liposomes made from all cellular membranes. Thus, it will be important to identify the specific organelles in which peptide loading occurs and to define the features of this local environment that influence the behavior of the class II molecule.

UNFOLDING AND CLEAVAGE OF ANTIGENS

The end product of the class II antigen processing pathway is a stable complex composed of class II MHC and peptides 14-25 amino acids in length (23,24). The bound peptides are largely derived from exogenous proteins that enter the endocytic pathway of APC and from endogenous proteins present in the secretory pathway. An accepted paradigm proposes that protein antigens are first cleaved into peptides by endopeptidases in the endosomal or lysosomal processing compartments of APC prior to binding class II MHC. However, one must also consider the possibility that unfolded proteins may bind class II prior to proteolytic cleavage. By contrast to native proteins, irreversibly denatured proteins have been shown to bind purified class II and, in some cases, are sufficient to stimulate specific T cells in the presence of fixed APC (25,26). The unfolding event itself can therefore be sufficient to allow class II binding and antigen presentation without the requirement for further processing (27). It is possible that proteolysis follows, rather than precedes, class II binding. Binding of unfolded proteins may precede proteolytic "trimming" of the portions of antigen that are not protected by intimate association with the peptide-binding site in class II (28). The physiological events responsible for unfolding and cleavage of antigen and their temporal sequence are likely to have a major effect on the ultimate repertoire of determinants that are made available for binding to class II and recognition by T cells.

Disulfide Reduction During Antigen Processing

Many potentially important protein antigens contain disulfide bonds. These covalent linkages serve to stabilize protein conformations and protect folded proteins from the action of proteases. It is not intrinsically obvious that a reducing environment should be present in organelles in the endosomal pathway of APC. The cytoplasm contains mM concentrations of reduced glutathione and no disulfide-containing proteins. By contrast, membrane and secretory glycoproteins present in vesicles generally contain disulfide bonds and it is logical to assume that an oxidative redox potential would be maintained to preserve protein structure. However, there is evidence that disulfide reduction occurs during antigen processing. We found that disulfide reduction occurred during processing of insulin for presentation to class II-restricted T cells (29). Disulfide reduction was found to be both necessary and sufficient for presentation of insulin by fixed APC. Cleavage by proteases was not required. Free cysteine thiol moieties generated by physiological reduction of antigen disulfides after endocytosis were found to be essential components of an immunodominant T cell determinant in the insulin A chain. Alkylation of the thiol groups disrupted T cell recognition with little effect on peptide binding to several mouse class II proteins. More recent studies with insulin peptides and DR1 have demonstrated that cysteine thiol groups can also be required for binding to the class II molecule. It is likely that disulfide reduction is an important step in the processing of many antigens yet only in a subset are thiol groups an essential feature of the core determinant.

The mechanism and subcellular site for disulfide reduction remains unknown. Collins et al. (30) have demonstrated that reduction of disulfides in a labeled protein conjugate taken up through receptor-mediated endocytosis in macrophages occurs in a dense, presumably lysosomal, compartment. It is possible that after disulfide reduction in lysosomes, partially degraded peptides are rescued prior to complete hydrolysis and transported to a compartment containing class II MHC. There is evidence for transport of reduced cysteine from the cytoplasm into lysosomes and this may directly reduce protein disulfide bonds (31). Alternatively, disulfides may be reduced in an earlier, less hydrolytic endosomal compartment. The potential role of protein disulfide isomerase or other enzymes in this process cannot be excluded at this time. The mechanism must account for the selective reduction of the disulfide bonds in antigen and not in the glycoproteins that reside in the appropriate vesicles. It is interesting to note that reducing agents have no detectable effect either on the structure or on peptide binding activity of solubilized or cell-surface class II molecules (29,32). Thus class II may be able to enter a specialized compartment in the endocytic pathway with reducing redox potential and bind polypeptides as they unfold.

Acidification And Disulfide Reduction Allow Intact Proteins To Bind Class II MHC.

It is evident that proteins encounter acidic and reducing environments during antigen processing. We tested the hypothesis that these conditions are sufficient to allow intact proteins to bind class II MHC without a requirement for prior cleavage into peptides (32). We found that fixed APC, pulsed with hen egg lysozyme at pH 5 in the presence of a reducing agent, could stimulate class II-restricted T cells. Both low pH and the reducing agent were required. Lysozyme was found to bind to purified IE^d under conditions where contaminating protease activity was highly unlikely. Binding was strictly dependent on the presence of reducing agent. Similar results were obtained with other disulfide-containing proteins. Lysozyme and other globular proteins are not denatured at pH 5. Thus complete unfolding is not required for class II binding. We have observed a conformational change in lysozyme by measuring intrinsic tryptophan fluorescence of protein solutions containing a reducing agent. The conformational change was not observed at neutral pH or in the absence of reducing agent, in agreement with the MHC binding experiments.

Fixed APC were also observed to stimulate IE^k-restricted T cells after exposure to cytochrome c at pH 5 but not pH 7. No reducing agent was required, consistent with the absence of disulfide bonds in this protein. Native cytochrome c and a number of other proteins were observed to bind specifically to certain purified class II proteins in competition assays at low pH. These results suggest that low pH, alone, is sufficient to allow some proteins to bind class II. Our theory is that low pH is sufficient to cause local destabilization of structure in some proteins so that selected regions have sufficient flexibility to interact with the class II binding site. This would obviously require that flexible domains have the appropriate sequence motif for binding to a given class II protein. It is possible that interaction with class II could catalyze further unfolding of the protein. We have found that a number of proteins bind class II at pH 5. A larger group of proteins can bind class II at low pH after reduction of disulfide bonds. Our findings support the hypothesis that low pH and disulfide reduction can be sufficient to partially unfold or structurally destabilize many proteins and allow them to directly bind class II without prior cleavage into short peptides. It is possible that this may be a major pathway for antigen processing. Subsequently, a variety of proteases could act on intermediates composed of class II molecules bearing large unfolded polypeptides and trim residues not protected by interaction with the peptide-binding site in class II. The repertoire of peptides available for recognition by T cells may be considerably different depending on which pathway dominates in Antigen processing (33,34). If proteins are first unfolded and cleaved into peptides before binding MHC then many potential determinants will be destroyed by proteases and the peptide repertoire will be influenced by the specificity of endopeptidases present in the relevant compartments. The remaining peptides would compete for binding MHC solely on the basis of relative affinity and concentration. By contrast, if partially unfolded proteins first interact with MHC then determinants that are made available first during the unfolding process will have a competitive advantage, over buried determinants, in binding class II (32).

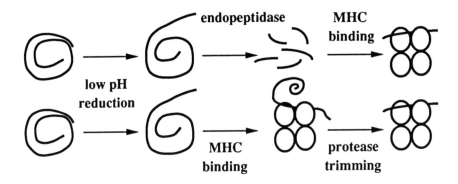

Pathways for antigen processing. Protein antigens may be cleaved into short peptides before binding class II MHC. Alternatively, partially unfolded proteins may bind class II prior to proteolytic trimming.

SUMMARY AND FUTURE DIRECTIONS

It is clear that local conditions present in selected endosomal compartments in APC regulate the formation of class II MHC/peptide complexes. H^+-ATPases in endosomes and lysosomes play a key role by regulating the pH in these compartments. Hydrogen ion concentrations regulate the activity of acid proteases required for release of Ii from newly synthesized class II molecules and for the cleavage of antigens into smaller peptides. The conformation and peptide binding activity of class II molecules is controlled by pH. Acidification may also play an important role in unfolding protein antigens after endocytosis. Small thiol-containing compounds, including reduced cysteine and glutathione, regulate local redox potentials and contribute toward antigen unfolding and cleavage. The potential role of enzymes with thioreductase activity has not been excluded. Interactions between components of cellular membranes and class II affect peptide-binding behavior and there is some evidence that specific co-factors may serve to catalyze peptide loading in the physiological environment. The intriguing possibility that molecular chaperones may rescue peptides from degradative compartments and deliver them to a different site for class II binding needs to be explored.

Much remains to be learned about the mechanism(s) responsible for the effect of pH on peptide binding by class II proteins. We do not know which amino acid residues in class II are sensitive to pH and regulate molecular flexibility and what specific conformational changes occur. Is peptide binding by all class II proteins favored at low pH or are there important exceptions among species, isotypes, or alleles? It is possible that different class II proteins have different, characteristic pH optima for peptide binding. This could limit the subcellular compartments in which they bind peptides and influence the repertoire of peptides presented by specific class II proteins. Are there pathological conditions where peptide loading occurs in a neutral pH environment that would allow subdominant or self determinants to become available for T cell recognition? In addition to hydrogen ions, other co-factors may also facilitate peptide binding by class II molecules. Before these can be identified, the hypothesis that peptide binding is facilitated in the cellular membrane environment must be critically analyzed.

Further investigation is needed to characterize the various steps involved in unfolding and cleavage of protein antigens and the sequence of events leading to the expression of peptide/MHC complexes on the cell surface. We have shown that acidification and disulfide reduction can be sufficient to allow intact proteins to bind to class II. Careful analysis of processing intermediates generated under physiological conditions will indicate whether this occurs *in vivo*. The mechanism and subcellular site for reduction of antigen disulfide bonds also needs to be defined. It will be important to identify possible antigen processing heterogeneity in different cell types and to determine how this affects the repertoire of determinants available for recognition by T cells.

ACKNOWLEDGMENTS

We thank Joseph Moore for excellent technical support and Dr. Dominique Weber for advice on the manuscript. This work was supported by the U.S. Public Health Service Grant AI30554 from the National Institutes of Health. MAS is a Howard Hughes Predoctoral Fellow.

REFERENCES

1. **Babbitt, B. P., P. M. Allen, G. Matsueda, E. Haber, and E. R. Unanue.** 1985. Binding of immunogenic peptides to Ia histocompatibility molecules. *Nature 317*:359.

2. **Watts, T. H. and H. M. McConnell.** 1986. High-affinity fluorescent peptide binding to I-A^d in lipid membranes. *Proc. Natl. Acad. Sci. USA 83*:9660.

3. **Buus, S., A. Sette, S. M. Colon, D. M. Jenis, and H. M. Grey.** 1986. Isolation and characterization of antigen-Ia complexes involved in T cell recognition. *Cell 47*:1071.

4. **Roosneck, E., S. Demotz, G. Corradin, and A. Lanzavecchia.** 1988. Kinetics of MHC-antigen complex formation on antigen-presenting cells. *J. Immunol. 140*:4079.

5. **Mellman, I., R. Fuchs, and A. Helenius.** 1986. Acidification of the endocytic and exocytic pathways. *Ann. Rev. Biochem. 55*:663.

6. **Jensen, P. E.** 1990. Regulation of antigen presentation by acidic pH. *J. Exp. Med. 171*:1779.

7. **Ceppellini, R., G. Frumento, G. B. Ferrara, R. Tosi, A. Chersi, and B. Pernis.** 1989. Binding of labelled influenza matrix peptide to HLA DR in living B lymphoid cells. *Nature 339*:392.

8. **Jensen, P. E.** 1991. Enhanced binding of peptide antigen to purified class II major histocompatibility glycoproteins at acidic pH. *J. Exp. Med. 174*:1111.

9. **Sadegh-Nasseri, S. and R. N. Germain.** 1991. A role for peptide in determining MHC class II structure. *Nature 353*:167.

10. **Jensen, P. E.** 1989. Stable association of processed antigen with antigen-presenting cell membranes. *J. Immunol. 143*:420.

11. **Jensen, P. E.** 1992. Long-lived complexes between peptide and class II MHC are formed at low pH with no requirement for pH neutralization. *J. Exp. Med. 176*:793.

12. **Lee, J. M. and T. H. Watts.** 1990. On the dissociation and reassociation of MHC class II-foreign peptide complexes. Evidence that brief transit through an acidic compartment is not sufficient for binding site regeneration. *J. Immunol. 144*:1829.

13. **Brown, J. H., T. S. Jardestzky, J. C. Gorga, L. J. Stern, R. G. Urban, J. L. Strominger, and D. C. Wiley.** 1993. Three-dimensional structure of the human class II histocompatibility antigen HLA-DR1. *Nature 364*:33.

14. **Harding, C. V., R. W. Roof, P. M. Allen, and E. R. Unanue.** 1991. Effects of pH and polysaccharides on peptide binding to class II major histocompatibility complex molecules. *Proc. Natl. Acad. Sci. U. S. A. 88*:2740.

15. **Reay, P. A., D. A. Wettstein, and M. M. Davis.** 1992. pH dependence and exchange of high and low responder peptides binding to a class II MHC molecule. *EMBO J. 11*:2829.

16. **Sette, A., S. Southwood, D. O'Sullivan, F. C. A. Gaeta, J. Sidney, and H. M. Grey.** 1992. Effect of pH on MHC class II-peptide interactions. *J. Immunol. 148*:844.

17. **Mouritsen, S., A. S. Hansen, B. L. Petersen, and S. Buus.** 1992. pH dependence of the interaction between immunogenic peptides and MHC class II molecules. Evidence for an acidic intracellular compartment being the organelle of interaction. *J. Immunol. 148*:1438.

18. **Scheirle, A., B. Takacs, L. Kremer, F. Marin, and F. Sinigaglia.** 1992. Peptide binding to soluble HLA-DR4 molecules produced by insect cells. *J. Immunol. 149*:1994.

19. **Tompkins, S. M., P. A. Rota, J. C. Moore, and P. E. Jensen.** 1993. A europium fluoroimmunoassay for measuring binding of antigen to class II MHC glycoproteins. *J. Immunol. Methods 163*:209.

20. **Sherman, M. A., H. A. Runnels, J. C. Moore, L. J. Stern, and P. E. Jensen.** 1993. Membrane interactions influence the peptide binding behavior of DR1. *J. Exp. Med. in press.*

21. **Lakey, E. K., E. Margoliash, and S. K. Pierce.** 1987. Identification of a peptide binding protein that plays a role in antigen presentation. *Proc. Natl. Acad. Sci. USA 84*:1659.

22. **VanBuskirk, A. M., D. C. DeNagel, L. E. Guagliardi, F. M. Brodsky, and S. K. Pierce.** 1991. Cellular and subcellular distribution of PBP72/74, a peptide-binding protein that plays a role in antigen processing. *J. Immunol. 146*:500.

23. **Buus, S., A. Sette, S. M. Colon, and H. M. Grey.** 1988. Autologous peptides constitutively occupy the antigen binding site on Ia. *Science 242*:1045.

24. **Rudensky, A. Y., P. Preston-Hurlburt, S. Hong, A. Barlow, and C. A. Janeway Jr..** 1991. Sequence analysis of peptides bound to MHC class II molecules. *Nature 353*:622.

25. **Streicher, H. Z., I. J. Berkower, M. Busch, F. R. N. Gurd, and J. A. Berzofsky.** 1984. Antigen conformation determines processing requirements for T cell activation. *Proc. Natl. Acad. Sci. USA 81*:6831.

26. **Allen, P. M. and E. R. Unanue.** 1984. Differential requirements for antigen processing by macrophages for lysozyme-specific T cell hybridomas. *J. Immunol. 132*:1077.

27. **Sette, A., L. Adorini, S. M. Colon, S. Buus, and H. M. Grey.** 1989. Capacity of intact proteins to bind to MHC class II molecules. *J. Immunol. 143*:1265.

28. **Donermeyer, D. L. and P. M. Allen.** 1989. Binding to Ia protects an immunogenic peptide from proteolytic degradation. *J. Immunol. 142*:1063.

29. **Jensen, P. E.** 1991. Reduction of disulfide bonds during antigen processing: Evidence from a thiol-dependent insulin determinant. *J. Exp. Med. 174*:1121.

30. **Collins, D. S., E. R. Unanue, and C. V. Harding.** 1991. Reduction of disulfide bonds within lysosomes is a key step in antigen processing. *J. Immunol. 147*:4054.

31. **Pisoni, R. L., T. L. Acker, K. M. Lisowski, R. M. Lemons, and J. G. Thoene.** 1990. A cysteine-specific lysosomal transport system provides a major route for the delivery of thiol to human fibroblast lysosomes: Possible role in supporting lysosomal proteolysis. *J. Cell Biol. 110*:327.

32. **Jensen, P. E.** 1993. Acidification and disulfide reduction can be sufficient to allow intact proteins to bind class II MHC. *J. Immunol. 150*:3347.

33. **Shivakumar, S., E. E. Sercarz, and U. Krzych.** 1989. The molecular context of determinants within the priming antigen establishes a hierarchy of T cell induction: T cell specificities induced by peptides of beta-galactosidase vs. the whole antigen. *Eur. J. Immunol. 19*:681.

34. **Gammon, G., E. E. Sercarz, and G. Benichou.** 1991. The dominant self and the cryptic self: shaping the autoreactive T-cell repertoire. *Immunol. Today. 12*:193.

12

PEPTIDE, INVARIANT CHAIN, OR MOLECULAR AGGREGATION PRESERVES CLASS II FROM FUNCTIONAL INACTIVATION

Scheherazade Sadegh-Nasseri

INTRODUCTION

T cells bearing $\alpha\beta$ T cell receptors recognize antigenic peptides bound to class I and class II glycoproteins encoded in the major histocompatibility complex (MHC). Assembly of newly synthesized class II α and β chains in the endoplasmic reticulum is assisted by invariant chain, leading to the formation of a large non-covalently linked molecular assembly composed of nine polypeptide chains (1). Details of the intracellular pathway taken by class II from synthesis to entry into the vesicular compartments, where invariant chain is replaced by antigenic peptides, have been discussed in another chapter in this book. In this chapter, I will focus on the mechanisms of peptide and class II interactions and how this knowledge lead to the understanding of some properties of class II pertinent to protein structure, folding, stability, pH sensitivity, and macromolecular assembly. Kinetic analysis has been utilized to reveal the complex series of events from the initial interaction of a peptide with an MHC class II molecule to the establishment of the thermodynamically favored peptide-class II complex as defined by the crystal structure (2, and L. Stern, personal communication). This knowledge can help in designing strategies to influence formation of peptide-class II complexes relevant to the prevention or cure of diseases.

UNUSUAL KINETICS OF THE MHC PEPTIDE-CLASS II INTERACTION

The peculiar biological properties of MHC proteins make studying the kinetics of their ligand binding revealing. Few distinct MHC molecules possessed by an organism must bind a variety of peptide sequences from different sources; making binding of peptide to MHC molecules rather nonspecific. Nevertheless, complexes of peptide-MHC molecules would be most efficient in triggering T cells if they are stable; as, specific T cells might not reach the antigen presenting cells soon after the complexes are expressed. Consistent with this requirement, several studies examining the kinetics of interaction of class II molecules and antigenic peptides have reported a remarkable slow dissociation rate for the bound peptide. Slow measured on rates led to an equilibrium binding constants of micromolar (3) to nanomolar range (4, 5). In none of these studies, occupancy by offered peptide exceeded 15-20 percent of the potentially available class II binding sites, even using high concentrations of peptide and long incubation times. In all these cases the reaction was assumed to be pseudo first-order.

Slow apparent association rates, i.e., k_{on} less than diffusion-limited reaction rates, can be due to numerous causes that fall into two general categories. First, heterogeneity of reactants can decrease the effective concentration of the ligand (peptide molecules) or the receptor (class II molecules) from the theoretical values. This heterogeneity might be due to the presence of nonreactive conformers among proteins of the same primary structure and a rare interconversion to the active conformers. Second, inadequate detection techniques could leave fast reactions unmeasured.

Cocrystalization of class I with self peptides (6), and elution of self peptides from affinity purified class II molecules (7), pointed to the occupancy of a large proportion of mature MHC molecules with peptide ligand (heterogeneity in receptor molecules capable of binding the ligand). Moreover, the techniques employed for isolation of peptide-class II complexes from unbound peptide in all reported cases were lengthy and took over 30 minutes. Interestingly, using a fast fluorescence microscopic detection system to study the kinetics of interaction of purified class II inserted in planar membrane bilayers and a peptide of pigeon cytochrome c (PCC), Sadegh-Nasseri and McConnell (8) reported a two step consecutive reactions. In addition to the commonly slow forming (accumulating) complex, a specific fast forming and fast dissociating complex of PCC peptide-I-E^k was noted. Consistent with the findings of other investigators, stable complex was biologically active and could stimulate specific T cell hybridomas to produce IL-2. A kinetic model was proposed to suggest fast dissociating complex was a kinetic intermediate in the generation of stable complex and that its conversion to the long lived complex involved a conformational change in the structure of the class II MHC molecules. Evaluation of this model required determining conformational changes in the structure of MHC that were peptide dependent.

SDS-PAGE AS A MEANS OF SEPARATING DIFFERENT CONFORMERS OF CLASS II HETERODIMERS

Due to the slow (rare) conversion of low affinity complex to the stable peptide-class II complex and the peptide occupancy of most of the affinity purified I-E^k, physicochemical demonstration of a peptide-induced conformational change in MHC class II molecules was a technical challenge. Ideal strategies would have focused on increasing the signal to noise ratio, either by enhancing the rate of conformational change or by reducing the background; i.e., starting with samples containing a markedly increased proportion of empty class II molecules. A key contribution towards solving this problem came from Dornmair et.al. (9) who demonstrated that exposure of class II heterodimers to either acid or high temperature led to dissociation of the self peptides that typically co-purified with the class II. Dissociation of peptides correlated with changes in the structure of class II that were detected by differences in migration pattern in an SDS gel. When samples were not boiled, dissociation of peptides correlated with generation of floppy dimer (F) migrating at a slower rate than the peptide occupied form, compact dimer (C). Further extremes of pH or temperature, led to the conversion of F structure into separate α and ß chains. The authors suggested that SDS stable floppy dimer was an structural intermediate between compact and dissociated heterodimer.

KINETIC STUDIES AT ACIDIC pH

Jensen (10) first reported an enhanced antigen presentation to specific T cell hybridomas by pulsing of either fixed cells or cell membrane preparations with peptides at acidic as compared to neutral pH. Thus, acidic pH might have increased the rate of peptide loading on class II.

The first direct demonstration of a positive role for peptide in determination of the structure of class II came from the work of Sadegh-Nasseri and Germain (11) who combined SDS-PAGE analysis (9) with the findings of Jensen. Affinity purified I-E^k molecules, either inserted in lipid vesicles or in detergent micelles, were found almost exclusively in the C state. Treatment at pH 4.5 and 37°C for 30 minutes followed by neutralization led to loss of this structure but the appearance of F dimer as well as separated α and β chains. Under these experimental conditions, the separated chains derived from loosely associated dimer (L) that fell apart in SDS-PAGE and migrated as the individual α and β chains. Inclusion of peptides that bound I-E^k during the pH transit reconstituted the C structure. This reversal involved stable binding of offered peptide to the I-E^k as judged by western blotting where biotin labeled peptide primarily associated with C dimers. Peptides known not to bind I-E^k such as peptide of hen egg lysozyme, HEL 46-61 (interacts with I-A^k), or ovalbumin peptide, OVA 323-339 (interacts with I-A^d), did not induce C dimers. Thus, it was clear that peptide binding in some way induced structural changes in class II molecules that correlated with their stability and mobility in an SDS-PAGE.

Although C dimer undoubtedly contains peptide, certain combinations of class II alleles and peptides may migrate as individual chains in standard SDS-PAGE without sample boiling (Sadegh-Nasseri, unpublished, and 12, 13). In these cases, lowering the SDS concentration, and keeping the temperature of the buffers low might help in preventing dissociation of peptide-class II complexes (Sadegh-Nasseri, unpublished).

The formation of C dimer then, could be used as an assay for peptide binding. For instance, biosynthesis, maturation and peptide loading of class II molecules within antigen presenting cells can be traced by following the changes in their SDS-PAGE revealed conformation (14, 15). *Ex vivo* (cell culture) experiments demonstrated that newly synthesized class II molecules when associated with the invariant chain were exclusively unstable in SDS-PAGE and migrated as individual α and β chains. Compact dimer formation was only achieved when class II molecules had passed through the ER and Golgi and had entered acidic endosomal compartments where invariant chain could dissociate (14). Thus, the conformational transition seen *in vitro* precisely correlated with the expected *in vivo* acquisition of antigen for presentation to T cells.

Additional support for the physiological relevance of the *in vitro* structural transitions observed for class II upon binding peptide, comes from the analysis of class II in the mice that lacked invariant chain (16, 17). A combination of lack of invariant chain, defective class II assembly, transport, and peptide acquisition led to surface expression of class II molecules that had the phenotype of empty molecules, i.e., migrated as F dimer in unboiled SDS-PAGE. Incubation with peptide at acidic pH followed by neutralization gave rise to the C conformation (16).

Effects of Low pH on Class II Binding To and Dissociation From Specific Peptide

One explanation for the apparent pH dependence of k_{on} was the generation of more free sites under these conditions. Thus, class II molecules that copurified with self peptides might have lost the bound peptides more rapidly under these conditions. To test this hypothesis, preformed complexes of biotinylated PCC-I-E^k were incubated at pH 4.5 and the kinetic rate for dissociation of the labeled peptide were determined. A significant increase in the dissociation rate of the biotinylated peptide at pH 4.5 v.s. pH 7.2 was observed (11). Dissociation half time for I-E^k and PCC peptide at neutral pH is ~30 hours (8), whereas at pH 4.5 $t_{1/2}$ reduces to 30 minutes-- a difference of 60 fold! Thus, pH of 4.5 might help class II peptide loading by removing the previously bound peptides from class II binding cleft.

If the slow forward rate of formation of stable peptide-MHC class II complex was only due to prior occupancy of the peptide binding site, then F dimer should readily bind specific peptide at neutral pH. Yet this does not seem to be the case. Similar to pH 4.5 (11), a brief exposure to pH 11.3 (the pH used for elution of I-E^k from anti-I-E^k antibody columns) generates a fraction of I-E^k that is largely in F and L dimer forms (Fig 1, lane 1). Incubation with PCC peptide at pH 7.2 for 5 minutes does not induce readily detectable compact dimers. However, after eighteen hours of incubation, a small signal associates with generation of C dimer (lane 2). In contrast, five minutes of incubation at pH 4.5 (lane 3) promotes a significant cohort of class II to convert to C dimers. Incubation with peptide for further 18 hours after neutralization does not result in additional C dimer signal (lane 4). Therefore, at neutral pH class II molecules that are free of stably bound peptide do form C dimers in the presence of peptide but the extent of the conversion is considerably less in comparison to acidic pH.

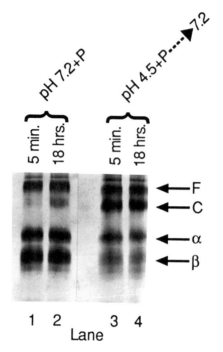

Fig. 1. *Empty peptide binding cleft and acidic pH together promote efficient and fast C structure formation induced by the offered peptide.*

To examine the kinetic rate for formation of C dimer from empty class II at acidic pH, I-E^k molecules, existing mainly as the F and L conformers, were incubated with biotinylated PCC at pH 4.5 for different length of time and then neutralized. Figure 2 depicts formation of C dimer as a function of time for I-E^k that is free of self peptides. PCC peptide was present at pH 4.5 for the length of time indicated and then samples were brought to pH 7.2. The data show that biotin signal is associated exclusively with the C structure and a laser densitometric scanning of the biotin signal indicates maximal peptide signal is reached with a $t_{1/2}$ of 3-5 minutes. This formation rate is in striking contrast to binding rates for the peptide-occupied class II molecules. These observations establish an important role for the extremes of pH in dissociation of stably bound peptides and in facilitation of compact folding of I-E^k in the presence of PCC.

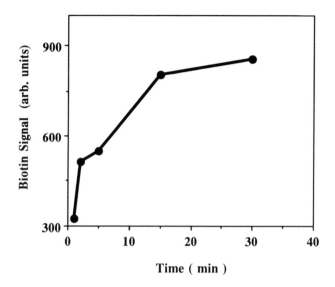

Fig. 2.   *Empty I-E^k binds PCC with a half time of 3-5 minutes at pH 4.5 (~200 times faster than at neutral pH).*

Thus, the same pH that results in enhanced dissociation of peptide from class II also allows for efficient peptide binding. This can be understood by considering that the conformational changes that lead to dissociation of self peptides from compact dimer might be related to the opening of the cleft. Binding of the labeled peptide to the empty and open cleft can readily take place (Fig.1 and 2). The observation that empty F dimer does not bind peptide efficiently at neutral pH, may hint closing of the peptide binding cleft upon increasing the pH in the absence of new peptide. The acidic pH, then, might impose its effect by retaining the peptide binding site open. A need for high concentrations of peptide of 100 µM at pH 4.5 for effective C dimer formation is consistent with formation of a low affinity intermediate complex prior to complex stabilization by neutral pH trapping.

Stabilization of the low affinity complex does not require a rise in the pH value to 7.2. This can be understood considering that dissociation of previously bound peptides from class II is highly sensitive to pH changes ≤ 4.5 (18, 19). Very little or no effect on the dissociation rate of the bound peptide is seen at pH > 4.5, while at pH 3.5 all the bound peptide is dissociated. Thus, increasing the pH of the reaction to pH ≥ 5 is more than sufficient to stabilize the weakly interacting peptide-class II complexes (20). In fact, this notion further strengthens the *in vitro* low pH transition protocol as a model system for simulation of the cellular events leading to peptide loading of class II. Existing data support a prolonged stay of class II in endosomal/lysosomal compartments prior to cell surface expression (21, 22).

THE CORRELATION BETWEEN THE KINETIC STATE AND A
PARTICULAR CONFORMATION OF CLASS II SEEN USING EMPTY
SOLUBLE DR1

pH 4.5 for I-E^k is the most efficient pH in loading I-E^k with peptide and variations of even one tenth of pH unit results in significant changes in proportion of C dimer formed (Sadegh-Nasseri, unpublished). However, at this pH, longer incubations in the absence of added peptide causes irreversible inactivation of I-E^k. After 3.5 hours incubation at pH 4.5, followed by addition of labeled peptide, only 30 % of the molecules can reconstitute into C dimers. This ratio is further decreased to 10 % by 5 hours of incubation at pH 4.5 (11, and Sadegh-Nasseri unpublished). Thus, extended incubation at acidic pH, induce irreversible inactivation of empty class II molecules.

The SDS-PAGE is not a suitable assay for the analysis of structure of MHC at acidic pH. Since, the combined effect of low pH and SDS, cause even the compact dimer to migrate as F and L structures. To address the interrelation between different kinetic forms of peptide-class II and different structures, it is ideal to start with a population of class II that is empty of peptide and mostly active.

Human class II, HLA-DR1 molecules produced as a soluble, secreted glycoprotein in insect cell infected with baculoviruses carrying truncated α and ß subunit genes, were shown to be free of peptides. Soluble DR1 (sDR1) was shown to exist mainly in molecular aggregates but bound peptide with the 1:1 stoichiometry. Stable peptide binding resulted in generation of single αß heterodimer from the large aggregates (23). Lack of bound peptide, absence of detergents, and their potential in stoichiometric binding to specific peptide are the desireable criteria in augmenting signal to background ratio regardless of a need for acidic pH.

Thus, sDR1 molecules have many advantages for kinetic analyses of peptide binding to class II and the conformational changes that might accompany. A peptide of hemagglutinin, HA 306-318 and its ^{125}I radiolabeled analogue were used for this study (24, and Sadegh-Nasseri, Stern, Wiley and Germain, in preparation). Spun columns served as a simple and rather fast (4 minutes) separation medium for ^{125}I HA peptide-DR1 complexes from free radiolabeled peptide.

Soluble DR1 Forms Complexes With HA Peptide Within Five Minutes

An incubation time of 5 minutes at neutral pH yields a stoichiometric binding of 1:1 of ^{125}I-HA to sDR1. These fast forming complexes also dissociate rapidly with a $t_{1/2}$ of 10 minutes. This is similar to the half time for dissociation of the kinetic intermediate formed by PCC peptide and I-E[k]. Fast complexes are peptide and class II specific because sDR1 does not bind ^{125}I labeled moth cytochrome c peptide 88-103 although I-E[k] does. Excess unlabeled HA peptide inhibits the formation of fast dissociating complexes containing radioactive peptide. Formation of the fast forming complex varies with the concentration of peptide and the binding curve resembles a sigmoidal function. Little signal is detected at peptide concentrations below 50 μM, but plateau binding is reached at 100 μM, consistent with the notion that fast forming complexes also dissociate fast-- 10 minutes $t_{1/2}$ v.s. 4 minutes separation time for the spun column. Effective isolation of fast dissociating complexes using the spun column assay, may be possible at higher concentrations of peptide at which the equilibrium of the pseudo-first order reaction between peptide and sDR1 has shifted towards the product.

An incubation of 48 hours or longer lead to the appearance of complexes that dissociate slowly. Differences in dissociation rates for the two kinetic forms of peptide-class II complexes could be observed using a native gel. Stable ^{125}I HA-DR1 complexes appear as a well delineated spot whereas a trail of labeled peptide migrating towards the cathode could be seen in lanes with fast dissociating complexes. The protein stain of an SDS gel analyzing sDR1 after a 5 minutes incubation with peptide does not indicate any detectable changes in the migration pattern of the sDR1. Nevertheless, after 48 hours or longer incubation with peptide, large amounts of material consisting of C dimers could be found. The dissociation half time for stable complex is ~140 hours at 37°C.

Direct Demonstration that Stable Complex and Compact Dimer are Identical

The key experiment addressing the relationship between the affinity state of peptide binding and the structure of sDR1 involved comparing the kinetic rates for formation of C dimers versus that of stable complexes. The latter are identified by two consecutive rounds of spun column separations. A first spun column separates the sDR1 complexes from the unbound excess peptide in the incubation mixture. A second spun column separates peptide molecules that dissociate over a 3 hour incubation at the room temperature. It is crucial to point out that the absolute signal from 125I HA-DR1 complexes, i.e., fast dissociating + slow dissociating, remained constant for every sDR1 sample prepared, including the shortest incubation time of 15 minutes. The results indicate that the formation of C dimers parallels that of the stable complexes, strongly suggesting that C dimer and the stable complex are equivalent.

These observations clearly demonstrate that sDR1, even if initially in molecular aggregates can rapidly bind specific peptide, but C dimers accumulate slowly. The conformational transition necessary for stable/C complex formation is a rare event and reaches its maximum with a $t_{1/2}$ of ~8 hours. This formation rate is about 10 fold faster than the rate constant for peptide binding to native DR1(5, 23).

Relationship Between Formation of sDR1 Compact Dimer and Peptide Concentration

The experiments described above utilized a 100 micromolar peptide concentration and led to the full occupancy of sDR1. To reevaluate the two-step sequential kinetic model, it was necessary to analyze forward rates for formation of the compact/stable complex at different concentrations of peptide. The slope of a plot of half times obtained at different concentrations of peptide against inverse of peptide concentration, provides a means of distinguishing complex kinetic mechanisms as reviewed by Witt and McConnell (25). A pseudo-first order reaction with a very slow on rate would yield a positive slope, whereas a two-step sequential reaction that involves a fast forming intermediate would exhibit a slope of zero. A complex reaction in which a proportion of one reactant or product undergoes denaturation would be distinguished by a negative slope.

Various concentrations of radiolabeled peptide were used ranging between 10-200 μM. The measured $t_{1/2}$ to reach equilibrium for formation of compact dimer at all peptide concentrations tested were virtually identical, ~8 hours. However, significant differences in the plateau levels were observed. The plot of different half times versus the inverse of peptide concentration gave a slope of zero, consistent with the two step sequential kinetic model. Furthermore, once plateau levels were established, adjusting the concentration of peptide to the highest concentration used did not elevate the maximum stable binding. Maximal levels of C dimer was achieved only at concentrations near or above 100 μM, although even the lowest peptide concentration used was at least 10-20 fold in molar excess over sDR1. This experiment is consistent with a proportional irreversible inactivation of sDR1 at subsaturating concentrations of peptide. Overall, the results support a complex kinetic mechanism wherein rare conformational changes among fast dissociating peptide-sDR1 complexes cause slow accumulation of the stable complexes. In the absence of saturating peptide concentrations to support a steady state for low affinity complex, a proportion of sDR1 undergos inactivation. Forming a low affinity complex with the peptide then, might rescue class II from denaturation.

Similarities Between the Kinetic Behavior of Empty sDR1 and Peptide Occupied Class II

By using peptide free sDR1 for kinetic analysis, it was expected to circumvent the complexities that rise from self peptide dissociation. Surprisingly, the kinetic data using sDR1 were in close aggreement with those of Witt & McConnell (26) and Tempè & McConnell (27) using preparations of I-E^k or I-A^d that mostly $\underline{did}$ contain endogenous peptides. First, the half time for binding of labeled peptide was independent of peptide concentration. Second, they observed irreversible denaturation of a proportion of the αβ heterodimer at subsaturating peptide concentrations. Thus, they interpreted the data to be consistent with a reaction mechanism that included a peptide exchange step (26). Supported by direct experimental evidence they explained that peptide binding site of occupied class II is in a closed form and can become open when peptide dissociates and leaves the site. Open class II can bind a new peptide and become closed again. If the new peptide is not readily available, the open site undergoes chain dissociation (irreversible inactivation).

AGGREGATION OF EMPTY sDR1 AS A SUBSTITUTE FOR SELF PEPTIDES

As empty sDR1 forms molecular aggregates in the absence of peptide, and the kinetic findings for sDR1 resembles that of peptide occupied class II, it is tempting to consider aggregation of empty sDR1 molecules as a molecular analog for peptide occupancy.

There are several lines of evidence to support the notion that the peptide binding cleft of MHC class II may bind to any compatible peptide sequences either free or in a large protein (28, 29, and Sadegh-Nasseri unpublished). The peptide sequence may be derived from a partially unfolded proteins in the endosomal compartments, or perhaps be a partially folded self (molecular aggregation). A physiological example for this might be the prevalence of immunological dominance as a result of binding of class II to the dominant epitopes in a partially unfolded protein. The protection of the dominant epitope leaves the adjacent subdominant epitopes exposed for proteolysis as postulated by Sercarz and colleagues (30). As such, it is also reasonable to acknowledge that empty class II molecules are structurally incomplete and stable binding to peptides helps them in conformational adjustments necessary in reaching their minimum energy levels (31).

A MODEL FOR THE MECHANISM OF PEPTIDE BINDING TO CLASS II

Here I propose that aggregation as induced by mutual recognition of peptide binding sites of class II heterodimer, is a mechanism for protection of the empty peptide binding cleft of sDR1. Class II in the absence of peptide is fragile and vulnerable to denaturation by adverse environmental effects. In solution, and in the absence of peptides, larger molecular aggregates composed of different numbers of $\alpha\beta$ heterodimers are in equilibrium with smaller aggregates of two heterodimers (double dimer).

Large Aggregates $\rightleftharpoons$ Double Dimers

Double dimers are analogous to class II occupied with peptides but with a low affinity. The interaction of two $\alpha\beta$ heterodimers assembled through their peptide binding clefts can be inhibited by peptide molecules. In the presence of high concentrations of peptide, a competitive inhibition reaction takes place. Complexes formed with the peptide prevent *p /$\alpha\beta$ heterodimer from reentry into the dimer assembly,

Double Dimer + *p $\Rightarrow$ *p /$\alpha\beta$ heterodimer (low affinity)

Conformational change that induce stable complex might take place for a small percent of the molecules (slow accumulation). Meanwhile, those *p /$\alpha\beta$ heterodimers that did not become stable, dissociate from *p, and the empty $\alpha\beta$ heterodimers become prone to denaturation. Thus, at subsaturating peptide concentration;

1) $*p/\alpha\beta$ *Heterodimer* $\rightleftharpoons$ $\alpha\beta$ *Heterodimer* $+ *p$
$$\Downarrow$$
Denature

This situation can be largely rescued in the presence of excess concentrations of peptide to shift the equilibrium of the low affinty interaction between peptide and class II towards the class II occupancy;

2) $*p/\alpha\beta$ *Heterodimer* $\Rightarrow$ *Compact dimer*

Thus, maximal numbers of compact dimer would accumulate when reaction (1) is minimized.

CONCLUSION

Kinetic studies of binding of antigenic peptides with class II have demonstrated a consistent picture with the biological role for MHC molecules. An initial low affinity intermediate complex (selection process) is followed by a conformational change in the structure of class II leading to the stabilization of the intermediate complex. Stable complex has long life and is capable of stimulating T cells. Class II molecules that are free of peptide are unstable and susceptible to denaturation (inactivation) at the physiologic temperature. Thus, peptide binding to mature class II can occur as a result of a peptide exchange mechanism in which a self peptide can be replaced by a new peptide. The susceptibility of class II to denaturation in the absence of bound peptide promotes its binding to available protein ligands. Depending upon circumstances, empty MHC can bind invariant chain at the time of synthesis, or the unfolding proteins in the endosome/lysosome, and in the case of sDR1, lack of peptide and invariant chain forces the empty sDR1 to form molecular aggregates. A low affinity complexes of peptide-sDR1, first free the single peptide-$\alpha\beta$ heterodimer from the multimeric assemblies. Next, stable/compact folded complexes accumulate as individual rare events. The tendency for aggregation of the peptide free class II might perhaps be the driving force in formation of the nine chain molecular assembly observed for class II and the invariant chain

(1). Absence of peptide causes the open binding cleft to form floppy dimer with a closed binding cleft that eventually denature to individual α and β chains.

FUTURE EXPERIMENTS

Future experiments will focus on examining the validity of the above model. Different size aggregates would physically be separated and their interconversion from one size to the other, as influenced by peptide or temperature, will be monitored. It is possible that certain sequences of peptides would only form fast dissociating complexes and fail to convert to the stable complexes. The criteria to distinguish such peptides, and their effects on the organization of the molecular assemblies would be a rewarding research effort.

Do different sets of chemical bonds distinguish the distinct kinetic and structural forms of peptide-class II complexes? For instance, peptide main chain atoms forming hydrogen bonds with class II could be the initial force in formation of the kinetic intermediates. Whereas, compact dimer forms as the interactions between the peptide side chains with multiple peptide binding pockets within the class II binding cleft are established (L. Stern personal communication).

ACKNOWLEDGEMENTS

I wish to thank Dr. Harden McConnell for his guidance of the initial part of this work, Dr. Ronald Germain for support and discussions, Dr. Lawrence Stern for collaboration, discussions, communication of the crystal structure of peptide-class II prior to publication, and Drs. R. Germain, F. Nasseri, L. Stern and D. Margulies for the critical reading of the manuscript. Special thanks to Cancer Research Institute for a fellowship Grant during the initial part, and OraVax Inc. as my employer during the recent part of this work.

REFERENCES

1. **Marks, M. S., J. S. Blum and P. Cresswell.** 1990. Invariant chain trimers are sequestered in the rough endoplasmic reticulum in the absence of association with HLA class II antigens. *J Cell Biol 111:839.*

2. **Brown, J. H., T. S. Jardetzky, J. C. Gorga, L. J. Stern, R. G. Urban, J. L. Strominger and D. C. Wiley.** 1993. The three dimensional structure of the human class II histocompatibility antigen HLA-DR1. *Nature 364:33.*

3. **Buus, S., A. Sette, S. M. Colon, D. M. Jenis and H. M. Grey.** 1986. Isolation and characterization of antigen-Ia complexes involved in T cell recognition. *Cell 47:1071.*

4. **Roche, P. A. and P. Cresswell.** 1990. High-affinity binding of an influenza hemagglutinin-derived peptide to purified HLA-DR. *J Immunol 144:1849.*

5. **Jardetzky, T. S., J. C. Gorga, R. Busch, J. Rothbard, J. L. Strominger and D. C. Wiley.** 1990. Peptide binding to HLA-DR1: a peptide with most residues substituted to alanine retains MHC binding. *Embo J 9:1797.*

6. **Bjorkman, P. J., M. A. Saper, B. Samraoui, W. S. Bennett, J. L. Strominger and D. C. Wiley.** 1987. The foreign antigen binding site and T cell recognition regions of class I histocompatibility antigens. *Nature 329:512.*

7. **Buus, S., A. Sette, S. M. Colon and H. M. Grey.** 1988. Autologous peptides constitutively occupy the antigen binding site on Ia. *Science 242:1045.*

8. **Sadegh-Nasseri, S. and H. M. McConnell.** 1989. A kinetic intermediate in the reaction of an antigenic peptide and I-Ek. *Nature 337:274.*

9. **Dornmair, K., B. Rothenh:ausler and H. M. McConnell.** 1989. Structural intermediates in the reactions of antigenic peptides with MHC molecules. *Cold Spring Harb Symp Quant Biol 1:409.*

10. **Jensen, P. E.** 1990. Regulation of antigen presentation by acidic pH. *J Exp Med 171:1779.*

11. **Sadegh-Nasseri, S. and R. N. Germain.** 1991. A role for peptide in determining MHC class II structure. *Nature 353:167.*

12. **Nelson, C. A., S. J. Petzold and E. R. Unanue.** 1993. Identification of two distinct properties of class II major histocompatibility complex-associated peptides. *Proc. Natl. Acad. Sci. USA 90:1227.*

13. **Germain, R. N. and A. G. Rinker Jr.** 1993. Peptide binding inhibits protein aggregation of invariant-chain free class II dimers and promotes selective cell surface expression of occupied molecules. *Nature 363:725.*

14. **Germain, R. N. and L. R. Hendrix.** 1991. MHC class II structure, occupancy and surface expression determined by post-endoplasmic reticulum antigen binding. *Nature 353:134.*

15. **Davidson, H. W., P. A. Reid, A. Lanzavecchia and C. Watts.** 1991. Processed antigen binds to newly synthesized MHC class II molecules in antigen-specific B lymphocytes. *Cell 67:105.*

16. **Viville, S., J. Neefjes, V. Lotteau, A. Dierich, M. Lemeur, H. Ploegh, C. Benoist and D. Mathis.** 1993. Mice lacking the MHC class II-associated invariant chain. *Cell 72:635.*

17. **Bikoff, E. K., L.-Y. Huang, V. Episkopou, J. Van Meerwijk, R. N. Germain and E. J. Robertson.** 1993. Defective major histocompatibility complex class II assembly, transport, peptide acquisition, and CD4+ T cell selection in mice lacking invariant chain expression. *J Exp Med 177:1699.*

18. **Harding, C. V., R. W. Roof, P. M. Allen and E. R. Unanue.** 1991. Effects of pH and polysaccharides on peptide binding to class II major histocompatibility complex molecules. *Proc Natl Acad Sci U S A 88:2740.*

19. **Reay, P. A., D. A. Wettstein and M. M. Davis.** 1992. pH dependence and exchange of high and low responder peptides binding to a class II MHC molecule. *EMBO J. 2829.*

20. **Jensen, P. E.** 1992. Long-lived complexes between peptide and class II major histocompatibility complex are formed at low pH with no requirement for pH neutralization. *J Exp Med 176:793.*

21. **Guagliardi, L. E., B. Koppelman, J. S. Blum, M. S. Marks, P. Cresswell and F. M. Brodsky.** 1990. Co-localization of molecules involved in antigen processing and presentation in an early endocytic compartment. *Nature 343:133.*

22. **Neefjes, J. J., V. Stollorz, P. J. Peters, H. J. Geuze and H. L. Ploegh.** 1990. The biosynthetic pathway of MHC class II but not class I molecules intersects the endocytic route. *Cell 61:171.*

23. **Stern, L. J. and D. C. Wiley.** 1992. The human class II MHC protein HLA-DR1 assembles as empty alpha beta heterodimers in the absence of antigenic peptide. *Cell 68:465.*

24. **Sadegh-Nasseri, S., L. J. Stern, D. C. Wiley and R. N. Germain.** 1993. Distinct kinetic states of MHC class II peptide interaction correlate with different class II structures. *Journal of Cellular Biochemistry 17C:73.*

25. **Witt, S. N. and H. M. McConnell.** 1993. The kinetics of peptide reactions with class II major histocompatibility complex membrane proteins. *Accounts in Chemical Research 26:442.*

26. **Witt, S. N. and H. M. McConnell.** 1991. A first-order reaction controls the binding of antigenic peptides to major histocompatibility complex class II molecules. *Proc Natl Acad Sci U S A 88:8164.*

27. **Tempe, R. and H. M. McConnell.** 1991. Kinetics of antigenic peptide binding to the class II major histocompatibility molecule I-Ad . *Proc. Natl. Acad. Sci. USA 88:4661.*

28. **Sette, A., L. Adorini, S. M. Colon, S. Buus and H. M. Grey.** 1989. Capacity of intact proteins to bind to MHC class II molecules. *J Immunol 143:1265.*

29. **Jensen, P. E.** 1993. Acidification and disulfide reduction can be sufficient to allow intact proteins to bind class II MHC. *Journal of Immunology 150:3347.*

30. **Sercarz, E. E., P. V. Lehmann, A. Ametani, G. Benichou, A. Miller and K. Moudgil.** 1993. Dominance and crypticity of T cell antigenic determinants. *Annu Rev Immunol 11:729.*

31. **Sadegh-Nasseri, S. and R. N. Germain.** 1992. How MHC class II molecules work: peptide- dependent completion of protein folding. *Immunol Today 13:43.*

13

INTERACTION OF MHC CLASS II MOLECULES WITH ENDOPLASMIC RETICULUM RESIDENT STRESS PROTEINS

W. Timothy Schaiff and Benjamin D. Schwartz

RETENTION OF CLASS II MOLECULES IN THE ENDOPLASMIC RETICULUM IN THE ABSENCE OF INVARIANT CHAIN

Major histocompatibility complex (MHC) class II molecules are highly polymorphic molecules consisting of an α-chain and a β-chain which, soon after synthesis, associate with a third, nonpolymorphic molecule, the invariant chain (Ii) (1). Initially it was thought that Ii did not influence the ability of class II molecules to be expressed on the cell surface. Indeed, in early experiments using transfected cells expressing class II in the absence of Ii, the class II molecules appeared to be efficiently expressed on the cell surface (2,3). However, when total cellular class II molecules were examined in subsequent experiments, the "efficiency" of surface expression of class II molecules in the absence of Ii was shown to be not as high as originally thought. Experiments in our laboratory (4,5) and by Anderson and Miller (6) demonstrated that although class II molecules were expressed on the cell surface of Ii-negative transfected cell lines, the majority of the class II molecules remained in the endoplasmic reticulum (ER). These results were subsequently confirmed using class II$^+$/Ii$^-$ spleen cells from mutant mice which no longer posses functional Ii genes (7,8).

By studying the proteins that coimmunoprecipitate with class II molecules synthesized in the absence of Ii, several candidate proteins were identified that could be responsible for the ER retention of class II molecules (5). The identification of the molecules, possible models for the interaction of ER resident stress proteins with class II molecules, and the possible significance of class II-stress protein interactions are discussed below.

ASSOCIATION OF CLASS II WITH GRP94, ERp72, AND p74 IN THE ABSENCE OF INVARIANT CHAIN

Initial experiments analyzed immunoprecipitates of class II molecules prepared from lysates of metabolically radiolabeled fibroblastoid cell lines expressing class II molecules in the absence or presence of Ii. These experiments from our laboratory demonstrated that a prominent, radiolabeled 74-kDa protein coimmunoprecipitated with class II molecules from cells expressing class II in the absence of Ii, but did not coimmuno-precipitate with class II molecules from cells coexpressing Ii (Figure 1A, lane 2 vs. 4). In addition, a faint 97-kDa band was also occasionally seen to coimmunoprecipitate under the same conditions. In order to identify these proteins, class II molecules were immunoprecipitated from non-radiolabeled cells so that N-terminal sequences could be obtained from the 74- and 97-kDa proteins.

When these nonradiolabeled immunoprecipitates were analyzed it was found that the pattern of high molecular weight proteins associated with class II molecules in the absence of Ii was somewhat different from that observed using metabolically radiolabeled cells. To further explore the spectrum of proteins coimmunoprecipitating with class II in the absence of Ii, class II molecules were immunoprecipitated from unlabeled cell lysates, electrophoresed on 8% gels, transferred to nitrocellulose, and stained with colloidal gold to visualize all proteins which coimmunoprecipitated with the class II molecules. Using this protocol, three prominent proteins of 94-, 74-, and 72-kDa were found to be associated with class II molecules in the absence of Ii (Figure 1B, lane 1). In immunoprecipitates of radiolabeled class II molecules, the 94-kDa band was only faintly visible and the 72-kDa protein was not observed. These results indicate that the 74-kDa molecule associated with class II in the absence of Ii is from a pool of molecules synthesized during the metabolic labeling period. However, the 94- and 72-kDa molecules associated with the class II molecules must be from a pool of molecules synthesized prior to the labeling period.

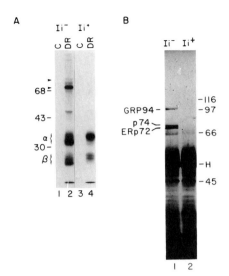

Fig. 1. Coimmunoprecipitation of 94-, 74-, and 72-kDa proteins with MHC class II molecules in the absence of invariant chain. Class II molecules were immunoprecipitated from metabolically radiolabeled cells (A) or from unlabeled cells (B) and run on 10% or 8% SDS-PAGE gels, respectively. (A) Fluorograph of class II molecules immunoprecipitated from cells transfected with class II alone (Ii⁻), or from cells cotransfected with Ii (Ii⁺). Positions of 74- and 94-kDa molecules are indicated by arrow heads. α: DRα-chains, β: DRβ-chains, C: control antibody, DR: anti-DR antibody. (B) Colloidal gold stained blot of class II molecules immunoprecipitated from unlabeled transfectants expressing class II alone (Ii⁻) or coexpressing Ii (Ii⁺).

Identification Of 94- And 72-kDa Proteins As GRP94 And ERp72

GRP94. N-terminal amino acid sequence analysis of the 97-kDa protein revealed identity with the 97-kDa ER resident glucose regulated protein GRP94 (9,10). The identity was confirmed by western blot analysis using antisera specific for GRP94 (Figure 1A, lane 2). GRP94 was not coimmunoprecipitated with class II molecules from cells co-transfected with Ii (Figure 1A, lane 4). These data indicate that GRP94 stably associates with class II molecules only in the absence of Ii. However the data do not preclude the possibility that there is a brief, transient association of GRP94 with class II molecules in cell coexpressing Ii. If such a transient association does occur, there may be too few GRP94-class II complexes to detect using the methods described in these experiments.

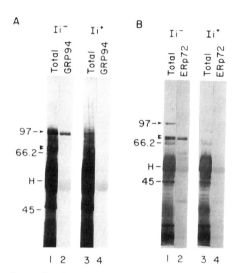

Fig. 2. Identification of coimmunoprecipitating 97- and 72-kDa proteins as GRP94 and ERp72, respectively. Class II molecules were immuno-precipitated from unlabeled transfected cells expressing either class II alone (Ii⁻) or cotransfected with Ii (Ii⁺). After electrophoresis on 8% SDS-PAGE gels, the gels were transferred to nitrocellulose and duplicate lanes either stained for total proteins immunoprecipitated (lanes 1 and 3) or probed with antisera specific for either GRP94 (A, lanes 2 and 4) or ERp72 (B, lanes 2 and 4). Positions of 94-, 74-, and 72-kDa proteins are indicated by arrowheads.

GRP94 is a 97-kDa member of the hsp90 family of heat shock proteins (10) and has been implicated as a chaperone protein involved in the folding of newly synthesized proteins in the ER. GRP94 has been found to be associated with several newly synthesized proteins in the ER such as mutant viral proteins (11) and unassembled immunoglobulin chains (12). Accumulation of misfolded proteins in the ER can lead to increased expression of GRP94 (13), however this increase was not observed in the transfectants expressing class II in the absence of Ii (5). It has been reported that GRP94 binds ATP (14) and has ATPase activity that, unlike GRP78/BiP, is not induced by peptides, but instead is stimulated to hydrolyse ATP by the presence of partially folded proteins (15). These data suggest that GRP94 plays an important chaperone role in the folding of class II molecules and that the lack of Ii prevents the dissociation of the GRP94-class II complex.

ERp72. N-terminal sequence analysis of the 72-kDa protein coimmunoprecipitating with class II molecules in cells lacking Ii revealed that the protein had a blocked N-terminus so that no sequence could be obtained. However, since it was known that the ER resident protein ERp72 also had a blocked N-terminus, antisera specific for ERp72 was used in western blot assays. As shown in Figure 2B, the 72-kDa protein was recognized by the anti-ERp72 antisera, indicating that this protein was ERp72. The fact that ERp72 is the lower band of the 72-/74-kDa doublet is demonstrated by side to side comparison of directly immunoprecipitated ERp72 and immunoprecipitated class II molecules from labeled Ii⁻ cells. As can be seen in Figure 3, ERp72 migrates to a position slightly below that of the 74-kDa protein coimmunoprecipitating with the class II molecules, indicating that the lower band of the doublet seen with un-labeled immunoprecipitates (Fig. 1B, lane 1, and Fig. 2, lanes 1 and 3) is ERp72 and the upper band is p74.

ERp72 is an ER resident stress protein with homology to another ER resident protein, protein disulfide isomerase (PDI) (16), which is involved in the correct formation of disulfide bonds of nascent proteins (Reviewed in refs. 17,18). Although ERp72 contains three of the active sites found in PDI (compared to two in PDI itself), ERp72 has not been shown to have PDI activity. However, it is likely that ERp72 is involved in maturation of nascent proteins in the ER. As with GRP94, it is possible that ERp72 transiently associates with and assists in proper folding and/or assembly of class II molecules in cells coexpressing Ii, but the association is too brief to detect using the methods described here.

p74. To date the identity of p74 is unknown. N-terminal sequence analysis failed to give sequence and no antibody tested has recognized the protein on western blots. In particular, one antibody that recognizes

denatured hsp70 proteins, including BiP, did not react with this molecule
on western blots. However, the possibility that p74 could be BiP should
not be definitively excluded at this point.

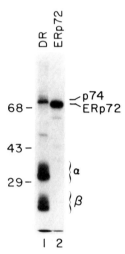

Fig. 3. Non-identity of ERp72 and p74. Class II molecules (lane 1) or
ERp72 (lane 2) were immunoprecipitated from the cell lysate of
metabolically radiolabeled transfectants expressing class II in the absence
of Ii, and the immunoprecipitates analyzed by 10% SDS-PAGE. The
positions of p74, ERp72, and the DRα- and β-chains are indicated at the
right of the figure.

POSSIBLE MODELS OF ER RESIDENT STRESS PROTEIN
INTERACTION WITH CLASS II MOLECULES IN NORMAL CELLS

Because cells which normally express class II molecules also express Ii,
it is necessary to envision models of how the data obtained in the Ii⁻
transfected cell lines reflect how ER resident stress proteins interact with
class II molecules in cells expressing Ii. Three models are proposed
below. It should be noted that for brevity the models are highly simplified
and do not take into account oligomerization of class II/Ii complexes in the
ER or the role of antigenic peptides in folding of class II molecules in
post-Golgi compartments.

In the first model (Fig. 4A), the ER resident stress proteins act as general chaperone molecules and are required for the proper folding of class II molecules. Once the class II chains have achieved the proper conformation, Ii induces the release of the class II molecules from the

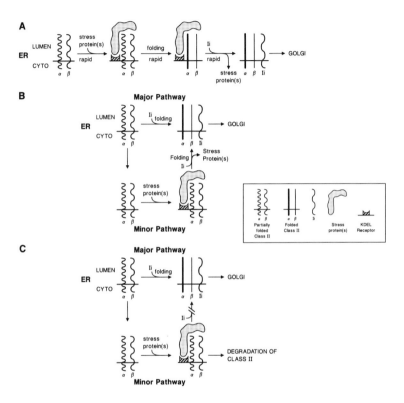

Fig. 4. *Models of ER resident stress protein-class II molecule interaction in cells naturally expressing class II and Ii*

chaperone protein(s), possibly by inducing a further conformational change. In the absence of Ii, the chaperone proteins remain associated with the class II molecules and can be coimmunoprecipitated as described above. If this model is correct, the association of the chaperones with class II molecules, the folding of the class II molecules, and the displacement of the chaperone molecules by Ii would have to be a very rapid process since the class II-chaperone complex has not yet been detected in cells naturally expressing class II and Ii.

In the second and third models (Fig. 4, B and C), the majority of class II molecules associate with Ii immediately after synthesis of the class II chains. The association with Ii then prevents the association of class II molecules with the stress proteins. In these models the stress proteins act as "scavengers" which bind to and retain the few class II molecules which are not associated with Ii. In the second model (Fig. 4B), Ii is able to displace the stress proteins allowing the class II-Ii complex to exit the ER. In the third model (Fig. 4C), Ii does not displace the stress proteins from the class II molecules and the stress proteins target the class II molecules for degradation.

Again, it should be pointed out that these models are highly simplified and that it is possible that all three models could be involved in the proper assembly of class II-Ii complexes, with different ER resident stress proteins involved in folding, retaining, and/or "tagging" class II molecules not associated with Ii. To date, there is no direct evidence to support or rule out any of the proposed models. Some indirect evidence supporting the first model (Fig. 4A), is the proposed role of GRP94 and ERp72 in the proper folding of nascent proteins, as discussed above. Such a role for these ER resident stress proteins would indicate that they are likely involved in folding of the nascent class II chains in all cells expressing class II molecules.

Indirect evidence against GRP94, ERp72, or p74 acting as general chaperone molecules in the synthesis of class II molecules is that these stress proteins are not found coimmunoprecipitated with class II molecules from cells which coexpress Ii. If these stress proteins were required for the proper folding of class II molecules regardless of the presence of Ii, it would be predicted that some of these molecules should be immuno-precipitated along with nascent class II chains and/or molecules not yet associated with Ii. If the first model is correct, then, as stated above, the process of association, folding , and displacement must be extremely rapid so that there are very few, and therefore undetectable levels of, class II-stress protein complexes in normal cells. More sensitive detection methods are required to determine if in fact there is a transient association of nascent class II molecules with GRP94, ERp72, or p74.

The lack of detectable class II-stress protein complexes in cells expressing Ii does give some indirect support to the second and/or third models of stress protein interaction with class II molecules. Because there is an excess of Ii relative to class II molecules in the ER (19), there are very few class II molecules present in the ER which are not associated with

Ii. Therefore, the number of class II-stress protein complexes would be very low, and therefore undetectable at the sensitivity of the assays used.

FUTURE DIRECTIONS

In this review two of the proteins likely to be involved in the folding and retention of class II molecules in the ER have been discussed. It is likely that, in addition to these proteins and possibly p74, there are numerous other proteins involved in these processes. In fact, Schreiber, et.al. (20), have recently described the association of the transmembrane ER protein calnexin with class II molecules. Additional studies need to be performed to determine what other proteins are involved in the folding and assembly of class II molecules in the ER and how these proteins interact with class II molecules and with each other during this process.

Three models were proposed as to how GRP94, ERp72, and/or p74 might interact with class II molecules, assist in the proper folding of the class II molecules, and/or retain in the ER class II molecules not associated with Ii. Experiments need to be performed to determine which model(s) is correct. In order to test the first model, it is necessary to develop more sensitive assays and techniques capable of detecting the transient association of class II molecules with stress proteins. Experiments also need to be performed to determine if Ii can displace these stress proteins once they have bound to class II molecules. If any of the stress proteins described is not displaced by Ii, then this protein is likely to be marking the class II molecule for degradation as described in the model in Fig. 4C. Displacement of stress proteins from class II molecules by Ii would support a role for these proteins as described in models in Fig. 4A or 4B.

Perhaps, most importantly, the functional significance of the association and retention of class II molecules in the ER by resident stress proteins must be determined. It is clear that the immune system has developed intricate safety mechanisms to ensure that class II molecules or their individual chains are not inappropriately expressed on the cell surface. It would be of interest to determine the consequences of altering the levels of the ER resident stress proteins described such that more of the class II molecules are allowed to escape the ER in the absence of Ii. In such a case, would the separation between the endogenous and exogenous pathways of antigen processing and presentation be further blurred by class II presentation of endogenous peptides which normally do not associate with class II molecules? Another fail safe mechanism for the system must also be taken into consideration in such a scenario. If class II molecules which escape the ER in the absence of Ii are capable of binding peptides in the ER, then SDS stable dimers should be present in cells expressing class II in the absence of Ii. In Ii knockout mice, it was demonstrated that the class II molecules were in fact not stable in the presence of SDS and were functionally empty (7,8). These data indicate that a mechanism other than retention must also be in place to prevent the improper association of peptides with class II molecules in the ER in the absence of Ii. One possible explanation is that class II molecules which escape the ER in the absence of Ii do so by exiting the ER before associating with ER resident stress proteins. If the stress proteins are required for proper folding, these class II molecules might not be folded properly, and thus, might not be able to functionally bind peptides in the ER. In addition, this lack of folding could also account for the altered conformation of surface class II molecules expressed in the absence of Ii (21).

In conclusion, although several ER resident proteins have been identified that are likely to play a role in the biosynthesis of class II molecules, the delineation of this role and the consequences of the interaction of class II molecules with ER stress proteins have yet to be determined.

REFERENCES

1. **Kvist, S., K. Wiman, L. Claesson, P.A. Peterson, and B. Dobberstein.** 1982. Membrane insertion and oligomeric assembly of HLA-DR histocompatibility antigens. *Cell 29*:61.

2. **Miller, J., and R.N. Germain.** 1986. Efficient cell surface expression of class II MHC molecules in the absence of associated invariant chain. *J. Exp. Med. 164*:1478.

3. **Sekaly, R.P., C. Tonnelle, M. Strubin, B. Mach, and E.O. Long.** 1986. Cell surface expression of class II histocompatibility antigens occurs in the absence of the invariant chain. *J. Exp. Med. 164*:1490.

4. **Schaiff, W.T., K.A. Hruska, Jr., C. Bono, S. Shuman, and B.D. Schwartz.** 1991. Invariant chain influences post-translational processing of HLA-DR molecules. *J. Immunol. 147*:603.

5. **Schaiff, W.T., K.A. Hruska, Jr., D.W. McCourt, M. Green, and B.D. Schwartz.** 1992. HLA-DR associates with specific stress proteins and is retained in the endoplasmic reticulum in invariant chain negative cells. *J. Exp. Med. 176*:657.

6. **Anderson, M.S., and J. Miller.** 1992. Invariant chain can function as a chaperone protein for class II major histocompatibility complex molecules. *Proc. Natl. Acad. Sci. USA. 89*:2282.

7. **Viville, S., J. Neefjes. V. Lotteau, A. Dierich, M. Lemeur, H. Ploegh, C. Benoist, and D. Mathis.** 1993. Mice lacking the MHC class II-associated invariant chain. *Cell 72*:635.

8. **Bikoff, E.K., L-Y. Huang, V. Episkopou, J. van Meerwijk, R.N. Germain, and E.J. Robertson.** 1993. Defective major histocompatibility complex class II assembly, transport, peptide acquisition, and CD4[+] T cell selection in mice lacking invariant chain expression. *J. Exp. Med. 177*:1699.

9. **Lee, A.S., J. Bell, and J. Ting.** 1984. Biochemical characterization of the 94- and 78-kilodalton glucose-regulated proteins in hamster fibroblasts. *J. Biol. Chem. 259*:4616.

10. **Mazzarella, B.A., and M. Green.** 1987. ERp99, an abundant, conserved glycoprotein of the endoplasmic reticulum, is homologous to the 90-kDa hear shock protein (hsp90) and the 94-kDa glucose regulated protein (GRP94). *J. Biol. Chem. 262*:8875.

11. **Navarro, D., I. Qadri, and L. Pereira.** 1991. A mutation in the ectodomain of herpes simplex virus 1 glycoprotein B causes defective processing and retention in the endoplasmic reticulum. *Virology 184*:253.

12. **Melnick, J., S. Aviel, and Y. Argon.** 1992. The endoplasmic reticulum stress protein GRP94, in addition to BiP, associates with unassembled immunoglobulin chains. *J. Biol. Chem. 267*:21303.

13. **Kozutsumi, Y., M. Segal, K. Normington, M-J. Gething, and J. Sambrook.** 1988. The presence of malfolded proteins in the endoplasmic reticulum signals the induction of glucose-regulated proteins. *Nature 332*:462.

14. **Clairmont, C.A., A. De Maio, and C.B. Hirschberg.** 1992. Translocation of ATP into the lumen of rough endoplasmic reticulum-derived vesicles and its binding to luminal proteins including BiP (GRP78) and GRP94. *J. Biol. Chem. 267*:3983.

15. **Li, Z., and P.K. Srivastava.** 1993. Tumor rejection antigen gp96/grp94 is an ATPase: implications for protein folding and antigen presentation. *EMBO J. 12*:3143.

16. **Mazzarella, R.A., M. Srinivasen, S.M. Haugejordan, and M. Green.** 1990. ERp72, an abundant luminal endoplasmic reticulum protein, contains three copies of the active site sequences of protein disulfide isomerase. *J. Biol. Chem. 265*:1094.

17. **Freedman, R.B., N.J. Bulleid, H.C. Hawkins, and J.L. Paver.** 1989. Role of protein disulphide-isomerase in the expression of native proteins. *Biochem. Soc. Symp. 55*:167.

18. **Freedman, R.B.** 1989. Protein disulfide isomerase: multiple roles in the modification of nascent secretory proteins. *Cell 57*:1069.

19. **Marks, M.S., J.S. Blum, and P. Cresswell.** 1990. Invariant chain trimers are sequestered in the rough endoplasmic reticulum in the absence of association with HLA class II antigens. *J. Cell Biol. 111*:839.

20. **Schreiber, K.L., M.P. Bell, C.J. Huntoon, S. Rajagopalan, M.B. Brenner, and D.J. McKean.** Class II histocompatibility molecules associated with calnexin during assembly in the endoplasmic reticulum. *Int. Immunol. In Press.*

21. **Peterson, M., and J. Miller.** 1990. Invariant chain influences the immunological recognition of MHC class II molecules. *Nature 345*:172.

14

INVARIANT CHAIN: VARIATIONS IN FORM AND FUNCTION

*Jim Miller, Mark Anderson, Lynne Arneson, Beatrice Fineschi,
Michelle Morin, Marisa Naujokas, Mary Peterson, Jon Schnorr,
Kevin Swier, and Linda Zuckerman*

INTRODUCTION

Invariant chain (Ii) is a nonpolymorphic glycoprotein that associates with class II rapidly after synthesis in the endoplasmic reticulum (ER) and remains associated throughout Golgi transport. In the trans-Golgi, this complex is diverted from the bulk flow of proteins to the cell surface and is transported to and retained within an endosomal compartment. It is this endosomal compartment where class II interacts with internalized antigens. The same proteases that are thought to generate antigenic peptides are thought to degrade Ii, allowing for class II-Ii dissociation. Class II free of Ii and loaded with peptide is exported to the plasma membrane where it is available for recognition by T cells. Although Ii is nonpolymorphic, it is not homogeneous and several different molecular isoforms have been identified (Table 1). Recently, several different activities have been attributed to Ii that are thought to modulate the intracellular transport and function of class II molecules (for review see (1)). Interestingly, some of the functions associated with Ii have been attributed to distinct molecular forms of Ii. In this review, we will summarize the functions of Ii, focusing on the specific roles of alternate forms of Ii, and suggest key unanswered questions that will need to be solved in order to understand how the generation of different Ii isoforms are regulated and how these various forms interact in mediating class II-restricted antigen presentation to T cells.

ROLE OF INVARIANT CHAIN IN CLASS II BIOSYNTHESIS

Assembly And Folding In The Endoplasmic Reticulum

Table I. Molecular Isoforms Of Invariant Chain[a]

Isoform	Characteristics
p31, p41[b]	Alternative splicing
	p41 contains an additional 64 aa segment between residues 191 and 192 in p31
p33, p43	Alternative initiation in human, but not mouse, Ii
p25	Proteolytic processing in ER (aa 98-215)
p28, p21, p18, p12, p10	Proteolytic processing in endosome
Ii-CS	Alternative glycosylation at is aa 201 represents ~2-5% of total Ii

[a]Numbering corresponds to the murine p31 sequence. Murine p31 is 215 aa and human Ii is 216 aa; human Ii has an aa insertion after positions 25 and 136 and a deletion of position 184, relative to murine Ii. For historical reasons, human p31 is often referred to as p33 and p33 as p35.
[b]The p31:p41 ratio is ~10:1 in B cells and L cell transfectants, ~4:1 in dendritic cells, ~3:1 in macrophages, ~1:1 in Langerhans cells (2), and ~1:1 in thymus (3).

Shortly after biosynthesis in the ER, Ii assembles into trimers (4), that can contain both p31 and p41 (5) and in human cells can also contain p33 and/or p43 (6). This Ii trimer is then sequentially loaded with three class II dimers, resulting in the generation of a nine chain complex (6,7). This nine chain complex remains intact throughout Golgi transport and targeting into endosomal compartments.

Although class II α and β chains still assemble in the absence of Ii, they are misfolded in the ER, resulting in the loss of certain mAb epitopes (8-10), decreased efficiency of egress from the ER to the Golgi (10,11), and stable association with resident ER heat shock proteins (12). Class II does eventually arrive at the plasma membrane (PM) in Ii-negative cell lines, however it can be more extensively glycosylated during transit through the Golgi (10) and it remains in an altered conformation as detected by mAb epitopes (8,9).

In transfection studies Ii can have dramatic effects on the rate of transport of class II out of the ER, although it has little effect on the steady state level of class II expressed at the cell surface (8,9,13,14). In contrast, in Ii-negative mice, the level of class II at the cell surface is significantly diminished (3,15). One possible interpretation of this apparent discrepancy is that the quality control system that retains misfolded proteins in the ER is less efficient in cultured cell lines than it is in normal cells in vivo. Thus, in the Ii-negative mice the class II that assembles in an altered conformation in the absence of Ii is efficiently retained, rarely gaining access to the PM, while in the transfectants, these molecules are temporally retained in the ER, but eventually slip through to the PM.

Invariant Chain Interferes With Class II-Peptide Association

One function that has been proposed for Ii is that it can protect the class II peptide binding site early in biosynthesis allowing for association of class II with exogenous peptide encountered later in biosynthesis (16,17). This hypothesis is based in the observation that class II-Ii complexes do not contain any endogenous peptides (18), yet class II associated with Ii is not available for binding to soluble peptides (16,17). Once Ii is dissociated, the class II peptide binding site is now free for loading with peptides (16,19). Furthermore, soluble forms of Ii can compete for class II-peptide binding (17). Thus, it is thought that Ii may block the class II peptide binding site early in biosynthesis, allowing class II to bind peptides only after Ii dissociation in endosomes.

Recently a fragment of Ii (CLIP) has been found stably associated with class II (18,20-23). CLIP is an overlapping set of peptides that span the region 80-104 of Ii. Synthetic peptides corresponding to this region bind with high avidity to class II and can compete with antigenic peptides for class II binding (22,24,25). In cell lines that are defective in antigen presentation, CLIP-associated class II can represent as much as 50% of the total class II molecules (24,25). From the available data it is not clear whether the ability of CLIP to compete is steric or allosteric. However, one intriguing possibility is that the CLIP region of Ii actually occupies the class II peptide binding site in the native Ii molecule as well. This model would account for the ability of Ii to block class II-peptide association and may provide a mechanism whereby CLIP-containing Ii fragments actually enhance class II-peptide association (1,26).

Ii may influence class II folding in the ER by a number of mechanisms, but one interesting possibility is that its principal function is to act as a surrogate peptide for class II. For class I molecules, peptide association in the ER is required for efficient folding, assembly with β2-microglobulin, and transport to the cell surface (27). Likewise, class II association with Ii is required for proper class II folding and efficient transport out of the ER. The ability of Ii to function as a surrogate peptide may account both for the ability of Ii to inhibit class II-peptide association and to promote correct folding of class II. Interestingly, the inability of class II to fold properly in the absence of Ii suggests that either Ii must have additional effects on class II folding or that analogous class II binding peptides are simply not available in the ER.

Post-Golgi Transport Of Class II

After maturation through the Golgi the class II-Ii complex is transported into endosomal compartments where Ii undergoes proteolysis and dissociation from class II and class II encounters antigenic peptides. Although this phenomenon is well established, the route by which the complex enters the endosomal compartment, the signals that determine this transport, and the exact localization of this compartment in the endosomal pathway are still matters of much discussion.

Golgi To Endosomal Transport. Although the current bias is that class II-Ii complexes are transported directly from the trans-Golgi to endosomes, this route has not been formally documented. The most compelling argument for direct Golgi-endosomal transport is that in the trans-Golgi class II appears to segregate from proteins that follow the constitutive route to the cell surface (28). Alternatively, class II-Ii complexes may follow the default pathway to the PM and enter endosomal compartments after rapid internalization (29). This is a difficult issue to resolve, as is evident from the continued debate on the transport pathway of lysosomal membrane proteins (30). Failure to detect the cell surface intermediate in many experiments could simply reflect a very efficient and rapid internalization step. Detection of a cell surface intermediate in cells expressing high levels of class II-Ii complexes could simply allow for more efficient detection of a naturally occurring intermediate or could reflect saturation of the trans-Golgi sorting step allowing the excess complexes to follow the default pathway to the PM (31). Although this distinction may seem somewhat trivial, it is important for understanding intracellular sorting pathways and immunologically relevant because direct Golgi to endosome transport and PM internalization may provide access of class II to different endosomal compartments and to the peptides generated within them.

Endosomal Localization Signals In The Class II-Invariant Chain Complex. Although class II in the absence of Ii can gain access to endosomes (32-34), coexpression of Ii can dramatically enhance the localization of class II to endocytic compartments (6,32,34-36). This signal is encoded within the cytosolic tail of Ii and deletion of this sequence results in transport of Ii alone (35,37) and of class II-Ii complexes (29,38) to the PM. Interestingly, we have recently found that the ability of Ii to localize to endosomal compartments requires multimerization, and class II-Ii complexes that only contain a single intact Ii cytosolic tail are found at the cell surface (5). Endosomal localization can be mediated either by direct targeting or by retention. There is clear evidence that the cytosolic tail of Ii encodes an endosomal retention signal (39,40). The best evidence that Ii also contains an endosomal targeting signal is that in the absence of class II Ii can gain access to endosomal compartments (35-37). However, it is not yet resolved if Ii molecules follow the same route or end up in the same compartments as do class II-Ii complexes (35,41).

There is also growing evidence that class II itself contains an endosomal localization signal, possibly encoded within aa 80-82 of the class II β chain (32). The ability of Ii-negative cells to efficiently present even some antigens suggests that class II can gain access to endosomal compartments (9,15,42-44). This has been confirmed by immunolocalization studies (32-34). Resolution of this controversy will require additional studies. None of the models are mutually exclusive, and the redundancy in intracellular transports signals observed in other systems (30,45) suggests that both class II and Ii may have independent endosomal targeting signals that work in concert to direct the transport of class II into the appropriate antigen processing and peptide association compartment(s).

Class II-Peptide Loading Compartment. Once the class II-Ii complex arrives in endosomes, Ii is degraded and class II associates with peptides. Disruption of Ii degradation results in the intracellular retention of class II-Ii complexes and the failure to generate class II-peptide complexes (39,40,46). Ii proteolysis results in the generation of a nested set of Ii fragments that contains the amino terminal endosomal retention signal and remain associated with class II (47-49). In addition some of these proteolytic intermediates differ by the presence or absence of the CLIP region that is thought to interfere with class II-peptide binding. This regulated proteolysis of Ii suggests a model (1) whereby proteolytic processing intermediates of Ii may regulate class II-peptide loading and complete dissociation from Ii would be required to release class II to the PM. The observation that the majority of class II remains associated with CLIP peptides in cell lines that are defective in antigen processing (24,25) suggests that at least some additional co-factors are necessary to promote class II-Ii dissociation. In support of this possibility, we have found that proteolytic cleavage of Ii from the cell surface of transfected cells expressing class II and cell surface forms of Ii does not free up the class II peptide binding site (50). However, it is not clear whether these putative co-factors are involved in Ii dissociation, peptide generation, peptide transport, or class II-peptide association. In addition, it is not clear whether Ii degradation, antigen processing, and peptide loading all take place within the same endosomal compartment. Recent data suggest that the class II localization compartment is a specialized compartment called MHC_{II}, that is similar to late endosomes (28), and may in fact be created by the expression of class II and/or Ii (36).

ROLE OF p41 IN ANTIGEN PRESENTATION

Many of the functions of Ii discussed above, in particular the ability of Ii to protect the class II peptide binding site early in biosynthesis and the ability of Ii to enhance the localization of class II within the endosomal compartments, predict that Ii would be critical for efficient antigen presentation to T cells. Surprisingly, although Ii can enhance antigen presentation, it does so in a very restricted fashion.

1. Ii can only enhance presentation of a subset of antigens (15,43,44). The nature of the antigens that are and are not influenced by Ii is not understood.

2. The co-expression of the p31 form of Ii, although it can effectively impart the class II folding, inhibition of peptide loading, and endosomal localization functions of Ii, has little effect on antigen presentation (9,44).

3. The co-expression of the p41 form of Ii can dramatically enhance antigen presentation (38,44), indicating that p41 must have a unique effect on class II.

4. The p41 form facilitates antigen presentation equivalently whether it is expressed as ~10% or as 100% of total Ii (44), indicating that the low level of p41 expressed by alternative splicing of the Ii gene may be sufficient for this function.

Both p31 And p41 Can Facilitate Class II Folding And Endosomal Localization

Although p31 and p41 can be expressed at widely different ratios in different cell types (2,3), both forms associate equally well with class II. Pulse-chase analyses of cells transfected either with p31 or p41 alone confirmed that both forms of Ii rapidly associate with class II early in biosynthesis and can increase the efficiency of class II transport from the ER to the Golgi (5,38,51). In addition, expression of either p31 or p41 alone is able to confer the conformational effects on class II necessary to generate the Ii-dependent mAb epitopes (44). These data indicate that both p31 and p41 associate with class II in a manner that can modify class II folding and early transport events. Furthermore, both p31 and p41 can enhance the localization of class II to endosomal compartments (36) and can effectively retain class II within these compartments (40).

If p31 can efficiently associate with class II, protect the class II peptide binding site, mediate class II folding and intracellular transport events, and retain class II within endosomal compartments, it is not clear why p31 does not enhance antigen presentation. One issue that we do not understand is what is limiting during class II-restricted antigen presentation events. Most of the functions of Ii listed above would increase the concentration of class II within endosomal compartments. In the case of antigen presentation by B cells, where antigen may be internalized as a single bolus following Ig-mediated endocytosis, the relative concentration of class II within the endosomal compartment within a given period of time may be the most critical factor. In this case p31 may have a dramatic effect on the efficiency of antigen presentation. In contrast, antigen presentation by L cell transfectants, where the p31 experiments have been performed, is mediated by fluid phase uptake of antigen over the course of a 24 hour assay. In this case intracellular proteases or chaperones may be the limiting factor and so the relative concentration of class II within endosomes (modulated by coexpression of p31) may have little discernable effect.

Intracellular Transport Of p41 Containing Class II-Invariant Chain Multimers

In contrast to p31, p41 does have a distinct effect on antigen presentation indicating that it does impart a unique effect above that of endosomal localization. Recent evidence in our lab indicates a biochemical correlate to this function of p41. We have found that dissociation of p41 from class II is delayed in comparison to p31 (51). In pulse chase experiments, loss of intact p31 and p41 proteins from class II immunoprecipitates occurs at a similar rate. However, amino terminal fragments derived from p41, but not p31, remain associated with class II for an extended period of time. We do not think that this difference in degradation results from an alteration in the structure and protease sensitivity of p41, because in mixed trimers that contain both p31 and p41, p41 can impart the delayed degradation onto the associated p31 chains (51). Although we cannot exclude a change in conformation that is transduced through all the chains in the trimer, these data are most consistent with the model that p41 may contain an independent transport signal that directs the class II-Ii complex to an endosomal compartment with altered proteolytic capacity. In this model, p41-dependent antigen presentation would represent the selective generation of a subset of antigenic peptides within this endosomal compartment.

The ability of p41 to confer altered proteolysis onto associated p31 molecules suggests that a single p41 molecule within an Ii trimer may be sufficient to direct the transport of the entire complex. If this is the case, in B cells where the ratio of p31:p41 is about 10:1, ~25% of the trimers would have at least one p41 and so the majority of class II would not gain access to this putative p41-dependent compartment. In contrast, in Langerhans cells, where the ratio is 1:1, ~90% of the trimers would contain at least one p41 molecule and the majority of class II would enter the p41-dependent compartment. This difference in localization of class II to different compartments may enable different APC to present different subsets of antigenic peptides.

At present this is the only biochemical difference between p41 and p31 that we have detected and it is not clear whether this accounts for the ability of p41 to enhance antigen presentation. Surprisingly, we have found that p41 can still enhance antigen presentation even in the absence of the endosomal localization signal in the cytosolic tail of Ii (38). These studies indicate that the putative transport signal encoded within p41 can function independently from the cytosolic tail signal and that this signal can function following internalization of class II-Ii complexes from the cell surface. Furthermore, these results suggest that the prolonged association of class II with the endosomal localization signal derived from the amino terminus of p41 is not the mechanism whereby p41 can enhance antigen presentation. Rather, other components of the putative p41-dependent endosomal compartment may be important in peptide generation or class II-peptide association.

THE CHONDROITIN SULFATE FORM OF INVARIANT CHAIN CAN FUNCTION AS AN ACCESSORY MOLECULE AT THE CELL SURFACE

We have recently found that coexpression of Ii with class II in EL4 cells can enhance the ability of these cells to stimulate primary allogeneic and SEB T cell responses (52). In contrast, coexpression of Ii was not required for efficient stimulation of peptide specific and allo-specific T cell hybridomas. Interestingly, we found that this function of Ii was provided by the chondroitin sulfate form of Ii (Ii-CS), a form that is coexpressed with class II at the cell surface (53). This finding in conjunction with the dependence of Ii on primary and not hybridoma T cell responses suggested that Ii-CS may function as an accessory molecule at the cell surface, either facilitating T cell/APC adhesion or directly signalling T cells. In either case, one could postulate the existence of a receptor expressed on the cell surface of T cells that binds Ii-CS. One potential candidate for this receptor is CD44. CD44 has been shown to function both in intracellular adhesion and costimulation of T cells (54). CD44 is expressed at high levels on thymocytes and memory T cells, and has been shown to bind hyaluronate and chondroitin sulfate (54). The ability to inhibit the Ii-dependent allogeneic response with both anti-CD44 antibodies and a soluble CD44-Ig fusion protein and the demonstration of direct binding between CD44 and Ii-CS strongly suggests that Ii-CS can function as an accessory molecule through interactions with CD44 (52).

In contrast to stimulation of bulk primary T cells, activation of established CD44-positive T cell clones is not enhanced by the presence of Ii-CS (55). Although we do not know whether this reflects the isoform of CD44 expressed by these T cells or the ability to signal appropriately through CD44, it does suggest that the ability of Ii-CS-CD44 interactions to enhance T cell activation may be restricted to subsets of T cells or to specific stages in T cell development. It is possible that Ii-CS may be most important in an early phase of T cell activation by an antigen presenting cell that either does not express other accessory molecules or is induced to express accessory molecules after an initial encounter with T cells. Alternatively, Ii-CS may be important in stimulating memory T cells, characterized in part by high expression of CD44. Ii-CS may allow stimulation of memory T cells by a broader array of antigen presenting cell types which lack other costimulatory molecules, thereby increasing the likelihood of stimulating these T cells during an immune response. The observation that CD44 is expressed at high levels in pro-thymocytes suggests that Ii-CS-CD44 interactions may be critical at some very early stage in T cell development.

Although antigen presenting cells isolated from peripheral lymph organs express very low levels of Ii-CS (53), it remains possible that subpopulations of APC may express more. CS addition occurs in two steps, xylose addition to a SerGly sequence in the ER and elongation of the CS side chain in the trans-Golgi. Which of these steps determines the amount of Ii-CS production is not known. Pulse chase analysis indicates that the subset of class II-Ii complexes that receive CS addition in the trans-Golgi are earmarked early in biosynthesis, as indicated by a very rapid transport from the ER to the Golgi and on to the cell surface (Ref 53 and L. Arneson and J. Miller, unpublished). Based on our observation that a single endosomal localization signal encoded by the Ii cytosolic tail is insufficient to target class II-Ii complexes to the endosome (5), we believe that class II-Ii-CS complexes may represent those molecules that did not form Ii trimers prior to association with class II. This monomeric class II-Ii complex would be available for CS addition and would be rapidly transported to the cell surface. The level of Ii-CS could also be regulated by modulating Ii-CS turnover. Ii-CS normally has a half life of about 60 minutes (53) that is dramatically extended by truncation of the cytosolic tail of Ii that contains the endosomal localization and internalization signals, but not by inhibition of lysosomal proteases (J. Schnorr and J. Miller, unpublished). Thus, turnover is likely to be mediated by internalization and CS removal. Regulation of Ii-CS internalization either intracellularly or by engagement with CD44 may dramatically extend the half life of this molecule and effectively increase the concentration at the cell surface.

MODEL FOR THE ROLE OF INVARIANT CHAIN IN CLASS II BIOSYNTHESIS AND ANTIGEN PRESENTATION

To provide a framework for designing additional experiments we have synthesized the available data into a model of the function of Ii at various stages of class II biosynthesis. This model does not reflect the uncertainties and controversies that are discussed in more detail above and is therefore highly speculative and highly subject to change.

Shortly after synthesis, Ii multimerizes into p31 homotrimers, p41 homotrimers, and p31/p41 mixed trimers, depending on the relative expression of p31 and p41 in a given cell type. In B cells only about 25% of the trimers will contain at least one p41 molecule, whereas in Langerhans cells p41 containing trimers will represent about 90% of the total. The preexisting Ii trimer then associates with class II molecules and when the nine chain complex is complete, it exits the ER and transits to the Golgi. Association of class II with either p31 or p41 can facilitate class II folding in the ER and transport to the Golgi. Ii may function as a surrogate peptide in the ER, interfering with peptide loading and enhancing folding. Although this function of Ii may be important for suppressing the presentation of peptides present in the ER, it may also reflect the necessity to protect class II from associating with newly synthesized and partially unfolded proteins in the ER. Nevertheless, the ability of Ii to protect class II from loading with peptides early in biosynthesis does not result in a generalized increase in the efficiency of peptide loading later in biosynthesis.

Once the class II-Ii complex arrives in the trans-Golgi, it is sorted from the bulk of secretory proteins that follow the constitutive route to the cell surface. The class II-Ii complex is transported directly from the Golgi to an endosomal compartment. The signals that regulate sorting and targeting and the endosomal compartment where the class II-Ii complex arrives are not fully understood. Once into the endosomal pathway, the cytosolic tail of Ii functions to retain the complex until Ii proteolysis and dissociation is complete, freeing class II for transport to the cell surface. Class II-Ii complexes that contain at least one p41 molecule sort into a different endosomal compartment from those that contain p31 homotrimers. Because the proteolytic machinery is different in these two compartments, degradation and dissociation of Ii in p41 containing complexes is delayed compared to those that only contain p31. In addition, class II localized in the p41 compartment gains access to a subset of antigenic peptides not available in the p31 compartment, accounting for the selective ability of p41 to enhance antigen presentation. Which of these compartments, if any, corresponds to the MHC_{II} compartment is not known. Class II in the absence of Ii probably gains access to the p31 compartment, accounting for antigen presentation by Ii negative cells. The retention function of Ii may account for the ability of p31 to facilitate the presentation of some antigens. In addition, the sequential proteolysis of Ii from the carboxy terminus into fragments that do and do not contain the CLIP region of Ii suggests the possibility that Ii degradation, peptide loading, and release of the endosomal retention signal may be an orchestrated series of events. Once class II has fully dissociated from Ii, it is transported to the cell surface and is available for recognition by T cells.

A subset of class II associates with Ii prior to Ii trimerization. Because no additional assembly is required, this three chain complex immediately exits the ER, transits through the Golgi, and becomes modified by the addition of CS to Ii. Trimerization of Ii may interfere with either xylose addition in the ER or CS elongation in the trans-Golgi. Because this complex only contains a single Ii cytosolic tail, the complex is either not transported to or not retained in the endosomal compartment and instead rapidly arrives intact at the cell surface. Once at the cell surface, Ii-CS can interact with CD44 and enhance activation of a subset of T cells. In the absence of CD44 engagement, this class II-Ii-CS complex is internalized, resulting in the dissociation of class II from Ii-CS, freeing the class II peptide binding site. This may provide a pool of class II that is available to associate with peptides in early endocytic compartments.

FUTURE DIRECTIONS

Although in the last few years we have learned much about the cell biology of class II restricted antigen presentation, there are still many unanswered questions. We have identified five key areas that need to be addressed in order to better understand the role of distinct molecular forms of Ii in the biosynthesis and function of class II.

1. What are the sorting signals that allow for segregation of Class II-Ii complexes in the trans-Golgi, targeting to endosomal compartments, and retention within these compartments? Are these signals encoded in class II, Ii, or both? How do these different signals interact and how are these intracellular targeting events mediated? What other cofactors are necessary?

2. How does CLIP regulate class II-peptide interactions? Does the CLIP region occupy the class II peptide binding site in the native Ii molecule? Do CLIP containing Ii fragments modulate the association of class II with antigenic peptides?

3. What is the compartment where class II-peptide association takes place? Is this the same compartment where Ii degradation and dissociation and/or antigen processing takes place? If not, what cofactors are required to facilitate class II-peptide interaction and binding?

4. How does the additional segment of p41 modulate the altered processing of Ii and enhance antigen presentation of class II?

5. What is the specific function of Ii-CS in T cell development and activation? How does Ii-CS-CD44 interaction differ from other potential costimulatory molecules? Are they redundant, independent, or synergistic? How is the production of Ii-CS regulated?

ACKNOWLEDGMENTS

We thank Dr. Andrea Sant and the members of her lab for extensive and fruitful discussions. We also thank all our colleagues for generously sharing reagents and techniques that have been instrumental in our experiments.

REFERENCES

1. Sant, A. J. and J. Miller. 1994. MHC class II antigen processing: Biology of invariant chain. *Curr. Opin. Immunol. 6:In press.*
2. Kampgen, E., N. Koch, F. Koch, P. Stoger, C. Heufler, G. Schuler and N. Romani. 1991. Class II MHC molecules of murine dendritic cells: synthesis, sialylation of invariant chain, and antigen processing capacity are down regulated upon culture. *Proc. Natl. Acad. Sci. USA. 88:3014.*
3. Bikoff, E., L. Huang, V. Episkopou, J. Meerwijk, R. Germain and E. Robertson. 1993. Defective major histocompatibility complex class II assembly, transport, peptide acquisition, and CD4+ T cell selection in mice lacking invariant chain. *J. Exp. Med. 177:1699.*
4. Marks, M., J. Blum and P. Cresswell. 1990. Invariant chain trimers are sequestered in the RER in the absence of association of HLA Class II antigens. *J. Cell Biol. 111:839.*
5. Arneson, L. and J. Miller. 1994. Invariant chain multimerization is necessary for endosomal localization of class-invariant chain complexes. *In preparation.*
6. Lamb, C. and P. Cresswell. 1992. Assembly and transport properties of invariant chain trimers and HLA-DR-invariant chain complexes. *J. Immunol. 148:3478.*
7. Roche, P., M. Marks and P. Cresswell. 1991. Formation of a nine subunit complex by HLA class II glycoproteins and the invariant chain. *Nature. 354:392.*
8. Rath, S., R. Lin, A. Rudensky and C. Janeway. 1992. T and B cell receptors discriminate major histocompatibility complex class II conformations influenced by the invariant chain. *Eur. J. Immunol. 22:2121.*
9. Peterson, M. and J. Miller. 1990. Invariant chain influences the immunological recognition of MHC Class II molecules. *Nature. 345:172.*

10. **Anderson, M. and J. Miller.** 1992. Invariant chain can function as a chaperone protein for class II major histocompatability complex molecules. *Proc. Natl. Acad. Sci. USA. 89:2282.*

11. **Schaiff, W., K. Hruska, C. Bono, S. Shuman and B. Schwartz.** 1991. Invariant chain influences post-translational processing of HLA-DR molecules. *J. Immunol. 147:603.*

12. **Schaiff, W., K. Hruska, D. McCourt, M. Green and B. Schwartz.** 1992. HLA-DR associates with specific stress proteins and is retained in the endoplasmic reticulum in invariant chain negative cells. *J. Exp. Med. 176:657.*

13. **Miller, J. and R. Germain.** 1986. Efficient cell surface expression of class II MHC molecules in the absence of associated invariant chain. *J. Exp. Med. 164:1478.*

14. **Sekaly, R., C. Tonnelle, M. Strubin, B. Mach and E. Long.** 1986. Cell surface expression of class II histocompatability antigens occurs in the absence of invariant chain. *J. Exp. Med. 164:1490.*

15. **Viville, S., J. Neefjes, V. Lotteau, A. Dierich, M. Lemeur, H. Ploegh, C. Benoist and D. Mathis.** 1993. Mice lacking the MHC class II-associated invariant chain. *Cell. 72:635.*

16. **Roche, P. and P. Cresswell.** 1990. Invariant chain association with HLA-DR molecules inhibits immunogenic peptide binding. *Nature. 345:615.*

17. **Teyton, L., D. O'Sullivan, P. Dickson, V. Lotteau, A. Sette, P. Fink and P. Peterson.** 1990. Invariant chain distinguishes between the exogenous and endogenous antigen presentation pathways. *Nature. 348:39.*

18. **Newcomb, J. and P. Cresswell.** 1993. Characterization of endogenous peptides bound to purified HLA-DR molecules and their absence from invariant chain-associated αβ dimers. *J. Immunol. 150:499.*

19. **Roche, P. and P. Cresswell.** 1991. Proteolysis of the class II-associated invariant chain generates a peptide binding site in intracellular HLA-DR molecules. *Proc. Natl. Acad. Sci. USA. 88:3150.*

20. **Rudensky, A., P. Preston-Hurlburt, S. Hong, A. Barlow and C. Janeway.** 1991. Sequence analysis of peptides bound to MHC class II molecules. *Nature. 353:622.*

21. **Hunt, D., H. Michel, T. Dickinson, J. Shabanowitz, A. Cox, K. Sakaguchi, E. Appella, H. Grey and A. Sette.** 1992. Peptides

presented to the immune system by the murine class II major histocompatibility complex molecule I-Ad. *Science. 256:1817.*
22. Chicz, R., R. Urban, W. Lane, J. Gorga, L. Stern, D. Vignali and J. Strominger. 1992. Predominant naturally processed peptides bound to HLA-DR1 are derived from MHC related molecules and are heterogeneous in size. *Nature. 358:764.*
23. Chicz, R., R. Urban, J. Gorga, D. Vignali, W. Lane and J. Strominger. 1993. Specificity and promiscuity among naturally processed peptides bound to HLA-DR alleles. *J. Exp. Med. 178:27.*
24. Sette, A., S. Ceman, R. Kubo, K. Sakaguchi, E. Apella, D. Hunt, T. Davis, H. Michel, J. Shabanowitz, R. Rudersdorf, H. Grey and R. DeMars. 1992. Invariant chain peptides in most HLA-DR molecules of an antigen processing mutant. *Science. 258:1801.*
25. Riberdy, J. and P. Cresswell. 1992. The antigen processing mutant T2 suggests a role for MHC-linked genes in class II antigen presentation. *J. Immunol. 148:2586.*
26. de Kroon, A. and H. McConnell. 1993. Enhancement of peptide antigen presentation by a second peptide. *Proc. Natl. Acad. Sci. USA. 90:8797.*
27. Bijlmakers, M. and H. Ploegh. 1993. Putting together an MHC class I molecule. *Curr. Opin. Immunol. 5:21.*
28. Peters, P., J. Neefjes, V. Oorschot, H. Ploegh and H. Geuze. 1991. Segregation of MHC class II molecules from MHC class I molecules in the Golgi complex for transport to lysosomal compartments. *Nature. 349:669.*
29. Roche, P., C. Teletski, E. Stang, O. Bakke and E. Long. 1993. Cell surface HLA-DR-invariant chain complexes are targeted to endosomes by rapid internalization. *Proc. Natl. Acad. Sci. USA. 90:8581.*
30. Trowbridge, I., J. Collawn and C. Hopkins. 1993. Signal-dependent membrane protein trafficking in the endocytic pathway. *Ann. Rev. Cell Biol. 9:129.*
31. Harter, C. and I. Mellman. 1992. Transport of the lysosomal membrane glycoprotein lgp120 to lysosomes does not require appearance on the plasma membrane. *J. Cell Biol. 117:311.*
32. Chervonsky, A., L. Gordon and A. Sant. 1994. A segment of the class II β chain modulates expression of MHC class II molecules in the endocytic pathway of antigen presenting cells. *Submitted.*

33. Salamero, J., M. Humbert, P. Cosson and J. Davoust. 1990.
    Mouse B lymphocyte specific endocytosis and recycling of MHC
    class II molecules. *EMBO J. 9:3489.*

34. Simonsen, A., F. Momburg, J. Drexler, G. Hammerling and O.
    Bakke. 1993. Intracellular distribution of the MHC class II
    molecules and the associated invariant chain in different cell
    lines. *Int. Immunol. 5:903.*

35. Lotteau, V., L. Teyton, A. Peleraux, T. Nilsson, Karlsson, S.
    Schmid, V. Quaranta and P. Peterson. 1990. Intracellular
    transport of class II MHC molecules directed by invariant chain.
    Nature. 348:600.

36. Romagnoli, P., C. Layet, J. Yewdell, O. Bakke and R. Germain.
    1993. Relationship between invariant chain expression and
    major histocompatibility complex class II transport into early
    and late endocytic compartments. *J. Exp. Med. 177:583.*

37. Bakke, O. and B. Dobberstein. 1990. MHC class II-associated
    invariant chain contains a sorting signal for endosomal
    compartments. *Cell. 63:707.*

38. Anderson, M., K. Swier, L. Arneson and J. Miller. 1993.
    Enhanced antigen presentation in the absence of the invariant
    chain endosomal localization signal. *J. Exp. Med. 178:1959.*

39. Neefjes, J. J. and H. L. Ploegh. 1992. Inhibition of endosomal
    proteolytic activity by leupeptin blocks surface expression of
    MHC class II molecules and their conversion to SDS resistant $\alpha\beta$
    heterodimers in endosomes. *EMBO J. 11:411.*

40. Loss, G. and A. Sant. 1993. Invariant chain retains MHC class II
    molecules in the endocytic pathway. *J. Immunol. 150:3187.*

41. Chervonsky, A. and A. J. Sant. 1994. In the absence of class II
    localization of Ii to endocytic compartments occurs by
    autophagy. *In preparation.*

42. Sekaly, R., S. Jacobson, J. Richert, C. Tonnelle, H. McFarland and
    E. Long. 1988. Antigen presentation to HLA Class II-restricted
    measles virus-specific T cell clones can occur in the absence of
    the invariant chain. *Proc. Natl. Acad. Sci. USA. 85:1209.*

43. Nadimi, F., J. Moreno, F. Momburg, A. Heuser, S. Fuchs, L.
    Adorini and G. Hammerling. 1991. Antigen presentation of
    hen egg lysozyme but not ribonuclease A is augmented by the
    MHC class II associated invariant chain. *Eur. J. Immunol.
    21:1255.*

44. **Peterson, M. and J. Miller.** 1992. Antigen presentation enhanced by the alternatively spliced invariant chain gene product p41. *Nature. 357:596.*
45. **Letourneur, F. and R. Klausner.** 1992. A novel di-leucine motif and a tyrosine-based motif independently mediate lysosomal targeting and endocytosis of CD3 chains. *Cell. 69:1143.*
46. **Mellins, E., L. Smith, B. Arp, T. Cotner, E. Celis and D. Pious.** 1990. Defective processing and presentation of exogenous antigens in mutants with normal HLA class II genes. *Nature. 343:71.*
47. **Nguyen, Q. and R. Humphreys.** 1989. Time course of intracellular associations, processing, and cleavages of Ii forms and class II MHC molecules. *J. Biol. Chem. 264:1631.*
48. **Pieters, J., H. Horstmann, O. Bakke, G. Griffiths and J. Lipp.** 1991. Intracellular transport and localization of major histocompatability complex class II molecules and associated invariant chain. *J. Cell Biol. 115:1213.*
49. **Maric, M., M. Taylor and J. Blum.** 1994. Endosomal aspartic proteinases as required for invariant chain processing. *Proc. Natl. Acad. Sci. USA. 91:In press.*
50. **Anderson, M., M. Peterson and J. Miller.** 1994. Proteolytic cleavage of cell surface invariant chain is insufficient to free the class II peptide binding site. *In preparation.*
51. **Fineschi, B., L. Arneson and J. Miller.** 1994. Differential proteolysis of invariant chain trimers containing a single p41 invariant chain component. *In preparation.*
52. **Naujokas, M., M. Morin, M. Anderson, M. Peterson and J. Miller.** 1993. The chondroitin sulfate form of invariant chain can enhance stimulation of T cell responses through interaction wiith CD44. *Cell. 74:257.*
53. **Sant, A., S. Cullen and B. Schwartz.** 1985. Biosynthetic relationships of the chondroitin sulfate proteoglycan with Ia and invariant chain glycoproteins. *J. Immunol. 135:416.*
54. **Lesley, J., R. Hyman and P. Kincade.** 1993. CD44 and its interaction with extracellular matrix. *Adv. Immunol. 54:271.*
55. **Zuckerman, L. and J. Miller.** 1994. Role of costimulatory molecules in T cell activation and anergy induction. *In preparation.*

15

MHC CLASS II PROCESSING PATHWAY AND A ROLE OF SURFACE INVARIANT CHAIN

Norbert Koch, Gerhard Moldenhauer, and Peter Möller

ANTIGEN PRESENTING CELLS

The presentation of antigen requires solely surface expression of class II molecules. This has been demonstrated by gene transfer and expression of class II in fibroblast cells that usually do not express these molecules nor present antigen (1). These class II transfected cells efficiently present antigenic peptides. Antigen presentation can also be achieved by soluble class II molecules which were immobilized on a solid surface, indicating that the presence of class II molecules is the minimal requirement to present antigen (2). The presentation of native antigen however, requires living cells. This emphasizes the importance of intracellular trafficking and metabolism of cells to exert antigen processing and presentation. Although the constitutive expression of MHC class II is restricted to cells of the immune system and some epithelial and endothelial cells, many different cell types can function as APC. Upon stimulation with cytokines class II is induced in a substantial number of cell types (3). These diverse APC add a great variety to the MHC class II processing pathway.

Degradation of antigen by APC requires intracellular proteases. The composition and the nature of these proteases may change according to cell type. Thus the presence of distinct sequence specific proteases has been implicated to generate disparate antigenic fragments from the same antigen (4). This could suggest that the nature of the APC could modulate the specificity of the immune response.

To operate the immune response T cells are strongly influenced by the mode of antigen presentation. Resting T cells require special signals to be activated. These signals are provided by subpopulations of APC. The

mode of administration of antigens determines selection of the appropriate
APC type. Processing and presentation by these specialized APC governs
the selective activation of T cell subsets with the release of certain
cytokines which for example regulate Ig class switching and other effector
functions.

In addition the efficiency of antigen presentation can vary with APC
types. For instance a few thousand dendritic cells can present antigen as
good as millions of B-cells (5). This could suggest that in vivo dendritic
cells are the most prominent APC. Biochemical analysis revealed that
biosynthesis of class II and associated invariant chain (Ii) polypeptides is
strongly increased in dendritic and langerhans cells (6). A high amount of
intracellular acidic compartments in these cells (putative processing
compartments) correlates to an unusual number of negatively charged sialic
acids attached to invariant chains. This modification increases resistance to
proteolysis and decelerates degradation of Ii in processing compartments.
A prolonged retention of class II in processing compartments could provide
time to saturate the class II cleft with peptides.

CLASS II ANTIGEN PRESENTATION PATHWAY

To achieve antigen specific recognition, the immune system
employs an ancient defense mechanism by which cells prevent infections,
namely by internalization and degradation of microbial antigens. This path
of nonspecific degradation was coupled to the intracellular MHC class II
pathway, which intersects the lysosomal route of antigen degradation. On
their way to the cell surface MHC class II molecules traverse endocytic
compartments, where they capture peptides derived from endocytosed
antigens and rescue them from further degradation. To gain access to
endosomes, class II molecules are associated with the invariant chain. Ii
directs the route of class II to endosomes and facilitates antigen
presentation (7). The cytoplasmic domain of Ii bears a signal which directs
transport of the class II/Ii complex to endosomes, where the oligomeric
complex is retained (8, 9). A second role of Ii is to prevent binding of
peptides and to keep class II in a virgin state until exogenous peptides are
available (10). In endocytic compartments Ii is degraded and the class II
cleft, which was protected by Ii, is available to peptides (11). Processed
antigen binds to the class II cleft and stabilizes the complex of α and ß
chains (12). Subsequently the class II/peptide complex appears at the cell
surface. It remains open how class II molecules enter the endocytic
pathway. Recent findings suggest that at least part of the class II/Ii
oligomers are exposed on the cell surface before they are internalized and
transfered to endosomes (13, 14). Here we present data that the various
forms of invariant chains are expressed on the cell membrane and there are

associated with class II molecules. The possible function of surface Ii will be discussed in this chapter.

SURFACE EXPRESSION OF THE MHC CLASS II-ASSOCIATED INVARIANT CHAINS

Recently we described several mAbs which recognize invariant chains on the surface of human B lymphoma cells (15). By Western blotting these mAbs were shown to bind to the C- terminal extracytoplasmic part of the invariant chain. Since several mAbs recognize surface determinants of Ii, these determinants were clustered and assigned to CD74 (16). Here we demonstrate surface expression of CD74 on primary B cells obtained from human tonsilles. By density centrifugation resting and pre-activated B cells were separated. Staining with CD74 and DR mAbs revealed that both B cell subsets express similar amounts of Ii and DR on the cell surface. Roughly estimated, on these B cell populations the CD74 epitope is found 10 times lower expressed than the DR epitope.

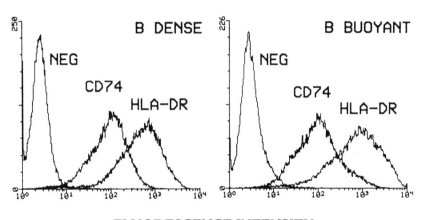

FLUORESCENCE INTENSITY

Fig.1 Surface expression of CD74 and HLA-DR antigens on dense and buoyant tonsillar B lymphocytes measured by flow cytometry.

In addition, surface expression of Ii could be demonstrated by immuno-electronmicroscopy and by immunoprecipitation of 125I surface labeled invariant chains (13).

Human invariant chain consists of at least 3 polypeptides, Ii33, Ii41 and Ii35 which are produced by differential splicing of RNA and by alternative initiation of translation. Here we demonstrate, that not only the most abundant Ii33, but also Ii41 and Ii35 are exposed on the cell surface. Fig.2 displays a two-dimensional separation of 125I surface labeled invariant chains.

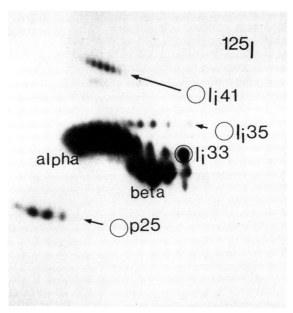

Fig.2 Surface expression of invariant chains detected by 125I labeling. JOK-1 B lymphoma cells were surface labeled with 125I. Cells were lysed and Ii was immunoprecipitated with Bu45, a mAb against human invariant chains. Subsequently the immunoprecipitate was separated in a two-dimensional gel electrophoresis. First dimension non-equilibrated-pH-gradient-electrophoresis and second dimension SDS PAGE. Three invariant chains Ii41, Ii35 and Ii33 and a degradation product of Ii33, p25 were identified. The circels indicate the position of high mannose type Ii chains which is known from metabolically labeled Ii. Surface forms of Ii chains are marked by arrows. HLA class II polypeptides are coprecipitated with Ii chains.

Almost exclusively the highly sialylated forms of invariant chains were labeled with 125I. Only small amounts of nonsialylated Ii33 are present (circle), and no precursor forms of 125I surface labeled Ii41 and Ii35 are detectable. This result indicates that upon passage of Ii through trans Golgi compartments, where sialyl-transferases attach sialic acids to the N- and O-linked carbohydrate side chains, invariant chains are expressed on the cell surface. Fig.2 additionally gives evidence, that a proportion of surface Ii chains are associated with class II polypeptides. HLA-D beta polypeptides are clearly visible and alpha chains partially superimpose with acidic invariant chains. The comigration of acidic Ii and HLA-D α chains could explain why in previous reports Ii was not identified in immunoprecipitates of 125I surface labeled class II molecules.

EXPORT OF INVARIANT CHAINS FROM THE ENDOPLASMIC RETICULUM AND TRANSPORT TO THE CELL SURFACE

Since the rate of biosynthesis of Ii is higher than that of class II α and ß chains , there is a significant amount of Ii that is not associated with class II. In pulse chase experiments it was demonstrated, that only a proportion of Ii polypeptides is exported from the ER and a considerable amount of Ii is retained in the ER. This may suggest that the ER retained Ii is disposed by a not yet defined mechanism, possibly on a branch path to lysosomes. The ER export signal of Ii has not yet been identified. It was suggested that association with class II facilitates transport of Ii from the ER to Golgi compartments (17). However, this seems to be no general mechanism because transfection of the murine Ii gene into rat fibroblast cells revealed highly sialylated Ii in the absense of class II molecules (18). Sialylation indicates transport of Ii to trans Golgi compartments.

THE BIOSYNTHETIC ROUTE OF INVARIANT CHAINS

On their route to the cell surface class II and Ii traverse compartments where their polypeptides are modified. Posttranslational modifications thus provide direct evidence for intracellular transport. Selective inhibition of transport and the presence or absence of modifications give insight into the pathway of class II/Ii. Fig.3 depicts the intracellular route of Ii and class II/Ii complexes.

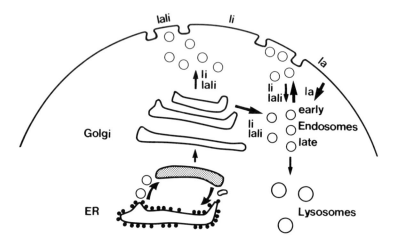

Fig. 3 The intracellular class II processing pathway. Model of intracellular transport of class II (Ia) from the endoplasmic reticulum (ER) to the cell surface. After membrane insertion in the ER Ia polypeptides associate with invariant chains. This oligomeric complex is exported to the Golgi complex. This exit can be blocked by brefeldin A. Upon passing the trans Golgi network the majority of Ia and Ii are sorted to endosomes. There Ii is degraded and class II is exposed on the cell surface. A proportion of Ii and class II chains is transported on the constitutive pathway to the cell membrane. The endocytic but not the constitutive pathway can be blocked by chloroquine. From the cell surface the IaIi complex and free Ii are rapidly internalized and directed to endosomes.

Towards their way to the Golgi complex Ii polypeptides are fatty acylated at a cysteine adjacent to the inner membrane (19). Palmitylation of Ii is a potential signal to leave the ER to a Golgi intermediate compartment. This can be suggested because non-acylated Ii chains do not carry sialic acids at their glycan side chains. Consistent with this result sialylation of Ii is blocked when fatty acylation is inhibited by cerulenin.

The export of newly synthesized Ii can also be blocked by brefeldin A, a drug that retrieves molecules in the ER (13).

Transport of membrane proteins to the cell surface is a continuous process which can be interrupted by signals that retain or sort the polypeptides apart from the bulk flow to the cell surface. After sialylation in the trans Golgi, Ii and Ia/Ii polypeptides appear in endosomes. The route taken to endosomes is either achieved by intracellular sorting of Ii chains to the endocytic route or by direct flow to the cell surface and internalization of the class II/Ii complex.

POSSIBLE ROLES OF SURFACE INVARIANT CHAINS

It is well known that class II molecules can recycle from the cell surface to endosomes. Like the transferrin receptor, they may enter early endosomes and subsequently return to the cell surface. It is not clear at present whether on this recycling pathway new peptides are accommodated in the antigen binding cleft. Surface Ii associated with class II molecules are internalized and could be targeted to late endosomes, where Ii is degraded and class II is charged with peptide. Similar routes have been described for mannose-6-phospate receptor, a polypeptide which delivers enzymes to lysosomes (20). Since the transport of Ii to the cell surface is rapid and the half life of surface Ii is low, a high turn over rate can be postulated. This cell surface pathway depends on the rate of biosynthesis of class II and Ii. Although the class II/Ii complex represents only a minor proportion of surface class II molecules in some cell types its internalization could be the major route for delivery to endosomes. At present a clear distinction between this rapid cell surface pathway and the intracellular sorting traffick of class II and Ii from the secretory pathway to endosomes cannot be made. In both cases the limitation of surface expression of Ii-free class II is its egress from endosomes.

Exposure of membrane proteins on the cell surface is a prerequisite to interact with other cell types. It was therefore tempting to speculate whether surface Ii has a cell adhesion function. Recent findings (21) suggest that a proportion of Ii that is modified to a chondroitin sulfate proteoglycan and expressed on the cell surface serves there as an adhesion molecule. It was shown that interaction to CD44 provides a signal capable to stimulate allogenic and mitogenic T cell responses.

ACKNOWLEDGMENTS

This work was supported by a grant from the Deutsche
Forschungsgemeinschaft Ko 810/4-3.

REFERENCES

1.   **Germain, R. N., and B. Malissen.** 1986. Analysis of the expression
     and function of class-II major histocompatibility complex-encoded
     molecules by DNA-mediated gene transfer. *Ann. Rev. Immunol.*
     4:281.

2.   **Wettstein, D. A., J. J. Boniface, P. A. Reay, H. Schild, and M.
     M. Davis.** 1991. Expression of a class II major histocompatibility
     complex (MHC) heterodimer in a lipid-linked form with enhanced
     peptide/soluble MHC complex formation at low pH. *J. Exp. Med.*
     174:219.

3.   **Momburg, F., N. Koch, P. Möller, G. Moldenhauer, G. W.
     Butcher, and G. J. Hämmerling.** 1986. Differential expression of
     Ia and Ia-associated invariant chain in mouse tissues after in vivo
     treatment with interferon gamma. *J. Immunol. 136*:940.

4.   **Vidard, L., K. L. Rock, and B. Benacerraf.** 1992. Diversity in
     MHC class II ovalbumin T cell epitopes generated by distinct
     proteases. *J. Immunol 149*:498.

5.   **Inaba, K., J. P. Metlay, M. T. Crowley, and R. M. Steinman.**
     1990. Dendritic cells pulsed with protein antigens in vitro can prime
     antigen-specific, MHC-restricted T cells in situ. *J. Exp. Med.*
     172:631.

6.   **Kämpgen, E., N. Koch, F. Koch, P. Stöger, C. Heufler, G.
     Schuler, and N. Romani.**1991. Class II major histocompatibility
     complex molecules of murine dendritic cells: synthesis, sialylation of
     invariant chain and antigen processing capacitiy are downregulated
     upon culture. *Proc. Natl. Acad. Sci. USA 88*:3114.

7. Stockinger, B., U. Pessara, R. H. Lin, J. Habicht, M. Grez, and
    N. Koch. 1989. A role of Ia-associated invariant chains in antigen
    processing and presentation. *Cell 56:*683.

8. Bakke, O., and B. Dobberstein. 1990. MHC class II-associated
    invariant chain contains a sorting signal for endosomal
    compartments. *Cell 63*:707.

9. Teyton, L., D. O'Sullivan, P. W. Dickson, V. Lotteau, A. Sette,
    P. Fink, and P.A. Peterson. 1990. Invariant chain distinguishes
    between the exogenous and endogenous antigen presentation
    pathways. *Nature 348*:39.

10. Roche, P.A., and P. Cresswell. 1990. Invariant chain association
    with HLA-DR molecules inhibits immunogenic peptide binding.
    Nature 345:615.

11. Riberdy, J. M., J. R. Newcomb, M. J. Surman, J. A. Barbosa,
    and P. Cresswell. 1992. HLA-DR molecules from an antigen-
    processing mutant cell line are associated with invariant chain
    peptides. *Nature 360*:474.

12. Germain, R. N., and L. R. Hendrix. 1991. MHC class II
    structure, occupancy and surface expression determined by post-
    endoplasmic reticulum antigen binding. *Nature 353*:134.

13. Koch, N., G. Moldenhauer, W. Hoffmann, and P. Möller. 1991.
    A rapid intracellular pathway gives rise to cell surface expression of
    the MHC class II associated invariant chain (CD74). *J. Immunol.
    147*:2643.

14. Roche, P. A., C. L. Teletski, E. Stang, O. Bakke, and E. O.
    Long. 1993. Cell surface HLA-DR-invariant-chain complexes are
    targeted to endosomes by rapid internalization. *Proc. Natl. Acad.
    Sci. USA* In press.

15. Wraight, C. J., P. van Endert, P. Möller, J. Lipp, R. Ling, I.
    C. M. MacLennan, N. Koch, and G. Moldenhauer. 1990.
    Human major histocompatibility complex class II invariant chain is
    expressed on the cell surface. *J. Biol. Chem. 265*:5787.

16. **Dörken, B., Möller, P., Pezzuto, A., Schwartz-Albiez, R. and Moldenhauer, G.** 1989. B cell antigens: CD74. In: Leukocyte Typing IV (ed. W. Knapp et al.), p. 106, Oxford University Press, Oxford.

17. **Marks, M. S., J. S. Blum, and P. Cresswell.** 1990. Invariant chain trimers are sequestered in the rough endoplasmic reticulum in the absence of association with HLA class II antigens. *J. Cell Biol.* *111*:839.

18. **Koch, N.** 1988. Posttranslational modifications of the Ia-associated invariant protein p41 after gene transfer. *Biochemistry 27*:4097.

19. **Koch, N., and G. J. Hämmerling.** 1986. The HLA-D-associated invariant chain binds palmitic acid at the cysteine adjacent to the membrane segment. *J. Biol. Chem. 261*:3434.

20. **Lobel, P., K. Fijimoto, R. D. Ye, G. Grifith, and S. Kornfeld.** 1989. Mutations in the cytoplasmic domain of the 275 kd mannose 6-phospate receptor differentially alter lysosomal enzyme sorting and endocytosis. *Cell 57*:787.

21. **Naujokas M. F., M. Morin, M. S. Anderson, M. Peterson, and J. Miller.** 1993. The chondroitin sulfate form of invariant chain can enhance stimulation of T cell responses through interaction with CD44. *Cell 74*:257.

16

CHARGING OF PEPTIDES TO MHC CLASS II MOLECULES DURING PROTEOLYSIS OF I_i

Minzhen Xu, Masanori Daibata, Sharlene Adams, Robert E. Humphreys, and Victor E. Reyes

The cleavage and release of I_i from MHC Class II molecules appears to catalyze the concurrent charging of the peptide binding site with T cell-presented peptides. Cathepsin B (CB) and cathepsin D (CD) each efficiently cleaves I_i without apparent damage to the MHC α,β chains. A photoactivated, radioiodinated, T cell-presented peptide is more efficiently bound to the MHC α,β molecules when present during CB cleavage than when added to I_i-freed α,β dimers. No peptide is bound during comparable exposure of I_i to CD, however, a trace amount of CD further enhances peptide binding during CB-mediated I_i release. Stages in the cleavage and release of I_i were characterized with I_i mutants at three CB and one CD potential proteolysis sites. CB or CD cleavage of these mutant I_i molecules yielded digestion patterns which, in their differences from wild type I_i, indicated specific, sequential steps in the cleavage/release pathway. These observations, with other control experiments, lead to a model that binding of a T cell-presented peptide occurs at some stage in the cleavage and release of I_i. Certain partially cleaved, mutant I_i fragments bound to MHC class II α,β chains might create relatively stable transition states to be used in screening for inhibitors of antigen presentation.

ROLE OF I_i IN THE REGULATION OF ANTIGEN PRESENTATION

The presence of the invariant chain I_i on class II α,β molecules prevents charging of T cell-presented peptides to their antigen binding sites (1). The upregulation of I_i on some leukemic cells might reflect blockage of endogenous tumor-associated antigens (2,3). Abnormalities in I_i expression or function might be associated with presentation of endogenous, autoimmune determinants.

The discrepancy between the slow rate of charging I_i-freed α,β chains with foreign peptide and the time required by living cells to process and present antigenic determinants led to hypotheses of several mechanisms to catalyze intracellular peptide binding to MHC class II α,β chains (reviewed in (4)). We have addressed the question whether peptide binding occurs during cleavage and release of I_i, perhaps in a concerted reaction. The peptide charging event occurs at or after cleavage and release of I_i in a post-Golgi/endosomal compartment (5-7). Intracellular proteases CB and CD, which can create functional fragments of some antigens (8,9), are found in intracellular compartments where peptide charging to MHC class II molecules is thought to occur (5). In those compartments cleavage of both antigen and I_i by CB and CD might take place.

Cathepsins B and D cleave and release I_i from solubilized MHC class II α,β chains without apparent damage to those chains, creating fragments of sizes which are observed in metabolically radiolabeled living cells (1,10). In living cells the CB inhibitor leupeptin limits proteolysis of I_i to the p21 and p10 fragments (11,12) and blocks antigen charging *in vitro* (13). Since CB but not CD is inhibited by leupeptin (14), the intracellular cleavage of I_i might involve the sequential action of CD and then CB. Treatment of antigen presenting cells with leupeptin blocked antigen presentation (13,15), reduced MHC class II surface expression (2,7), and led to the appearance of I_i-derived p21 and p14 which were presumably generated by a protease not blocked by leupeptin, *e.g.*, CD (11,12).

In order to establish the stages in cleavage and release of I_i by CB and CD, and to determine possible concurrent charging of the MHC class II antigen binding site, we have analyzed peptide charging during cleavage and release of I_i with CB, CD, and both proteases. We have also constructed a series of mutants in putative CB and CD cleavage sites of I_i. These mutants have significantly altered CB and CD cleavage patterns which support a detailed hypothesis about the staged cleavage and release of I_i fragments from MHC class II α,β chains.

CHARGING OF PEPTIDES TO MHC CLASS II MOLECULES DURING PROTEOLYSIS OF I_i

Peptide binding was enhanced when the indicator peptide was present during CB-mediated I_i release from detergent-solubilized MHC class II α,β,I_i complexes. The azidobenzoyl-coupled, radioiodinated, influenza virus MA(18-29) peptide was added to solubilized microsomal membranes from JESTHOM B cell line for digestions for varying times with three concentrations of CB. The photoactivated peptide was crosslinked where it became bound. Anti-MHC class II immunoprecipitates were analyzed by SDS-PAGE and autoradiography of the slab gels. Crosslinked peptide was detected after as early as 5 min of CB digestion with progressive increases in levels of peptide crosslinked to the class II MHC α and β chains as a function of digestion time.

The binding of peptide to MHC class II α,β, I_i trimers during CB release of I_i was greater than binding to class II α,β dimers from which I_i had been released and the proteases inactivated (Fig. 1). The peptide was bound much more efficiently to class II α,β chains when present during the CB digestion than when added afterwards, also at pH 5.0. The peptide was not bound to class II α,β,I_i complexes which were not treated with CB. CB and CD cleaved and released I_i from MHC class II molecules without cutting the α,β chains (10).

Peptide binding was not enhanced during I_i cleavage by CD. Under a wide range of concentrations of CD which cleaved I_i but did not lead promptly to complete dissociation of its fragments from class II α,β chains (10), no enhanced binding of the radioiodinated MA(18-29) peptide was seen (Fig. 1).

However, trace levels of CD enhanced further the CB-mediated peptide binding effect. When the lowest level of CD found to cleave I_i was added to the assay for peptide binding in the presence of varying concentrations of CB, the net level of peptide binding was enhanced about 3 times that seen without CD.

230 Minzhen Xu *et al.*

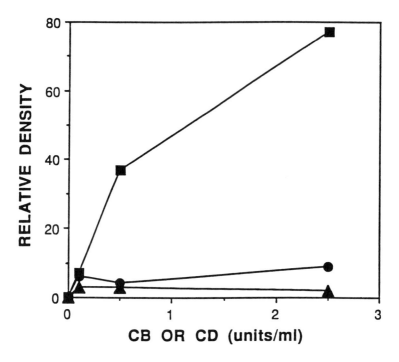

Fig. 1. Binding of the indicator peptide, radioiodinated azidobenzoyl-coupled influenza virus MA(18-29) peptide was detected by autoradiography of SDS electrophoretic gels of immunoprecipitated MHC class II molecules. Densitometry showed a dependence of binding on CB concentration during a 30 min interval and a greater degree of binding when the peptide was present during I_i cleavage and release by CB (■) than when the peptide was added afterward (●). CD treatment did not lead to enhanced binding of the concurrently added indicator peptide (▲).

Competition for indicator peptide binding at comparable concentrations of peptides was seen only with peptides presented by the same HLA-DR1 MHC class II allele. The specificity of [125I]MA(18-29) binding to MHC class II molecules of HLA-DR1-positive JESTHOM cells inhibited by HLA-DR1-restricted MA(18-29) and HLA-DR1-restricted influenza HA(306-318) peptides, but not by HLA-DP-restricted dengue NS3(251-265), and HLA-B37-restricted influenza NP(336-356) peptides.

HYPOTHESIS ON A CONCERTED EXCHANGE MECHANISM

These results support the hypothesis that peptide insertion actually occurs during some stage in the cleavage and release of I_i. It is true that I_i cleavage and removal makes available the antigenic peptide binding site on the MHC class II α,β chains. The greater binding which is observed when the peptide is present during protease cleavage and release of I_i than when the peptide is added after I_i release could reflect the fact that most sites are rapidly filled by ambient peptides by the time the control study is performed. However, two specific experiments support the more exacting view of a concerted exchange mechanism. Enhancement of peptide binding occurred only during CB cleavage and not during CD cleavage, which does efficiently cleave I_i. Secondly, the addition of CD to the CB-mediated experiment further enhances the level of peptide binding, possibly by enhancing cleavage at a CB-sensitive site. In order to define the molecular stages in the cleavage and release of I_i by CB and CD, we prepared mutations at putative CB and CD cleavage sites in I_i.

Design of I_i Mutants

Mutations were made in one putative CD cleavage site and three clusters of putative CB cleavage sites in I_i (Table I). These mutations were based on three principles. (1) Potential protease-specific cleavage motifs were identified. CB was thought to attack at the unusual frequency of clustered LK and LR residues. CD was reported to prefer paired hydrophobic residues except when the C-terminal one is Val, Gly or Ala. (2) Putative cleavage sites should lead to the MHC class II α,β chains-associated I_i fragments which are found both with CB or CD digestions of isolated proteins and naturally in pulse-chase metabolically radiolabeled cells. (3) The facts that the N-terminal CB-derived p21 fragment is bound to class II α,β chains and that naturally occurring I_i digestion products around L^{97}-G^{119} are also bound to class II α,β chains (16) point to the cleavage site producing p21 being C-terminal to L^{97}-G^{119}.

Table I. Potential cathepsin B and cathepsin D cleavage sites in I$_i$

```
      10        20        30        40        50
                AA
                 \
MDDQRDLISNNEQLPMLGRRPGAPESKCSRGALYTGFSILVTLLLAGQAT
 /                                           //  //


      60        70        80        90        100
                          A A   A   T
            \             \ \   \ \   \ \
TAYFLYQQQGRLDKLTVTSQNLQLENLRMKLPKPPKPVSKMRMATPLLMQ
  /                                              //


     110       120       130       140       150
                                   A       A
                                    \       \
ALPMGALPQGPMQNATKYGNMTEDHVMHLLQNADPLKVYPPLKGSFPENL


     160       170       180       190       200
A  A                      VA
 \  \         \         \         \
RHLKNTMETLDWKVFESWMHHWLLFEMSRHSLEQKPTDAPPKESLELEDP
        /          //


     210

SSGLGVTKQDLGPVPM
```

*Cathepsin B sites are indicated by \ above the sequence and cathepsin D sites are indicated by / below the sequence. The transmembranal region Gly[31]-Gln[57], the hexa-cationic, tetraprolyl kink Leu[77]-Met[93], and the α-helix Phe[146]-Val[164] are underlined.

CB Proteolysis of the I_i Mutants

CB digestions were carried out on detergent-solubilized, [^{35}S]methionine-labeled COS cell transfectants with the mutant I_i genes listed in Table I and normal HLA-DR1 α and ß genes. Anti-MHC class II α,ß immunoprecipitates of complexes digested with CB yielded different patterns of I_i cleavage (Table II). All of the mutants except [$R^{151}K^{154}$] had reduced levels or the absence of some principal fragments produced by CB digestion of WT I_i: p21, p14, p10, and p6.

Table II.   Appearance of fragments of wild type and mutant I_i chains after digestion with cathepsin B or cathepsin D.

Wild Types Mutant I_i	CB FRAGMENTS				CD FRAGMENTS			
	p21	p14	p10	p6	p21	p1	p10	p6
WT	+	+	+	+	+	+	+	+
M[R^{78}→A;K^{80}→A;K^{83}→A;K^{86}→T]	+	-	-	-	+	-	-	-
M[K^{137}→A;K^{143}→A]	±	+	+	+	+	±	+	±
M[R^{151}→A;K^{154}→A]	+	±	+	+	+	+	+	+
M[L^{174}→V;F^{175}→A]	-	+	+	+	-	+	+	+

Summary of fragments generated in CB digestion and CD digestions. The letters in parenthesis indicate where the mutations were made. For example, in M[K^{137}→A;K^{143}→A], or more simply [$K^{137}K^{143}$], the lysine in positions 137 and 143 were replaced by alanines. ± indicates that the band from the mutant I_i is weaker than the corresponding one from WT I_i. Fragments of comparable size from digestions with CB or CD are not implied to result from cleavage at identical sites.

Specifically, mutant [$L^{174}F^{175}$] had only a trace amount of p21 and mutant [$R^{78}K^{80}K^{83}K^{86}$] did not produce p14, p10, and p6. Mutants [$R^{19}R^{20}$] and [$K^{137}K^{143}$] produced only about one tenth as much p21 as did WT I_i but were equally efficient in removing intact WT and mutant I_i.

Since mutant [$R^{78}K^{80}K^{83}K^{86}$] accumulated p21 but did not produce p14, p10, and p6, these latter small fragments might be derived from p21 by cleavage about $R^{78}K^{80}K^{83}K^{86}$. To test that hypothesis, digestions were carried out with varying CB concentrations. At 30 U/ml CB all intermediate fragments were lost excepting p6 which still associated with MHC class II α,β chains in anti-class II mAb immunoprecipitates of complexes formed with WT I_i or mutant $[K^{13}K^{143}]$. In the case of mutant $[R^{78}K^{80}K^{83}K^{86}]$, neither p21 nor p6 was associated with class II molecules, indicating CB cleavage at one site of $[R^{78}K^{80}K^{83}K^{86}]$ might be needed to produce MHC class II molecule-associated p14, p10, and p6. These results and the densitometer finding that density of the high dose CB-generated p6 band equalled the sum of the densities of p14, p10 and p6 bands produced at lesser doses of CB supported the idea that p14, p10 and p6 were derived from p21 with p6 being the final MHC class II-bound product.

There was an increased rate of release from MHC class II α,β chains of CB-generated p21 from some I_i mutants. In order to determine in mutants $[K^{137}K^{143}]$ and $[L^{174}F^{175}]$ whether CB-derived p21 was not produced, or degraded or dissociated from HLA-DR1 α,β chains, class II α,β,I_i trimers were immunoprecipitated, digested with CB, and the entire reaction mixture was analyzed by SDS-PAGE. These experiments showed that CB generated p21 from mutant $[K^{137}K^{143}]$ as strongly as it did from WT I_i, indicating that p21 was promptly dissociated from DR1 α,β chains and remained in the reaction mixture. These experiments showed that CB cleavage about $[K^{137}K^{143}]$ preceded CB cleavage about $[R^{78}K^{80}K^{83}K^{86}]$.

CD Proteolysis of the I_i Mutants

CD digestions of mutant and WT I_i complexes were also performed. Upon CD digestion of α,β,I_i trimers, a p21 was produced from WT I_i and all of the mutants excepting mutant $[L^{174}F^{175}]$. In mutant $[L^{174}F^{175}]$ more of p14, p10, and p6 was actually produced than from WT I_i, indicating that although the $L^{173}L^{174}F^{175}$ site was altered, other sites could be attacked by CD, possibly more efficiently than they were in WT I_i. The sizes of I_i fragments seen after CD digestion approximated those seen after CB digestion possibly because each CD site was in the vicinity of a CB site (Table I). Nevertheless, there were distinct differences between CB and CD in the production of these fragments. (*1*) CD digestion did not lead to premature release of I_i fragments from class II molecules. Mutants $[R^{78}K^{80}K^{83}K^{86}]$ and $[K^{137}K^{143}]$ were somewhat resistant to CD digestion and had reduced levels of p21 and increased intact I_i relative to the observations with

CB. (2) With these latter two mutants, one sharp p21 band was formed upon CD digestion; but in the case of CB digestion, the p21 band appeared to be microheterogeneous.

Inhibition of Antigen Presentation by I_i Sequences

PH-1.0, the homolog of the $I_i(F^{148}-V^{164})$ peptide emphasizing the longitudinal hydrophobic strip, and $I_i(L^{81}-M^{104})$, the MHC class II-associated invariant chain peptide fragment of I_i (the 'CLIP' peptide) (17), inhibited antigen presentation by paraformaldehyde-fixed antigen presenting cells. These fragments might compete for sites in the MHC class II α,β chains where antigenic peptides bind, *i.e.* the peptide-binding groove, or they might otherwise influence the structure of the MHC class II α,β/antigenic peptide complex which is recognized by T cell receptors.

A MODEL FOR CONCERTED CHARGING OF PEPTIDE IN A TRANSITION STATE IN THE STAGED CLEAVAGE AND RELEASE OF I_i FROM MHC CLASS II α,β CHAINS

The staged cleavage and release of I_i appears to be related to catalysis or regulation of binding of antigenic peptides to MHC class II α,β chains. We have demonstrated that the binding of immunogenic peptides to MHC class II α,β chains occurs best as a concerted process with the release of I_i by CB and that a trace level of CD further enhances peptide binding found in the presence of CB. A model emphasizing secondary structural motifs in I_i, and possible CB and CD cleavage sites clustered in or around some of those motifs, can be proposed for further testing of the mechanism of I_i cleavage and release, which is possibly associated with peptide charging at one stage of I_i fragment release (Figure 2). An N-terminal cytoplasmic tail M^1-R^{30} contains sorting signals to direct the intracellular transportation of MHC class II molecules and I_i. A transmembranal helix runs from A^{32} to L^{55}. A tight tetraprolyl palindrome ($L^{81}-M^{93}$) includes potential CB cleavage sites LR^{78}, MK^{80}, PK^{83}, PK^{86}, VSK^{90}, MR^{92} and is followed by a CD site LLM^{99}. A loose tetraprolyl, kinked region occurs at $P^{96}-P^{111}$.

The 'perfect' amphipathic helix F^{146}-V^{164} (18) contains potential CB cleavage sites LR^{151}, LK^{154}, WK^{163}, and is preceded by CB cleavage sites LK^{137}, LK^{143}. A C-terminal tail contains a potential CD site LLF^{175}.

Cleavages at the CD sites and within clusters of the CB sites, in a staged fashion, might lead to release of I_i fragments with the insertion of foreign peptide into the MHC class II antigen binding site. CD might cleave first at LLF^{175} to permit more efficient cleavage at LK^{137} or LK^{143} thereby releasing putative amphiphilic α-helix F^{146}-L^{164}. That helix is then destroyed by CB digestion at the LR or LK positions in the helix. Cleavages in one of the $R^{78}K^{80}K^{83}K^{86}$ sites leads to p14, p10 and finally p6. Immunogenic peptides could be inserted after CB cleavage at $K^{137}K^{143}$ or $R^{78}K^{80}K^{83}K^{86}$, leading to the dissociation of p21 or p6. Perhaps p6 may be further degraded from the C-terminus by CB, leading to peptides such as $I_i(L^{81}$-$M^{104})$ and $I_i(L^{96}$-$G^{119})$ which remain associated with the α,β chains (16,17,19). Since the fragment $I_i(L^{81}$-$M^{104})$ has no stabilizing effect on MHC class II complexes at boiling temperatures (19), as described for T cell-presented peptides by Germain and Hendrix (20) and Neefjes and Ploegh (7), the p6 I_i fragment might allosterically regulate the locking in of a T cell-presented peptide. The fact that these peptides constitute about 80% of the peptide pool bound to MHC class II from an antigen presenting mutant cell line (19) also suggests that their binding structurally alters MHC class II molecules to affect binding of antigenic peptides.

FUTURE DIRECTIONS

This model can be tested further by analysis of the binding of T cell-presented peptides to complexes with these mutant I_i chains as a function of proteolysis. Also, the capacity of transfected cell lines expressing these I_i mutants and normal MHC class II α,β chains to present antigen should be assessed.

Some I_i mutants, especially $[R^{78}K^{80}K^{83}K^{86}]$, might create a 'frozen intermediate' at a stage of peptide charging, and thus be useful in assaying for allele-specific or -nonspecific inhibitors of antigen presentation by MHC class II molecules.

Mechanisms of antigen charging of MHC class II molecules in nonlymphoid cells might be differently regulated with respect to

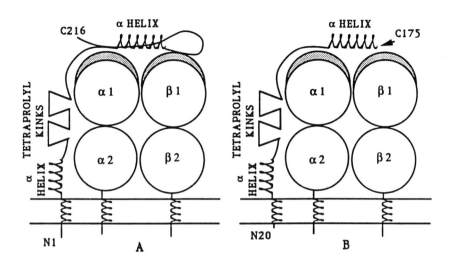

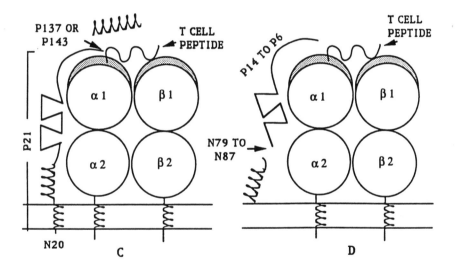

Fig. 2. Model of cleavage and release of I_i from MHC class II α,β chains as a concerted process with binding of a T cell-presented peptide. Shaded region is the binding site for antigenic peptides. Local secondary structures of I_i are described in the text. Binding of the T Cell-preesented peptide might occur after cleavage at I_i 137 or 143 or at I_i 79 to 87 prior to release of the remaining I_i fragments.

synthesis, transport and processing, and function of I_i and MHC class II α,β genes. For example mucosal epithelial cells, which support T cell differentiation (21) express MHC class II molecules (22) and I_i (Reyes *et al.*, unpublished observations). Structural variations in these molecules might relate to the tolerizing function of the mucosal environment and be reflected in the mechanisms of I_i cleavage/release and peptide charging.

The role of I_i in restricting the presentation of endogenous and exogenous antigens can be explored in the context of the origin of autoimmune diseases, tolerance induction, and the evolution of malignant potential in some tumors. Alterations in the synthesis, or cleavage and release of I_i from MHC class II α,β chains might lead to binding of self (endogenous) peptides to the MHC class II molecules and thus to autoimmune disease. Alternately, mutations in I_i leading to altered patterns of cleavage and release, or in promoters to enhance expression, might lead functionally to the suppression of tumor-associated (endogenous) antigenic determinants in MHC class II - positive malignancies. Leukemic cells of hairy cell leukemia, and some other B lineage malignancies, express very large amount of I_i or of a p15 which might be a fragment of I_i (23,24). Such an "over expression" of I_i might inhibit presentation of tumor-associated determinants in such cells. Therapies aimed at altering I_i promoter functions or mRNA levels could lead to an effective anti-leukemia immune response. Such questions can be addressed *in vitro* and in animal models of MHC class II-positive malignancies.

Immunosuppression by agents altering pH, or transport of proteins among intracellular vesicle might reflect perturbation of the I_i cleavage/release - peptide charging mechanism. Understanding this point and knowing the I_i structures involved at crucial catalytic steps, could lead to the design of drugs with more specific and potent therapeutic effects on peptide charging to MHC class II molecules.

ACKNOWLEDGMENTS

The current address of V.E.R. is the Department of Pediatrics, C-66, Children's Hospital, University of Texas Medical Branch, Galveston, TX 77555. This work has been supported by an ACS grant to R.E.H., by American Cancer Society, American Heart Association, and NIH grants to V.E.R., by an NIH fellowship to S.A., and by an Arthritis Foundation fellowship to M.X. We thank Patricia Downe for the preparation of this manuscript.

REFERENCES

1. **Roche, P. A. and P. Cresswell.** 1991. Proteolysis of the class II-associated invariant chain generates a peptide binding site in intracellular HLA-DR molecules. *Proc. Natl. Acad. Sci. USA 88*:3150.

2. **Loss, G. E.,Jr., C. G. Elias, P. E. Fields, R. K. Ribaudo, M. McKisic, and A. J. Sant.** 1993. MHC class II-restricted presentation of an internally synthesized antigen displays cell-type variability and segregates from the exogenous class II and endogenous class I presentation pathways. *to be named*

3. **Clements, V. K., S. Baskar, T. D. Armstrong, and S. Ostrand-Rosenberg.** 1992. Invariant chain alters the malignant phenotype of MHC class II$^+$ tumor cells. *J. Immunol. 149*:2391.

4. **Germain, R. M. and D. H. Margulies.** 1993. The biochemistry and cell biology of antigen processing and presentation. *Ann. Rev. Immunol. 11*:403.

5. **Guagliardi, L. E., B. Koppelman, J. S. Blum, M. S. Marks, P. Cresswell, and F. M. Brodsky.** 1990. Co-localization of molecules involved in antigen processing and presentation in an early endocytic compartment. *Nature 343*:133.

6. **Neefjes, J. J., V. Stollorz, P. J. Peters, H. J. Geuze, and H. L. Ploegh.** 1990. The biosynthetic pathway of MHC class II but not class I molecules intersects the endocytic route. *Cell 61*:171.

7. **Neefjes, J. J. and H. L. Ploegh.** 1992. Inhibition of endosomal proteolytic activity by leupeptin blocks surface expression of MHC class II molecules and their conversion to SDS resistant alpha beta

6666

heterodimers in endosomes. *EMBO J. 11*:411.

8. **Takahashi, H., K. B. Cease, and J. A. Berzofsky.** 1989. Identification of proteases that process distinct epitopes on the same protein. *J. Immunol. 142*:2221.

9. **Rodriguez, G. M. and S. Diment.** 1992. Role of cathepsin D in antigen presentation of ovalbumin. *J. Immunol. 149*:2894.

10. **Reyes, V. E., S. Lu, and R. E. Humphreys.** 1991. Cathepsin B cleavage of I_i from class II MHC alpha- and beta-chains. *J. Immunol. 146*:3877.

11. **Blum, J. S. and P. Cresswell.** 1988. Role for intracellular protease in the processing and transport of class II HLA antigens. *Proc. Natl. Acad. Sci. USA 85*:3975.

12. **Nguyen, Q. V., W. Knapp, and R. E. Humphreys.** 1989. Inhibition by leupeptin and antipain of the intracellular proteolysis of I_i. *Human Immunol. 24*:153.

13. **Puri, J. and Y. Factorovich.** 1988. Selective inhibition of antigen presentation to cloned T cells by protease inhibitors. *J. Immunol. 141*:3313.

14. **Bond, J. S.** 1989. Commercially Available Proteases. In *Proteolytic Enzymes: A Practical Approach*. R. J. Beynon and J. S. Bond, eds. IRL Press, New York, p.232.

15. **Streicher, H. Z., I. J. Berkower, M. Busch, F. R. N. Gurd, and J. A. Berzofsky.** 1984. Antigen conformation determines processing requirements for T-cell activation. *Proc. Natl. Acad. Sci. USA 81*:6831.

16. **Chicz, R. M., R. G. Urban, W. S. Lane, J. C. Gorga, L. J. Stern, D. A. A. Vignali, and J. L. Strominger.** 1992. Predominant naturally processed peptides bound to HLA-DR1 are derived from MHC-related molecules and are heterogeneous in size. *Nature 358*:764.

17. **Riberdy, J. M., J. R. Newcomb, M. J. Surman, J. A. Barbosa, and P. Cresswell.** 1992. HLA-DR molecules from an antigen-processing

mutant cell line are associated with invariant chain peptides. *Nature 360*:474.

18. **Elliott, W. L., C. J. Stille, L. J. Thomas, and R. E. Humphreys.** 1987. An hypothesis on the binding of an amphipathic, alpha helical sequence in Ii to the desetope of class II antigens. *J. Immunol. 138*:2949.

19. **Sette, A., S. Ceman, R. T. Kubo, K. Sakaguchi, E. Appella, D. F. Hunt, T. A. Davis, H. Michel, J. Shabanowitz, R. Rudersdorf, H. M. Grey, and R. DeMars.** 1992. Invariant chain peptides in most HLA-DR molecules of an antigen-processing mutant. *Science 258*:1801.

20. **Germain, R. N. and L. R. Hendrix.** 1991. MHC class II structure, occupancy and surface expression determined by post-endoplasmic reticulum antigen binding. *Nature 353*:134.

21. **Mosley, R. L. and J. R. Klein.** 1992. Peripheral engraftment of fetal intestine into athymic mice sponsors T cell development: Direct evidence for thymopoietic function of murine small intestine. *J. Exp. Med. 176*:1365.

22. **Sanderson, I. R., A. J. Ouellette, E. A. Carter, and P. R. Harmatz.** 1992. Ontogeny of class II MHC mRNA in the mouse small intestinal epithelium. *Molec. Immunol. 29*:1257.

23. **Spiro, R. C., T. Sairenji, and R. E. Humphreys.** 1984. Identification of hairy cell leukemia subset defining p35 as the human homologue of I_i. *Leuk. Res. 8*:55.

24. **Elliott, W. L., S. Lu, Q. Nguyen, P. S. Reisert, T. Sairenji, C. H. Sorli, C. J. Stille, L. J. Thomas, and R. E. Humphreys.** 1987. Hyperexpressed hairy leukemic cell I_i might bind to the antigen-presenting site of class II MHC molecules. *Leuk. Res. 1*:395.

17

RECOGNITION OF CLASS II MHC/PEPTIDE COMPLEXES BY T CELL RECEPTORS AND ANTIBODIES THAT MIMIC THEM

Philip A. Reay and Mark M. Davis

INTRODUCTION

Recent studies involving X-ray crystallography (1,2) and sequencing of peptides eluted from MHC molecules (3) have provided a wealth of information regarding the peptide features necessary for MHC interaction. However, such studies provide only limited information regarding T cell receptor (TCR) recognition, and cannot indicate the affinities with which the three molecules interact. Additionally, it is of interest to compare the different ligands recognized by T cells at different stages of their development - in particular those mediating positive and negative thymic selection.

Our approach to studying these issues has been to reconstruct and analyze the T cell recognition event *in vitro*. As a model system we have chosen the interaction of TCRs derived from either the 5C.C7 T cell or the 2B4 hybridoma with a COOH-terminal fragment of moth cytochrome c (MCC, residues 95-103) bound to the murine class II MHC molecule I-E^k (4).

SOLUBLE I-E^k AND TCR MOLECULES

Glycolipid forms of I-E^k (5) and a TCR heterodimer (6) were expressed by fusing the extracellular domains of the relevant α and β chains to a signal sequence specifying addition of a glyco-phosphatidyl inositol (GPI) anchor. These were expressed at high densities on CHO cells and released from the cell membrane by PI-PLC cleavage, allowing purification of large quantities of the pure proteins with specific monoclonal antibodies.

An unexpected property of soluble GPI-linked I-E^k is that it appears to be devoid of endogenous peptides (7), similar to class II MHC molecules expressed using baculovirus vectors (8). Using an assay that detects I-E^k-associated biotinylated peptide we have shown that such "empty" molecules bind MCC peptide 40-fold faster at the low pH values found in endosomes (7). In contrast, once peptide is bound its dissociation is relatively pH-independent (in the range 5.0-7.0), as is the structure of the complex as probed by NMR (9). This suggests that the empty I-E^k molecule is relatively mobile and adopts a peptide-receptive conformation at pH 5-6. Once bound the peptide may stabilize the complex into a pH-insensitive structure. These probably correspond to the forms of I-E^k that can be identified by gel electrophoresis (Sadegh-Nasseri, 1994 and Romagnoli et al, 1994, this volume).

A TOPOLOGY FOR T CELL RECOGNITION

Previous studies with peptide analogues have identified some of the residues of the MCC epitope that are important for I-E^k or TCR interaction (10,11). However, the problems of synthesizing and assaying a large number of peptides, coupled with the lack of an appropriate direct I-E^k binding assay, meant that only a small selection of analogues were studied, and the role of many positions therefore remained ambiguous. To circumvent this we have used multi-pin peptide synthesis to generate a "global" replacement set of analogues of the MCC (93-103) epitope, whereby the residue at each position was sequentially replaced by each of the other 19 naturally occurring L-amino acids (12). Using the ELISA-based binding assay described above, we found that I-E^k association is mostly affected by substitutions at three positions (95, 100 and 103). The substitutions at these residues which retain >50% binding are shown in figure 1. These data suggest that effective I-E^k binding requires a large hydrophobic amino acid at position 95 followed by either a small hydrophobic or polar residue at 100, and a positively charged or hydrophobic group at 103.

95	96	97	98	99	100	101	102	103
Ile	Ala	Tyr	Leu	Lys	Glu	Ala	Thr	Lys
Leu					Asn			Arg
Val					Glu			Tyr
Phe					Ser			Phe
Trp					Thr			Met
Thr					Met			Ile
					Val			Leu
					Ala			Val
					Gly			Ala
					Pro			Ser

Fig. 1. Monosubstitutions at positions 95, 100 and 103 which maintain I-E^k binding capacity.

Although positions 95 and 103 are likely to correspond to the pockets observed at either end of a collection of bound peptides in the recently solved structure of the HLA-DR1 molecule, no middle pocket corresponding to position 100 was observed (13). While this may reflect the heterogeneity of peptides bound to HLA-DR1, it is also possible that the structure of I-E^k differs significantly from this human molecule in this region.

In contrast to the relative degeneracy with respect to I-E^k association, T cells specific for the MCC(95-103) peptide were stimulated by only a limited number of substitutions at positions uninvolved with binding (Figure 2).

Positions 97, 99 and 102 seem highly specific for TCR contact as they are either intolerant to any replacement, or require a conservative substitution (e.g. aromatic at 97). In contrast, positions 96 and 101, initially both alanines, tolerated non-conservative replacement whose only obvious similarity is their relatively small side chains (glycine and cysteine). Interestingly, although substitution at positions 96 and 101 have no effect on MHC binding, 2-D NMR analysis of labelled MCC bound to I-E^k indicates that these alanine side chains are actually in close contact with the MHC (9). We have therefore proposed that positions 96 and 101 may act as "pivots", chiefly affecting the peptide conformation in the groove, while residues 97, 99 and 102 directly contact the TCR (12).

T cell recognition is also markedly affected by the nature of the residue interacting with I-E^k at position 100 or 103 (but not at 95). Thus, although analogues substituted with hydrophobic groups at position 103 all bind effectively, none of them are recognized. Similarly, recognition is ablated by substitution of Asn with Met at position 100. These results indicate that very similar peptides can have significantly altered conformations within the MHC groove. Analogous effects have also been observed with a class I MHC-restricted peptide complex (14).

95	96	97	98	99	100	101	102	103
Ile	Ala	Tyr	Leu	Lys	Gln	Ala	Thr	Lys
Leu	Gly	Phe	Ile		Asp	Gly		Arg
Val	Cys	Trp	Val		Glu	Cys		Ser
Ala			Ala		Ser			
Met			Gly		Thr			
Phe			Pro		Arg			
Trp			Lys		Pro			
Tyr			Gln		Ala			
Thr			Cys		Gly			
Asn			Ser					
Pro			Thr					

Fig. 2. Substitutions at positions 95-103 retaining T cell stimulatory capacity.

Further evidence that MCC positions 97, 99 and 102 interact directly with TCR is provided by experiments using peptides substituted with charged amino acids at these residues as immunogens in mice transgenic for either the TCR α or ß chain from the IE/MCC-reactive T cell clone 5C.C7. In each case, reciprocal charge changes are apparent in the junctional (CDR3) sequence of one of the 5C.C7 TCR chains expressed by T cells capable of recognizing the altered peptides (15 and Jorgensen et al, in preparation). In addition, TCR ß-chain transgenic mice respond to peptides altered at positions 97 and 99, but not to those substituted at 102. This suggests that the α-chain interacts with peptide positions 97 and 99, while the ß-chain interacts with position 102.

Taken together these results suggest a topology for the recognition of I-Ek/ MCC by the T cell clone 5C.C7 as indicated in figure 3.

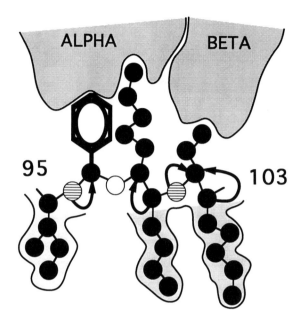

Fig. 3. Hypothetical model for the topology of I-E^k/MCC/TCR (5C.C7) interaction.

GENERATION OF TCR SURROGATES: ANTIBODIES SPECIFIC FOR I-E^k/MCC COMPLEXES

The ability to load >60% of I-E^k molecules with a given peptide *in vitro* (7,16) has allowed us to generate three independent antibodies (D-4, G-32 and G-35) specific for the I-E^k/MCC complex (17) as indicated in figure 4.

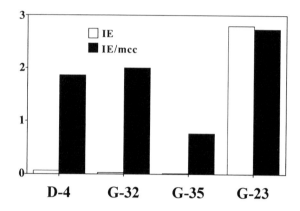

Fig. 4. I-E^k/MCC Specific Antibodies. Antibodies were tested by ELISA on soluble I-E^k either unbound or prebound to MCC.

Fig. 5 is a cartoon showing the effect of alterations in MCC and I-E^k sequence on the reactivity of one of these antibodies (G-35).

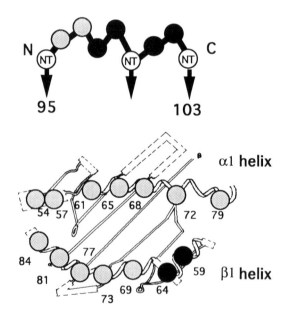

Fig. 5. An epitope map for antibody G-35. The location and effects of changes in sequence of either MCC peptide or the helices of I-E^k on G-35 reactivity are indicated.

G-35 reactivity is affected by changes at positions 102, 101, 99 and 98 (marked as •) but not 97 or 96 of the peptide, and by only 2 of 13 site directed mutants of the I-E^k-helices (18). This epitope map suggests that the MCC peptide is aligned in the binding groove as indicated, a result that is consistent with the effects of sequence changes within the groove as previously noted (19). Additionally, this is the same orientation as noted for class I MHC-associated peptides (1,2).

The availability of such "MHC-restricted" antibodies allows immuno-chemical analysis of the TCR ligand (i.e. I-E^k/MCC complexes) independent of T cell readouts. This uncoupling, in turn, permits the relationship between T cell responsiveness and the number of complexes per cell to be examined. By using radiolabelled D-4 antibody in binding experiments we have determined the number of complexes formed after pulsing a transfected cell expressing I-E^k with different peptide concentrations (Reay et al, in preparation). Figure 6 shows these results compared to the degree of T cell activation assessed by the amount of IL-3 released.

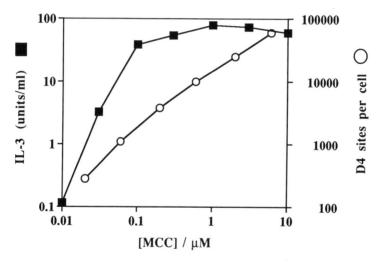

Fig. 6. Relationship between the average number of I-E^k/MCC complexes per cell and T cell stimulation capacity.

While the average number of I-E^k/MCC complexes formed per cell increases linearly over a wide range of peptide concentration, the degree of T cell activation reaches a plateau at 5000-10000 per cell. This direct analysis also demonstrates, in agreement with previous results (20,21), that an average of only 200 complexes per cell is sufficient to provoke a detectable T cell response in a bulk culture. However, preliminary results suggest that an individual presenting cell must express many more complexes (>2000) to stimulate a single T cell (data not shown). This has important consequences for the immunodominance of individual epitopes as it places more stringent requirements in terms of density and binding affinity on immunogenic peptides.

THE THYMIC LIGAND FOR POSITIVE SELECTION

Figure 7 demonstrates that G-35 inhibits the activation of the I-E^k/MCC-specific hybridoma 2B4, but not of hybridoma YO1 specific for a fragment of hemoglobin bound to I-E^k . In contrast, the peptide-independent anti-I-E^k antibody 14.4.4 inhibits the activation of both of these hybridomas.

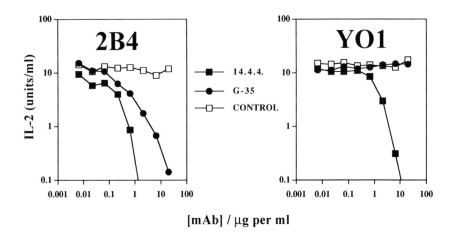

Fig. 7. Specific inhibition of in vitro T cell response by G-35.

This specific inhibition prompted us to use G-35 to probe the nature of the I-E^k/peptide molecules mediating selective events in T cell development (17). Accordingly, we examined the expression of CD4 and CD8 on developing transgenic thymocytes from a mouse expressing the I-E^k/MCC-specific TCR from the 5C.C7 clone (Fazekas de St. Groth, unpublished results) following serial injection of G-35 (figure 8). The injection regimen used resulted in mice having a circulatory concentration of 40μg/ml G-35.

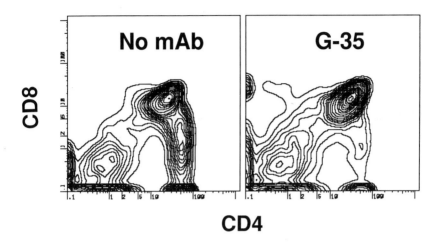

Fig. 8. G-35 blocks T cell positive selection. CD4/CD8 profiles of 5C.C7 transgenic thymocytes from mice injected with G-35.

In a mock injected mouse the transgenic thymocytes show positive selection into the CD4$^+$ subset, as noted previously in this system (22; Fazekas de St. Groth, pers. comm.). In contrast, such positive selection is severely impaired in mice injected with G-35, with the percentage of CD4$^+$8$^-$ and CD4$^+$8^{lo} cells reduced by 70%. Additionally, a significant subpopulation, unobserved in control mice, expressing the transgenic TCR on CD4$^-$8$^+$ cells is observed. No changes are seen in the CD4/8 profiles of the non-transgenic thymocytes from these same mice, or in those from non-transgenic littermate controls.

These results demonstrate that G-35 recognizes the I-E^k molecule(s) mediating positive selection of T cells carrying the receptor from the 5C.C7 clone. This implies that not only are such I-E^k molecules occupied with peptide(s), but also that these peptides must be structurally related to the nominal MCC antigen sequence (i.e.. they are mimics).

TCR- I-E^k/MCC AFFINITY

A major reason for expressing TCR and its I-E^k/peptide ligand as soluble forms was to investigate the physical parameters of their interaction. We have used antibody competition studies to determine the affinity between these macromolecules as outlined in figure 9.

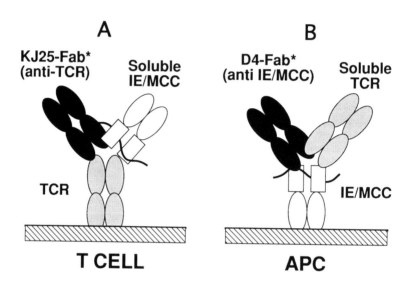

Fig.9. Summary of antibody competition studies used to measure the affinity of TCR-I-E^k/MCC Interaction.

The binding of an anti-TCR antibody (specific for the CDR2 region of Vβ3) to T cells was inhibited by very high concentrations (>10μM) of I-E^k bound to appropriate peptide, but not by complexes with substituted peptides not recognized by the T cells (16). In a reciprocal experiment (b) we have shown that 35μM soluble TCR will inhibit 50% of the binding of the I-E^k/MCC-reactive antibody D4 described above (23). In both of these protocols the calculated Kd value for the TCR-IE^k/MCC interaction is 40-60 μM. A similar value has been derived from blockade of T cell responses by soluble TCR (24), although the non-linearity of such responses (see above) means that it is difficult to estimate the percentage of complexes at the cell surface that are actually being blocked. More recently we have used a newly developed technique based on surface plasmon resonance (Pharmacia Biosensor) to confirm the Kd values of 40-50μM, and to show that this is a slow-on, fast-off interaction (23).

The very low affinity of TCR-I-E^k/peptide binding suggests that the primary function of this interaction is signalling rather than mediation of cell conjugate formation. Thus, it is likely to be the interactions of the large numbers of adhesion molecules which play the critical part of holding cells in apposition long enough for triggering to occur (25,26).

FUTURE DIRECTIONS

The I-E^k/MCC-specific antibodies described will provide an important tool with which to characterize the formation of this T cell determinant *in vivo*. In particular, immuno-histochemistry will be used to clarify the intracellular compartment(s) in which peptide binding occurs. Additionally, these reagents will be used to quantitate the rates of formation and dissociation of complexes in different types of antigen presenting cells.

A further use of the G-35 antibody will be to isolate the peptides associated with I-E^k in the thymus that mediate positive selection of the 5C.C7 TCR. Sequencing of these will identify the differences compared to the nominal MCC peptide antigen. Using the techniques described we will then be able to measure the TCR affinity for I-E^k complexed with these peptides, allowing testing of models of positive selection that have been proposed.

ACKNOWLEDGEMENTS

We would like to express our gratitude to all members of the lab who have contributed to the described work. In particular we thank Kiyoshi Matsui, Jay Boniface, Elliott Ehrich, Jeff Jorgensen and Barbara Fazekas for sharing their unpublished data, Bill Ho and Kristin Baldwin for mouse breeding and FACS analysis, and Suzanne Erickson with Yvonne Gautam for preparing soluble molecules. This work was supported by the Howard Hughes Medical Institute.

REFERENCES

1. **Madden, D. R., J. C. Gorga, J. L. Strominger, and D. C. Wiley.** 1992. The three dimensional structure of HLA-B27 at 2.1 Å resolution suggests a general mechanism for tight binding to MHC. *Cell 70*: 1035.

2. **Fremont, D. H., M. Matsumura, E. A. Stura, P. A. Peterson, and I. A. Wilson.** 1992. Crystal structures of two viral peptides in complex with murine MHC class I H-2K^b. *Science 257*: 919.

3. **Falk, K., O. Rötzschke, Stevanovíc, G. Jung, and H.-G. Rammensee.** 1991. Allele-specific motifs revealed by sequencing of self-peptides eluted from MHC molecules. *Nature 351*: 290.

4. **Schwartz, R. H.** 1985. T-lymphocyte recognition of antigen in association with gene products of the major histocompatibility complex. *Annu. Rev. Immunol. 3*: 237.

5. **Wettsein, D. A., J. J. Boniface, P. A. Reay, H. Schild, and M. M. Davis.** 1991. Expression of a class II major histocomptibility complex (MHC) heterodimer in a lipid-linked form with enhanced peptide/soluble MHC complex formation at low pH. *J. Exp. Med. 174*: 219.

6. **Lin, A. Y., B. Devaux, A. Green, C. Sagerstrom, J. F. Elliott, and M. M. Davis.** 1990. Expression of T cell antigen receptor heterodimers in a lipid-linked form. *Science 249*: 677.

7. **Reay, P. A., D. A. Wettstein, and M. M. Davis.** 1992. pH dependence and exchange of high and low responder peptides binding to a class II MHC molecule. *The EMBO J. 11*: 2829.

8. **Stern, L. J., and D. C. Wiley.** 1992. The human class II MHC protein HLA-DR1 assembles as empty ab heterodimers in the absence of antigenic peptide. *Cell 68*: 465.

9. **Driscoll, P.C., J. D. Altman, J. J. Boniface, K, Sakaguchi, P. A. Reay, J. G. Omichinski, E. Appella, and M. M. Davis.** 1993. Two dimensional nuclear magnetic resonance analysis of a labeled peptide bound to a class II major histocompatibility complex molecule. *J. Mol. Biol. 232*: 342.

10. **Fox, B. S., C. Chen, E. Fraga, C. A. French, B. Singh, and R. H. Schwartz.** 1987. Functionally distinct agretopic and epitopic sites. Analysis of the dominant T cell epitope of moth and pigeon cytochromes c with the use of synthetic peptide antigens. *J. Immunol. 139*: 1578.

11. **Bhayani, H. and Y. Paterson.** 1989. Anaysis of peptide binding patterns in different major histocompatibility complex/T cell receptor complexes using pigeon cytochrome c-specific T cell hybridomas. *J. Exp. Med. 170*: 1609.

12. **Reay, P. A., R. Kantor, and M. M. Davis.** 1994a. Use of global amino acid replacements to define the requirements for MHC binding and T cell recognition of moth cytochrome c (93-103). *J. Immunol.* (in press).

13. **Brown, J. J., T. S. Jardetzky, J. C. Gorga, L. J. Stern, R. G. Urban, J. L. Strominger, and D. C. Wiley.** 1993. Three dimensional strcture of the human class II histocompatibility antigen HLA-DR1. *Nature 364*:33.

14. **Chen, W., J. McCluskey, S. Rodda, and F. R. Carbone.** 1993. Changes at peptide residues buried in the major histocompatibility complex (MHC) class I binding cleft influence T cell recognition. A possible role for indirect conformational alterations in the MHC class I or bound peptide in determining T cell recognition. *J. Exp. Med. 177*: 869.

15. **Jorgensen, J. L., U. Esser, B. Fazekas de St. Groth, P. A. Reay, and M. M. Davis.** 1992. Mapping T-cell receptor-peptide contacts by variant peptide immunization of single-chain transgenics. *Nature 355*:224.

16. **Matsui, K., J. J. Boniface, P. A. Reay, H. Schild, B. Fazekas de St. Groth, and M. M. Davis.** 1991. Low affinity interaction of peptide-MHC complexes with T cell receptors. *Science 254*:1788.

17. **Reay, P. A., K. Baldwin, K. Haase, B. Devaux, E. Ehrich, and M. M. Davis.** 1994b. T cell positive selection is blocked by class II MHC/peptide specific antibodies. *Submitted.*

18. **Ehrich, E. E., B. Devaux, E. P. Rock, J. L. Jorgensen, M. M. Davis, and Y.-h. Chien.** 1993. T cell receptor interaction with peptide/MHC and superantigen/MHC ligands is dominated by antigen. *J. Exp. Med.* (in press).

19. **Ronchese, F., R. H. Schwartz, and R. N. Germain.** 1987. Functionally distinct subsites on a class II major histocompatibility complex molecule. *Nature 329*: 254.

20. **Harding, C. V., and E. R. Unanue.** 1990. Quantitation of antigen presenting cell MHC class II/peptide complexes necessary for T cell stimulation. *Nature 346*: 574.

21. **Demotz, S., H. M. Grey, and A. Sette.** 1990. The minimal number of class II MHC-antigen complexes needed for T cell stimulation. *Science 249*: 1028.

22. **Berg, L. J., A. M. Pullen, B. Fazekas de St. Groth, D. Mathis, C. Benoist, and M. M. Davis.** 1989. Antigen/MHC-specific T cells are preferentially exported from the thymus in the presence of their MHC ligand. *Cell 58*: 1035.

23. **Matsui, K., P. Steffner, J. J. Boniface, P. A. Reay, and M. M. Davis.** 1994. Kinetic analysis of T cell receptor binding to peptide-MHC complexes in a cell free system. *Proc. Natl. Acad. Sci. USA Submitted.*

24. **Weber, S., A. Traunecker, F. Oliveri, W. Gerhard, and K. Karjalainen.** 1992. Specific low-affinity recognition of major histocompatibility complex plus peptide by soluble T-cell receptor. *Nature 356*:793.

25. **Spits, H., W. van Shooyen, H. Keizer, G. Van eventer, M. Van de Rijn, C. Terhorst, and J. E. De Vries.** 1986. Alloantigen recognition is preceded by nonspecific adhesion of cytotoxic T cells and target cells. *Science 232*:403.

26. **Kupfer A and S. J. Singer.** 1989. Cell biology of cytotoxic and helper T cells functions. Immunofluorescence microscopic studies of single cells and cell couples. *Annu. Rev. Immunol. 7*: 309.

18

MOLECULAR ANALYSIS OF
MHC-PEPTIDE-TCR INTERACTIONS

Alessandro Sette, Jeff Alexander, Jörg Ruppert, Ken Snoke,
Alessandra Franco, Glenn Y. Ishioka, and Howard M. Grey

INTRODUCTION

T cell stimulation, leading to cell differentiation and proliferation, requires delivery of two distinct, independent signals. The first signal results from the interaction of the T cell receptor with antigen/major histocompatibility complex (MHC) molecules on the surface of antigen presenting cells. The nature of the second signal is less defined and may differ, depending upon the cell type (e.g., Th1 and Th2) and the stage of cell differentiation at the time of stimulation. These signals were regarded to be of the ON-OFF variety, i.e., signaling either occurs or does not occur. Therefore, the possibility that quantitative differences in the affinity of interaction between T cell receptor (TCR) and antigen/MHC may, in fact, lead to changes in the nature of the signaling itself has not been investigated in much detail.

In the past two years, however, several observations have been made which suggest that subtle changes in the structure of the T cell receptor ligands may change the nature of the signal(s) transmitted, and that signal transduction via the T cell receptor may vary qualitatively, depending upon the nature of the ligand. These studies led to the formulation of the concept that certain antigen analog/MHC complexes can act as specific antagonists of the TCR (1). The concept of MHC/antigen analog complexes acting as specific TCR antagonists has important implications in terms of the physiological events linked to signal transduction in T cells, and may also have therapeutic potential in the treatment of autoimmune disease or allergy. Accordingly, the potential of TCR antagonism as a therapeutic approach to autoimmune diseases is currently being evaluated.

THE PHENOMENON OF TCR ANTAGONISM

During certain studies on the capacity of MHC binding peptides to block antigen presentation to T cells, large numbers of peptides were screened in parallel for either their capacity to bind purified MHC molecules or to inhibit antigen presentation. Using influenza hemagglutinin (HA) specific, DR1 restricted cloned T cells, the at first puzzling observation was made that nonantigenic HA analogs were extremely powerful inhibitors of T cell activation, much more than peptides unrelated to the HA 307-319 antigen but with similar DR1 binding capacities.

A more in-depth examination of this phenomenon led to the discovery of a new mechanism for inhibition of T cell responses which we designated as TCR antagonism. More specifically, using both HA analogs and analogs of another DR1 restricted peptide [tetanus toxoid (TT) 830-843], we found that nonstimulatory HA analogs could much more efficiently inhibit antigen presentation to HA specific T cells, as compared to their capacity to inhibit TT specific, DR1 restricted T cells. Conversely, TT analogs were much more efficient in inhibiting TT specific responses than they were at inhibiting HA specific T cells. Direct MHC binding experiments and antigen presentation cellular experiments indicated that the mechanism responsible was distinct from competition for binding to the MHC molecules.

To test this hypothesis and directly differentiate MHC blockade from TCR antagonism, a "prepulse" assay was designed. APC were pulsed with a suboptimal dose of antigenic peptide for 2 hr, unbound antigen was removed by washing, and the APC were added with various inhibitory peptides, including the antigen analogs. The proliferation of subsequently added T cells was measured three days later. It should be noted that under these experimental conditions, competition for peptide binding to DR is prevented, since only a small number of the available DR molecules will be occupied by antigenic peptide.

When the capacity of antigen analogs to inhibit in the antigen prepulse assays was examined, it was found that antigen analogs retained their potent inhibitory capacity even in the presence of otherwise stimulatory amounts of antigen. The antigen specificity of the inhibition was indicated by the fact that HA analogs inhibited the HA specific clone but not the TT specific line, and conversely, the TT analogs were potent inhibitors of the TT specific line but not the HA specific clone. Having thus defined the phenomenon of TCR antagonism, additional experiments were performed to more precisely characterize this effect. It was determined that live APC were not required for TCR antagonism, thus ruling out active metabolic events such as recycling of MHC molecules being involved in the effect. Other explanations, such as binding of free antigen analogs to the TCR and induction of T cell tolerance, were likewise excluded (1). This series of experiments thus demonstrated for the first time that nonstimulatory antigen analogs could compete effectively with antigen for TCR engagement and thus act as antagonists of TCR mediated T cell stimulation.

Since the initial observations, the phenomenon of TCR antagonism has been extended to several other T cell types. TCR antagonism has been demonstrated for CD4+ T cells restricted to various class II molecules of both mouse and human origins (DR1, DR4, DR5, DQw6, IA^d, IA^s, and IE^k). Many different antigenic systems have also been used, including epitopes derived from HA, TT, proteolipid protein (PLP), ovalbumin (OVA), haemoglobin (HB), and cytochrome c (Cyt$_c$). Furthermore, it has also been shown that the phenomenon of TCR antagonism can be generalized to class I restricted T cells (2-4).

THE RELATIONSHIP BETWEEN STRUCTURE AND ACTIVITY OF TCR ANTAGONISTS

To characterize the molecular mechanisms involved in the TCR antagonism effect, a systematic analysis of the effect on TCR antagonism of different single amino acid substitutions in the antigen molecule have been performed. Alexander et al. (5) found that by modifying any of the five major T cell contact residues of HA 307-319 recognized by a DR1 restricted T cell clone, powerful TCR antagonists could be generated.

Although no general rules could be formulated, it was noted that peptides with conservative substitutions at a major TCR residue tended to either remain antigenic or were powerful antagonists. Semiconservative or nonconservative substitutions tended to yield less potent antagonists, or even analogs with no antagonist potential whatsoever. Taken together, these data suggest a model for TCR antagonism based on ligand affinities in which highest affinity interactions are stimulatory, lower affinity interactions are antagonistic, and even lower affinity interactions are neither antigenic nor antagonistic.

In the next series of experiments, a series of polyalanine-based analogs, in which an increasing number of the crucial T cell contact residues were inserted into the polyalanine backbone, were synthesized and tested for antagonism (5). Analogs containing a single TCR contact residue were neither antagonistic nor antigenic, while analogs containing two TCR contacts were weakly antagonistic. Analogs containing three or four TCR residues were strongly antagonistic, and finally, analogs containing four or five TCR residues lost all antagonistic potential and displayed antigenic capacity. Thus, these data illustrate how the similarity of each analog to the unmodified HA epitope dictates its capacity to be antigenic, antagonistic, or devoid of any detectable TCR binding activity.

In conclusion, data available to date support a model relating TCR antagonism to defined ligand affinities. According to this model, certain affinity thresholds are required for signaling, and the engagement of the TCR below this affinity threshold results in antagonism at the functional level. In this light, the relationship between partial agonism (6-8) and co-stimulatory activity could readily be explained if different and discrete affinity thresholds are required to trigger co-stimulatory activity. Thus, binding with a lower than optimal affinity may be sufficient to transmit certain signals. In order to trigger co-stimulatory signals, a higher affinity interaction that results in some TCR conformational change or cluster formation necessary for activation of the co-stimulatory pathway may, however, be required.

THE MOLECULAR MECHANISM OF TCR ANTAGONISTS: A MODEL

The data presented in the previous section suggest that affinity related mechanisms may be involved in the phenomena of TCR antagonism and partial agonism, but still do not offer any clear model of what specific mechanism might be operating at the molecular level. The simplest mechanism would involve simple receptor saturation. According to this model, antagonist/MHC complexes would directly compete with agonist/MHC complexes for binding to the TCR. This simplistic mechanism has to be discarded, however, on the basis of two arguments. First, it has been shown that as little as a threefold excess of antagonist peptides can effectively antagonize T cell responses. Since suboptimal antigen doses were used in these experiments, this leads to a calculation of only a few thousand antagonist/MHC complexes being required for antagonist activity to be displayed. This is about one-tenth the number of TCRs normally expressed by T cells. Thus, this situation is stoichiometrically incompatible with the receptor saturation model. Second, the identification of ligands acting as partial agonists/antagonists is diagnostic, in pharmacology, of mechanisms of action more complex than receptor saturation, and usually involves interference with receptor conformational changes or its interaction with co-receptors or allosteric ligands (5).

In search of an explanation for the TCR antagonist effect, we have studied the early events associated with T cell activation. We have found that TCR antagonists can inhibit intracellular events such as Ca^{++} influx and intraperitoneal (IP) turnover. By contrast, antagonist peptides had no effect on antigen induced APC-T cell conjugate formation (9).

This result has two different implications. First, it further illustrates, as already implied by the studies of Evavold and Allen (6), that the TCR is not an ON/OFF switch. In this context, O'Rourke and Mescher (10) have also demonstrated that cell adhesion can be achieved independently of the signals necessary for IP turnover. Thus, TCR engagement can result in either "horizontal" signaling (formation of APC/T cell conjugates via adhesion molecules and other specific membrane events) and/or "vertical" signaling (Ca^{++} influx, IP turnover, and other intracellular activation events). Second, the results described above imply that TCR antagonists could act at the membrane interface and uncouple the membrane "horizontal" activation events (which are not interfered with) from the intracellular "vertical" activation events (which are blocked by antagonists).

Thus, based on the above considerations, we can propose a speculative model of the molecular events that take place in the course of TCR antagonism (Fig. 1). According to this model, engagement of the TCR by wild type antigen/MHC complexes results in "horizontal" signaling, while engagement of the same TCR by antagonist/MHC complexes will not (Fig. 1a). Under normal circumstances, horizontal signaling is followed by TCR clustering associated with "vertical" signaling (Ca^{++} influx, IP turnover, etc.). When antigen/MHC and antagonist/MHC complexes are both present on the APC surface, however, mixed TCR oligomers will form, with some TCR engaging antigen/MHC complexes, while others engage antagonist/MHC complexes (Fig. 1b). These mixed TCR-MHC clusters are thought to be incapable of proper "vertical" signaling. It is noteworthy that, independent of these studies, solution by x-ray crystallography of the 3-D structure of DR1 has suggested that formation of complexes of TCR dimers engaging MHC/antigen dimers might be involved in signal transduction for T cell activation (11).

Figure 1.

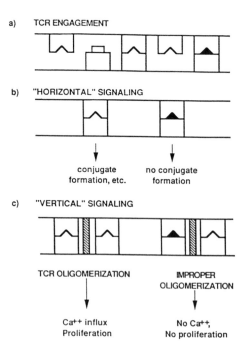

CAN A SINGLE TCR ANTAGONIST INHIBIT MULTIPLE TCR'S WITH DIFFERENT FINE SPECIFICITIES?

The data presented in the previous sections establish T cell receptor antagonism as a highly efficient method of inhibiting antigen induced T cell clonal responses. As such, it has potential for use in treatment of diseases in which suppression of a T cell immune response might have a beneficial effect, such as allergic and autoimmune diseases. A prerequisite for the use of this strategy is the demonstration of the capacity of a single or few analogs inhibiting most or all of the clones that are induced in vivo by a particular T cell epitope.

Data to support the possibility that TCR antagonists could inhibit polyclonal responses have been obtained in our laboratory in a human system. Specifically, we have examined the ability of TCR antagonist peptides to inhibit a panel of six different DR restricted, influenza hemagglutinin 307-319 specific T cell lines, all expressing different fine antigen specificities and TCR rearrangements. Antigen analogs that demonstrated no antigenicity were examined for their ability to act as TCR antagonists for these six T cell lines. Several different peptide analogs capable of acting as TCR antagonists on at least one T cell line were identified. Analogs capable of inhibiting all T cells have yet to be identified. However, some analogs were capable of inhibiting a majority (five of six) of the T cell specificities examined (3).

ANIMAL STUDIES

In recent years there have been several studies that have used antigen analogs to prevent or treat autoimmune disease. However, these studies were initiated to test the concept that addition of a peptide that could compete with a disease-inducing peptide for the same MHC restriction element could prevent disease (i.e., MHC blockade). Due to the short in vivo half-life of peptides and the rapid appearance of empty MHC molecules, this inhibition would be predicted to be transient. In fact, antigen given subsequent to the administration of the inhibitor will readily induce disease (12). In contrast to these findings, McDevitt's group (13, 14) has shown that pretreatment of animals with an analog of the encelphalitogenic peptide, followed one-two weeks later by immunization with the encephalitogenic peptide itself, led to experimental autoimmune encephalomyelitis (EAE) inhibition, i.e., the analog had a long-lasting effect that did not require its co-administration with antigen. Second, the analog could be administered after the encephalitogenic peptide at a time just prior to the disease onset, and the disease could still be ameliorated. In a second study by van Eden (15), it was shown that an antigen analog of myelin basic protein (MBP) could inhibit MBP-induced EAE, but not a mycobacterial heat shock protein (HSP) induced arthritis. Since both antigens were restricted by the same MHC specificity, it is clear that MHC blockade could not explain the specificity of inhibition.

At present it is difficult to evaluate whether or not any of these in vivo results might be wholly or partially due to TCR antagonist or partial agonist effects. Clearly, additional studies are required to fully understand these in vivo phenomena. Nevertheless, the bulk of the data suggest that antigen analogs, when given in vivo, might be much more efficient inhibitors of the immune response than unrelated peptides that block MHC interaction and that furthermore, they may have some capacity to induce long-lasting, antigen specific hyporesponsiveness.

In our laboratory we have examined whether TCR antagonism could be demonstrated in vivo by examining the capacity of TCR antagonist peptides to inhibit EAE in SJL (H-2^s) mice. This disease is induced in H-2^s mice by a well-defined T cell peptide epitope, the murine proteolipid protein (PLP) 139-151 peptide. Moreover, EAE in SJL mice represents an attractive model system for examining the potential for peptide antagonist based therapy, since the pathogenesis of the disease involves a highly heterogeneous population of T cells. Finally, previous work has shown that EAE could be ameliorated by in vivo treatment with antigen unrelated peptides with strong IA^s binding capacity, most likely through an MHC blockade mechanism (12).

In the first series of experiments, residues crucial for MHC binding and T cell activation with the PLP peptides were defined. It was found that residues L_{145} and P_{148} were important for IA^s binding, while residues G_{142}, K_{143}, W_{144}, G_{146}, and H_{147} were important for T cell activation. Residue W_{144} appeared to be especially critical, since substitutions at this residue led, in most instances, to complete loss of antigenicity for our panel of six different PLP 139-151 specific T cell clones.

Nonantigenic peptides that still retained good IA^s binding capacity were tested next for TCR antagonism. Following this strategy, several antagonistic peptides were identified. In general, similar to the study of Snoke et al. (3), most peptides were antagonistic for at least one T cell clone, but no peptide could be identified which simultaneously inhibited all of the different T cell clones. Because of the failure to identify a single analog capable of simultaneously inhibiting all of the different T cell clones, we decided to utilize a pool of the most effective PLP analogs for subsequent in vivo studies. When this TCR antagonist pool was utilized in vivo, it was found that: 1) The pool was much more effective than each of the peptides injected separately (Fig. 2); 2) The antagonist pool was approximately 10-fold more efficient than the best IA^s blocker we have identified to date. This finding correlated with in vitro data obtained in which the TCR antagonist peptides were found to be much more potent than antigen unrelated MHC blockers (Table 1); 3) The antagonist pool was effective at approximately equimolar amounts with the encephalitogenic PLP 139-151 peptide (Table 1); 4) The TCR antagonist pool was also moderately effective when administered before the antigenic challenge.

Figure 2.

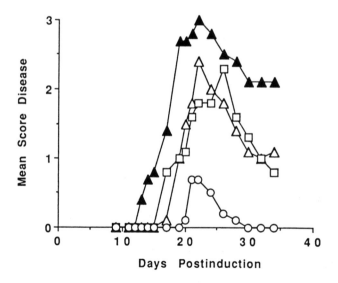

Table I. Relative In Vivo Efficacy of IA^s Blockers Versus TCR Antagonist Peptides in Inhibiting EAE Induction

Molar Excess [a]	ROIV Peptide	No. Experiments	TCR Antagonist Pool	No. Experiments
10x	44.7 ± 36.1[b]	3 [c]	92.0 ± 1.2	5
3.3x	27.0 ± 38.1	2	55.0 ± 2.8	3
1.1x	11.6 ± 20.6	2	49.6 ± 13.1	3
0.1x	ND [d]		2.7 + 11.3	2

The efficacy of preventing EAE induction in SJL mice following treatment with an unrelated IA^s blocking peptide, ROIV, was compared to treatment with a pool of Y_{144} and L_{144} analogs containing TCR antagonist activity. Animals were treated by co-immunizing them with the inhibitor peptide(s) and the encephalitogenic PLP 139-151 peptide.
[a] Molar excess of inhibitor peptide co-injected into mice relative to the PLP 139-151 peptide.
[b] Mean % inhibition of EAE ± SD.
[c] Number of independent experiments.
[d] Not done

CONCLUSIONS

The present report describes the use of TCR antagonist peptides to inhibit the experimental autoimmune disease, EAE. The results obtained suggest that TCR antagonists are highly effective inhibitors when administered together with antigen, and may also act as specific inhibitors when administered before the encephalitogenic challenge. Testing the therapeutic potential of the TCR antagonist and other antigen-related approaches in autoimmunity will require exact knowledge of the actual autoantigenic determinants recognized in the pathogenesis of human autoimmune diseases. In this regard, it is interesting to note that both MBP and PLP have been invoked as autoantigens involved in the pathogenesis of multiple sclerosis (MS) in humans, and that mapping of the MBP and PLP epitopes recognized by human autoreactive T cells has already been undertaken by several laboratories, including ours. In addition, recent developments in the field of sequencing naturally processed MHC bound peptide determinants (16) offer further hope that epitopes involved in the pathogenesis of other autoimmune diseases will also be identified. Accordingly, in the not too distant future, evaluation of antigen-specific strategies to immunomodulation may finally become possible.

ACKNOWLEDGMENTS

The authors thank Joyce Joseph for her assistance in preparing the manuscript. Partially supported by National Institutes of Health Grant AI18634 (H.G.)

REFERENCES

1. **De Magistris, M. T., J. Alexander, M. Coggeshall, A. Altman, F. C. A. Gaeta, H. M. Grey, and A. Sette**. 1992. Analog antigen/MHC complexes act as antagonists of the T cell receptor. *Cell 68*:625.

2. **Ostrov, D., J. Krieger, J. Sidney, A. Sette, and P. Concannon**. 1993. T cell receptor antagonism mediated by interaction between TCR junctional residues and peptide antigen analogues. *J. Immunol. 150*:4277.

3. **Snoke, K., J. Alexander, A. Franco, L. Smith, J. V. Brawley, P. Concannon, H. M. Grey, A. Sette, and P. Wentworth**. 1993. The inhibition of different T cell lines specific for the same antigen with TCR antagonist peptides. *J. Immunol. 151*:6815.

4. **Jameson, S. C., F. R. Carbone, and M. J. Bevan**. 1993. Clone-specific T cell receptor antagonists of major histocompatibility complex class I-restricted cytotoxic T cells. *J. Exp. Med. 177*:1541.

5. **Alexander, J., K. Snoke, J. Ruppert, J. Sidney, M. Wall, S. Southwood, C. Oseroff, T. Arrhenius, F. C. A. Gaeta, S. M. Colón, H. M. Grey, and A. Sette**. 1993. Functional consequences of engagement of the T cell receptor by low affinity ligands. *J. Immunol. 150*:1.

6. **Evavold, B. D. and P. M. Allen**. 1991. Separation of IL-4 production from Th cell proliferation by an altered T cell receptor ligand. *Science 252*:1308.

7. **Sloan-Lancaster, J., B. D. Evavold, and P. M. Allen**. 1993. Induction of T-cell anergy by altered T-cell-receptor ligand on live antigen-presenting cells. *Nature 363*:156.

8. **Racioppi, L., F. Ronchese, L. A. Matis, and R. N. Germain**. 1993. Peptide-major histocompatibility complex class II complexes with mixed agonist/antagonist properties provide evidence for ligand-related differences in T cell receptor-dependent intracellular signaling. *J. Exp. Med. 177*:1047.

9. **Ruppert, J., J. Alexander, K. Snoke, M. Coggeshall, E. Herbert, D. McKenzie, H. M. Grey, and A. Sette**. 1993. Effect of T-cell receptor antagonism on interaction between T cells and antigen-presenting cells and on T-cell signaling events. *Proc. Natl. Acad. Sci. USA 90*:2671.

10. **O'Rourke, A. M. and M. F. Mescher**. 1992. Cytotoxic T-lymphocyte activation involves a cascade of signalling and adhesion events. *Nature 358*:253.

11. **Brown, J. H., T. S. Jardetzky, J. C. Gorga, L. J. Stern, R. G. Urban, J. L. Strominger, and D. C. Wiley**. 1993. Three-dimensional structure of the human class II histocompatibility antigen HLA-DR1. *Nature 364*:33.

12. **Lamont, A. G., A. Sette, R. Fujinami, S. M. Colón, C. Miles, and H. M. Grey.** 1990. Inhibition of experimental autoimmune encephalomyelitis induction in SJL/J mice by using a peptide with high affinity for IA^s molecules. *J. Immunol. 145*:1687.

13. **Wraith, D. C., D. E. Smilek, D. J. Mitchell, L. Steinman, and H. O. McDevitt.** 1989. Antigen recognition in autoimmune encephalomyelitis and the potential for peptide-mediated immunotherapy. *Cell 59*:247.

14. **Smilek, D. E., D. C. Wraith, S. Hodgkinson, S. Dwivedy, L. Steinman, and H. O. McDevitt.** 1991. A single amino acid change in a myelin basic protein peptide confers the capacity to prevent rather than induce experimental autoimmune encephalomyelitis. *Proc. Natl. Acad. Sci. USA 88*:9633.

15. **Wauben, M. H. M., C. J. P. Boog, R. van der Zee, I. Joosten, A. Schlief, and W. van Eden.** 1992. Disease inhibition by major histocompatibility complex binding peptide analogues of disease-associated epitopes: more than blocking alone. *J. Exp. Med. 176*:667.

16. **Hunt, D. F., H. Michel, T. A. Dickinson, J. Shabanowitz, A. L. Cox, K. Sakaguchi, E. Appella, H. M. Grey, and A. Sette.** 1992. Peptides presented to the immune system by the murine class II major histocompatibility complex molecule I-A^d. *Science 256*:1817.

19

SELECTIVE IMMUNOSUPPRESSION BY MHC CLASS II BLOCKADE

Luciano Adorini and Jean-Charles Guéry

OVERVIEW

Administration of major histocompatibility complex (MHC) class II-binding synthetic peptides can induce selective immunosuppression via different mechanisms of action. MHC class II antagonists can block the MHC binding site, preventing binding of any antigenic peptide to MHC class II molecules and thereby precluding activation of class II-restricted T cells. Competition between self peptides for binding to class II molecules plays an important role in the regulation of antigen presentation to class II-restricted T cells, with implications for the positive and negative selection of developing T cells. This form of immunosuppression could, in principle, represent an approach towards selective immunointervention in autoimmune diseases, allograft rejection, and allergy.

INTERACTIONS OF PEPTIDES WITH MHC CLASS II MOLECULES

Antigen processing and presentation is a complex set of events leading to the activation of specific T cells by a binary ligand formed by antigenic peptides bound to MHC-encoded molecules. It is now well established that peptides bound to class I or class II MHC molecules on the surface of APC serve as ligands for the T cell receptors (TCR) of $CD8^+$ or $CD4^+$ T cells, respectively (1). The essential steps involved in peptide-MHC complex formation: antigen proteolysis in antigen-presenting cells (APC), binding of peptides to MHC molecules, and expression on the APC surface of the peptide-MHC complexes have been clarified in considerable detail (2). Following the pioneering work of Bjorkman, Wiley and Strominger (3), a more refined structural analysis of peptide interactions with MHC class I molecules has become available (4). The structure of MHC class II molecules has also been elucidated, demonstrating that, unlike the class I binding site, which is closed at both ends, the groove of class II molecules allows peptides to protrude from both sides (5). In addition, only a major deep pocket (and three or four more shallow ones) capable of accommodating bulky peptide side chains is present in class II DR molecules, against the six found in class I molecules, suggesting different rules governing peptide binding. Interestingly, class II, unlike class I molecules, occur as dimers oriented to allow simultaneous interaction with two TCR complexes, suggesting implications for T cell triggering.

The isolation and characterization of naturally processed peptides has permitted to determine the basic requirements for peptide binding to class I molecules. The majority of peptides bound to MHC class I molecules are 8-10 residues long and possess 1-3 anchor positions accounting for allotype-specific binding motifs (6), consistent with the structure of the binding site in MHC class I molecules, comprising multiple pockets and close-ended at both sides. In most peptide-class I complexes the majority of the bonding energy is provided by conserved peptide termini and extensive hydrogen bonding along the peptide backbone, without involvement of peptide side chains.

The structural requirements for peptide-class II interactions have been defined first by peptide analogues (7, 8) and by site-directed mutagenesis of class II molecules (9, 10), and later confirmed by isolation of natural ligands (11-14). Peptide libraries have also been successfully used to identify critical amino acid residues involved in peptide binding to class II MHC molecules (15). In contrast to peptides bound to class I molecules, which are predominantly octamers or nonamers, naturally processed peptides bound to class II molecules are longer and display a higher degree of heterogeneity both in length and in the site of terminal truncations, implying that the mechanisms of processing for peptides binding to class I and class II molecules are substantially different. The considerable variation in length among class II-binding peptides (10-30 resides) is consistent with the structure of the class II binding site, open at both ends.

This body of knowledge represents the current structural basis for designing peptides able to selectively interfere with the activation of pathogenic T cells.

IMMUNOSUPPRESSION BY PEPTIDES RELATED OR UNRELATED TO THE ANTIGEN

In vivo inhibition of T cell activation by administration of MHC-class II-binding peptides has been demonstrated in a variety of systems by analysing responses to peptides or to protein antigens either in normal mice(16-18) or in autoimmune disease models (19-25). Based on the structural relationship between competitor and antigenic peptides, these experimental systems can be divided in two groups.

When the inhibitory peptide is unrelated to the antigenic epitopes, but is immunogenic, the inhibition observed could be due to a clonal dominance of competitor-specific over antigen-specific T cells. Moreover, competitor-specific T cells might mediate a suppressive effect by secreting inhibitory cytokines preventing priming of antigen-specific T cells. As a result, the observed inhibition of T cell activation may not be merely due to MHC antagonism and other immunoregulatory mechanisms cannot be excluded. The situation is simplified, from a mechanistic point of view, by administration of non immunogenic competitor peptides unrelated to the antigen: if inhibition of T cell priming is observed it can be ascribed to MHC blockade (26).

In models where the inhibitor peptide is structurally homologous to the antigenic peptide, in addition to the above mentioned mechanisms, antigen-specific mechanisms may be involved. These include induction of cross-reactive T cell tolerance (27), induction of regulatory suppressor T cells (28), alteration in signalling pathway through TCR occupancy leading to anergy (29), change in lymphokine production (30) or induction of apoptosis (31). In addition, a structurally related epitope can engage the TCR of antigen-specific T cells without leading to any signal transduction, a concept known as TCR antagonism (32).

MHC BLOCKADE

Peptides binding to the same class II molecule can compete with each other, in vitro, for presentation to T cells (33). In addition, priming for T cell responses can be inhibited in vivo by coadministration of antigen and a molar excess of MHC class II-binding competitor peptides (16). This raises the possibility of inducing selective immunosuppression by blocking the binding site of class II molecules associated to autoimmune diseases, thus preventing their capacity to bind any antigen, including autoantigens. Following this observation, several groups have used a similar strategy to prevent induction of T cell-mediated autoimmune diseases like experimental allergic encephalomyelitis (EAE) in mice (21, 23) and rats (24), autoimmune carditis (22), and insulin-dependent diabetes mellitus (25).

Direct Evidence For MHC Blockade

To monitor directly APC-associated class II-antigen complexes in a functional assay, we established an ex vivo system able to detect on the surface of lymph node APC complexes of naturally processed antigenic peptides and MHC class II molecules generated in vivo (26). This was accomplished by showing that lymph node cells (LNC) from mice immunized with hen egg-white lysozyme (HEL) display at their surface HEL peptide-class II complexes able to stimulate, in the absence of any further antigen addition, HEL peptide-specific, class II-restricted T cell hybridomas. HEL peptide 46-61/I-A^k class II complexes were readily detectable on LN APC from mice primed either with intact HEL protein or HEL46-61 in adjuvant, and the degree of 46-61-specific T cell hybridoma activation was dependent on the amount of antigen injected. The lysosomotropic agent chloroquine failed to interfere with 46-61/A^k presentation by LNC from HEL-primed mice, indicating that these antigenic complexes were formed in vivo rather than by carry over and processing of HEL in vitro. To analyse the phenotype of the APC expressing these complexes, we depleted the LNC population of T and B cells. The remaining cells, mostly macrophages and dendritic cells, were 10-fold more potent in antigen-presenting activity as compared to unseparated LNC. Dendritic cells have been shown in vivo to be the most potent APC in their ability to present exogenous antigen as well as to prime naive T cells (34, 35) and it is likely, although not formally proven, that in our experimental system they are most relevant. In time course experiments, monitoring both antigen-presenting activity and T cell proliferative responses in the same LNC population, we found that antigenic complexes were readily detectable from day 2 to day 16 after immunization and peaked at day 2, at a time where no HEL-specific T cells were detectable. Conversely, T cell proliferative responses to HEL peptides were observed from day 5, when antigen presentation started to decrease. Thus, as expected, antigen presentation precedes priming of HEL-specific bulk T cells in situ (J-C Guéry, unpublished). Therefore, coadministration of antigen and MHC class II-binding peptide competitors should inhibit T cell priming by blocking the initial formation of antigenic complexes. Indeed, coinjection of HEL or HEL peptide 46-61 and the I-A^k-binding peptide HEL112-129 inhibits the ability of LNC to stimulate I-A^k-restricted T cell hybridomas. Under these conditions, we showed that inhibition of HEL46-61-specific T cell proliferative response is associated with in vivo inhibition of antigen presentation, and found a correlation between increased formation of competitor/I-A^k complexes and inhibition of I-A^k-restricted antigen presentation, indicating competitive blockade of class II molecules in vivo (26).

Exogenous Competitors Can Inhibit The Presentation Of Endogenous Antigen To Class II-Restricted T Cells

An important issue, from a theoretical and practical point of view, is whether exogenous competitors can inhibit the presentation of endogenous, as well as exogenous, antigens to class II-restricted T cells. Although the nature of autoantigens is still undefined in most autoimmune diseases, it is likely that endogenous antigens are the most relevant in the induction of autoreactive T cells leading to MHC class II-associated autoimmune diseases. Interestingly, an exogenous competitor can inhibit equally well class II-restricted T cell activation induced by exogenous or by transfected endogenously synthesized antigen, suggesting the feasibility of inhibiting, by MHC blockade, the presentation of self antigens synthesized by APC (36).

To address this question in a more physiological situation, we recently developed a novel experimental system to analyse processing and presentation to class II restricted T cells of endogenously synthesized naturally processed self antigens constitutively expressed by mouse APC (Guéry et al., submitted). First, binding to mouse class II molecules of synthetic peptides from the self antigen, in this case ß2 microglobulin (ß2M), is determined. Then, high affinity binding peptides are injected into mice expressing appropriate class II molecules but lacking ß2M, and peptide-specific class II-restricted T cell hybridomas are generated. These hybridomas are then used to read-out the formation of self ß2M peptide-class II complexes in APC from mice expressing ß2M. The mouse ß2M peptide 25-40 binds to I-A^d (37) and it is immunogenic in BALB/c mice with a knocked-out ß2M gene while non immunogenic in normal BALB/c mice. A panel of ß2M25-40-specific T cell hybridomas generated in ß2M-negative mice was used as readout for the formation of complexes between self ß2M peptides and class II molecules. The results show that: 1) processing of endogenous ß2M is sensitive to protease inhibitors and lysosomotropic amines, and is not due to reuptake of shed or released protein, 2) the ß2M sequence 25-40 contains two partially overlapping determinants presented by either I-A^d or I-E^d class II-molecules, 3) these self ß2M peptide-class II complexes are detected on any class II$^+$ APC examined in different tissues, including the thymus, and therefore represent naturally processed epitopes able to induce T cell tolerance as indicated by the lack of immunogenicity of this peptide in BALB/c mice, 4) splenic dendritic cells are extremely efficient in processing and presenting this self epitope to class II-restricted T cells.

Using this system, we assessed whether an exogenous competitor peptide added in vitro may inhibit presentation of a naturally processed endogenous antigen. An I-E^d-restricted T cell hybridoma, specific for the sequence 26-35 of mß2M, readily detects constitutive antigenic complexes expressed by L cell fibroblasts transfected with I-E^d genes, or by normal spleen cells from BALB/c mice. Preincubation of these APC with the I-E^d-binding peptide HEL105-120 or 107-129 inhibits, dose-dependently, activation of the ß2M-specific, E^d-restricted T cell hybridoma. This inhibition is specific for the class II molecules to which the competitor peptide binds, as demonstrated by the lack of inhibition of I-A^d-restricted T cell specific for the overlapping ß2M peptide. Therefore, exogenous MHC antagonists can inhibit constitutive presentation by splenic APC of an endogenous, naturally processed self epitope.

We also wished to determine if naturally processed self peptides could behave as MHC antagonists. Among naturally processed self peptides bound to I-A^d molecules, the E$_\alpha$ sequence 52-68 has been found to be a strong binder (13). Interestingly, this sequence had been previously eluted from I-A^b molecules (11) and found to represent a major self peptide, expressed by about 10% of class II molecules present on the APC surface (38). The E$_\alpha$ 52-68 peptide is able to block efficiently presentation of exogenous mouse ß2M25-40 by BALB/c ß2M-negative APC to A^d- but not to E^d-restricted T cells. We next analyzed the capacity of this peptide to inhibit A^d-restricted presentation of the naturally processed endogenous ß2M peptide by IFN$_\gamma$-stimulated macrophages from BALB/c mice. Preincubation of BALB/c macrophages with the A^d-binding peptide Eα52-68, but not with the E^d-binding peptide HEL105-120, inhibits dose-dependently presentation of self ß2M peptides by A^d molecules. These results indicate that self peptides eluted from MHC class II molecules can block presentation of both exogenous and endogenous self peptides. Blocking self peptides naturally occupying the class II binding site are probably selected for their high binding capacity , as indicated by the presence of specific binding motives, and may represent powerful MHC antagonists in vivo. In addition these peptides, which are likely to induce tolerance during development, are devoid of immunogenicity, potentially a major problem in immunotherapy based on chronic administration of MHC antagonists.

The capacity of an exogenous competitor to inhibit in vitro presentation to T cells of endogenous, as well as exogenous self antigens suggests that in vivo MHC blockade by administration of MHC antagonists could inhibit presentation to class II-restricted T cells of endogenous cellular antigens, certainly relevant, at least in some cases, in the induction of autoreactive T cells leading to HLA-associated autoimmune diseases.

MHC Blockade Does Not Inhibit Ongoing Antibody Responses

MHC blockade has mostly been evaluated by inhibition of T cell proliferation, which does not reflect the entire immune response (30). Therefore, we have examined the effect of MHC class II blockade on the in vivo antibody responses to T cell-dependent antigens. In these experiments, the MHC class II antagonist used was the self peptide corresponding to mouse lysozyme residues 46-62 (ML46-62) which binds to I-A^k but not to Ia^d class II-molecules and selectively inhibits, when injected together with antigen, priming of I-A^k-restricted T cells (16, 26). Injection of this non immunogenic MHC class II-binding self peptide at the time of antigen priming can inhibit both primary and secondary antibody responses to protein antigens related (HEL) or not (bovine ribonuclease) to the competitor peptide (39). By monitoring complexes formed in vivo between two different epitopes derived from HEL processing and I-A^k molecules, we have shown that ML46-62 prevents equally well presentation of both the competitor-related epitope HEL46-61 and of the unrelated sequence HEL34-45, also binding to I-A^k molecules. This inhibition of I-A^k-restricted antigen presentation is associated with the inhibition of T cell proliferation to determinants restricted by the blocked I-A^k molecules, regardless of their sequence homology with the competitor peptide. The specificity of this inhibitory effect was indicated by the lack of inhibition of HEL108-116/I-E^d complexes in H-2^d mice, according to the binding specificity of ML46-62. All antibody isotypes are equally inhibited, suggesting that MHC blockade inhibits activation of both Th1- and Th2-type helper T cells. Therefore, the inhibition of clonal expansion of Th cells by blocking the initial antigen-TCR interaction seems to affect both types of Th cells, preventing B cell proliferation and differentiation as well as generation of antigen-specific memory cells. This is exemplified by the inhibition of the secondary antibody response, where the unresponsiveness of B cells to soluble antigen is likely to arise both from lack of antigen-specific T cell help, mainly of Th-2 type, and lack of antigen-specific memory B cells able to endocytose soluble antigen and to present efficiently antigenic peptide-MHC complexes to primed T cells. As for T cell activation, the inhibition of antibody responses is selective for the MHC class II molecules binding the competitor peptide, and its extent depends on the molar ratio between antigen and competitor. However, administration of MHC antagonists after antigen priming does not affect ongoing antibody responses, as expected from MHC blockade. These results are consistent with the observation that MHC blockade can prevent, but not treat autoimmune diseases (21). Therefore, in established autoimmune diseases MHC blockade should be associated to treatments able to eliminate or inhibit activated T cells.

Prevention Of Autoimmune Diseases By MHC Blockade

MHC blockade in vivo is capable of preventing autoimmune diseases, for example coadministration of an encephalitogenic peptide together with an unrelated non immunogenic peptide binding to the same MHC class II molecule prevents induction of experimental allergic encephalomyelitis (EAE) (21, 23), and administration of the self lysozyme peptide ML46-62, binding to I-A^k, prevents myosin-induced autoimmune carditis in H-2^k mice (22). Administration of class II-binding peptide antagonists can prevent not only induced but also spontaneous autoimmune diseases. Using the non-obese diabetic (NOD) mouse as an autoimmune disease model it has been demonstrated that a peptide capable of blocking antigen presentation by the NOD class II molecule I-A^{g7} can also prevent the spontaneous development of insulin-dependent diabetes mellitus (IDDM) (25). This indicates that presentation of autoantigens by the I-A^{g7} molecule plays a critical role in induction of IDDM, and that interference with its ability to present antigen can prevent the onset of a spontaneous autoimmune disease. However, because in this case the MHC antagonist used was itself immunogenic, other mechanisms able to inhibit class II-restricted T cell activation, such as immune deviation, could not be excluded. In any case, prevention of disease required continuous administration of MHC antagonist, and after cessation of treatment progression to IDDM continued as in untreated NOD mice. This observation emphasizes the passive nature of MHC blockade, and demonstrates that in a spontaneous autoimmune disease prevention of pathogenic T cell activation depends on the continuous presence of a sustained level of MHC antagonist.

Prospects For Clinical Applicability Of MHC Blockade

Administration of MHC antagonists could thus represent an approach to prevent HLA-associated autoimmune diseases, assuming that the association between class II molecules and disease reflects their capacity to present autoantigens to autoreactive T cells. However, several points need still to be addressed to evaluate the practical feasibility of this form of immunointervention. Since MHC blockade is a passive type of treatment, appropriate delivery systems (40), instrumental to achieve the sustained plasma level of soluble MHC blockers necessary for effective inhibition of T cell activation, need to be developed. The use of proteins and peptides as therapeutic agents is presently limited by poor oral bioavailability, although in some cases orally delivered, peptide-based drugs can show bioavailability and efficacy. Even after parenteral administration, peptides have usually a very short plasma half-life, and they are certainly not the ideal drug to induce MHC blockade in vivo. MHC blockade in vivo could only be achieved when high concentrations of peptide antagonist were present in the fluid phase, as decrease in the plasma concentration of competitor peptide results in immediate re-acquisition of antigen presenting capacity in peripheral blood APC (L. Adorini, unpublished). These results indicate that saturation of class II molecule for any appreciable time does not occur in vivo. Therefore, prevention of T cell activation by MHC blockade may not be of therapeutic value unless sustained high concentrations of antagonist can be maintained in extracellular fluids, and this will be very difficult to achieve by systemic administration of synthetic peptides. It is hoped that non-peptidic MHC blockers may be more effective. In this respect, it appears that small organic compounds may not be suitable, because – unlike peptides – they have a very fast off-rate, whereas peptidomimetic antagonists could exhibit adequate pharmacokinetic and pharmacodynamic characteristics.
Strong and weak points of MHC antagonists as immunosuppressive drugs in the treatment of autoimmune diseases could be summarized as follows:

Strong points
- Information about the autoantigen/s are not required
- A degree of selectivity is afforded by targeting MHC class II molecules associated to disease
- The rapid reversal of blockade upon suspension of treatment should avoid undesired side effects of immunosuppression

Weak points
- A sustained level of antagonist must be maintained for a long period
- Ongoing immune responses are not inhibited

REFERENCES

1. **Sette, A., and H. M. Grey.** 1992. Chemistry of peptide interaction with MHC proteins. *Curr. Opin. Immunol. 4:*79.

2. **Germain, R. N., and D. H. Margulies.** 1993. The biochemistry and cell biology of antigen processing and presentation. *Annu. Rev. Immunol. 11:*403.

3. **Bjorkman, P. J., M. A. Saper, B. Samraoui, W. S. Bennet, J. L. Strominger, and D. C. Wiley.** 1987. Structure of the human class I histocompatibility antigen, HLA-A2. *Nature. 329:*506.

4. **Fremont, D. H., M. Matsumura, E. A. Stura, P. A. Peterson, and I. A. Wilson.** 1992. Crystal structures of two viral peptides in complex with murine MHC class I H-2K^b. *Science. 257:*919.

5. **Brown, J. H., T. S. Jardetzky, J. C. Gorga, L. J. Stern, R. G. Urban, J. L. Strominger, and D. C. Wiley.** 1993. Three-dimensional structure of the human class II histocompatibility antigen HLA-DR1. *Nature. 364:*33.

6. **Falk, K., O. Rotzsche, S. Stevanovic, G. Jung, and H. G. Rammensee.** 1991. Allele-specific motifs revealed by sequencing of self-peptides eluted from MHC molecules. *Nature. 351:*290.

7. **Babbitt, B. P., G. Matsueda, E. Haber, E. R. Unanue, and P. M. Allen.** 1986. Antigenic competition at the level of peptide-Ia binding. *Proc. Natl. Acad. Sci. USA. 83:*4509.

8. **Sette, A., S. Buus, S. Colon, J. A. Smith, C. Miles, and H. M. Grey.** 1987. Structural characteristics of an antigen required for its interaction with Ia and recognition by T cells. *Nature. 328:*395.

9. **Peccoud, J., P. Dellabona, P. Allen, C. Benoist, and D. Mathis.** 1990. Delineation of antigen contact residues on an MHC class II molecule. *EMBO J. 9:*4215.

10. **Krieger, J. I., R. W. Karr, H. M. Grey, W. Y. Yu, D. O'Sullivan, L. Batovsky, Z. L. Zheng, S. M. Colòn, F. C. A. Gaeta, J. Sidney, M. Albertson, M.-F. Del Guercio, R. W. Chesnut, and A. Sette.** 1991. Single amino acid changes in DR and antigen define residues critical for peptide-MHC binding and T cell recognition. *J. Immunol. 146:*2331.

11. **Rudensky, A. Y., P. Preston-Hurlburt, S.-C. Hong, A. Barlow, and C. A. Janeway.** 1991. Sequence analysis of peptides bound to MHC class II molecules. *Nature. 353:*622.

12. **Rudensky, A. Y., P. Preston-Hurlburt, B. K. Al-Ramadi, J. Rothbard, and C. A. Janeway.** 1992. Truncation variants of peptides isolated from MHC class II molecules suggest sequence motifs. *Nature 359:*429.

13. **Hunt, D. F., H. Michel, T. A. Dickinson, J. Shabanowitz, A. L. Cox, K. Sakaguchi, E. Appella, H. W. Grey, and A. Sette.** 1992. Peptides presented to the immune system by the murine class II major histocompatibility complex molecule I-A^d. *Science. 256:*1817.

14. **Chicz, R. M., R. G. Urban, W. S. Lane, J. C. Gorga, L. J. Stern, D. A. A. Vignali, and J. L. Strominger.** 1992. The predominant naturally processed peptides bound to HLA-DR1 are derived from MHC-related molecules and are heterogenous in size. *Nature. 358:*764.

15. **Hammer, J., P. Valsasnini, K. Tolba, D. Bolin, J. Higelin, B. Takacs, and F. Sinigaglia.** 1993. Promiscuous and allele-specific anchors in HLA-DR-binding peptides. *Cell. 74:*197.

16. **Adorini, L., S. Muller, F. Cardinaux, P. V. Lehmann, F. Falcioni, and Z. A. Nagy.** 1988. In vivo competition between self peptides and foreign antigens in T cell activation. *Nature. 334:*623.

17. **Adorini, L., E. Appella, G. Doria, and Z. A. Nagy.** 1988. Mechanisms influencing the immunodominance of T cell determinants. *J. Exp. Med. 168:*2091.

18. **Lamont, A. G., M. F. Powell, S. M. Colon, G. Miles, H. M. Grey, and A. Sette.** 1990. The use of peptide analogs with improved stability and MHC binding capacity to inhibit antigen presentation in vitro and in vivo. *J. Immunol. 144:*2493.

19. **Wraith, D. C., D. E. Smilek, D. J. Mitchell, L. Steinman, and H. O. McDevitt.** 1989. Antigen recognition in autoimmune encephalomyelitis and the potential for peptide-mediated immunotherapy. *Cell. 59:*247.

20. **Sakai, K., S. S. Zamvil, D. J. Mitchell, S. Hodgkinson, J. B. Rothbard, and L. Steinman.** 1989. Prevention of experimental encephalomyelitis with peptides that block interaction of T

cells with major histocompatibility complex protein. *Proc. Natl. Acad. Sci. USA. 86:*9470.

21. **Lamont, A. G., A. Sette, R. Fujinami, S. M. Colon, G. Miles, and H. M. Grey.** 1990. Inhibition of experimental autoimmune encephalomyelitis induction in SJL/J mice by using a peptide with high affinity for IA^s molecules. *J. Immunol. 145:*1687.

22. **Smith, S. C., and P. M. Allen.** 1991. Myosin-induced acute myocarditis is a T cell-mediated disease. *J. Immunol. 147:*2141.

23. **Gautam, A. M., C. I. Pearson, A. A. Sinha, D. E. Smilek, L. Steinman, and H. O. McDevitt.** 1992. Inhibition of experimental autoimmune encephalomyelitis by a nonimmunogenic non-self peptide that binds to I-A^u. *J. Immunol. 148:*3049.

24. **Wauben, M. H. M., C. J. P. Boog, R. van der Zee, I. Joosten, A. Schlief, and W. van Eden.** 1992. Disease inhibition by major histocompatibility complex binding peptide analogues of disease-associated epitopes: more than blocking alone. *J. Exp. Med. 176:*667.

25. **Hurtenbach, U., E. Lier, L. Adorini, and Z. A. Nagy.** 1993. Prevention of autoimmune diabetes in non-obese diabetic mice by treatment with a class II major histocompatibility cpmplex-blocking peptide. *J. Exp. Med. 177:*1499.

26. **Guéry, J.-C., A. Sette, J. Leighton, A. Dragomir, and L. Adorini.** 1992. Selective immunosuppression by administration of MHC class II-binding peptides. I. Evidence for in vivo MHC blockade preventing T cell activation. *J. Exp. Med. 175:*1345.

27. **Ria, F., B. M. C. Chan, M. T. Scherer, J. A. Smith, and M. L. Gefter.** 1990. Immunological activity of covalently linked T-cell epitopes. *Nature. 343:*381.

28. **Sercarz, E. E., and U. Krzych.** 1991. The distinct specificity of antigen-specific suppressor T cells. *Immunol. Today. 12:*111.

29. **Sloan-Lancaster, J., B. D. Evavold, and P. M. Allen.** 1993. Induction of T cell anergy by altered T cell receptor ligand on live antigen-presenting cells. *Nature. 363:*156.

30. **Evavold, B. D., and P. M. Allen.** 1991. Separation of IL-4 production from Th cell proliferation by an altered T cell receptor ligand. *Science. 252:*1308.

31. **Iwabuchi, K., K. Nakayama, R. L. McCoy, F. Wang, T. Nishimura, S. Habu, K. M. Murphy, and D. Y. Loh.** 1992. Cellular and peptide requirements for in vitro clonal deletion of immature thymocytes. *Proc. Natl. Acad. Sci. USA. 88:*9000.

32. **De Magistris, T. M., J. Alexander, M. Coggeshall, A. Altman, F. C. A. Gaeta, H. M. Grey, and A. Sette.** 1992. Antigen analog-major histocompatibility complexes act as antagonists of the T cell receptor. *Cell. 68:*625.

33. **Werdelin, O.** 1982. Chemically related antigens compete for presentation by accessory cells to T cells. *J. Immunol. 129:*1883.

34. **Crowley, M. T., K. Inaba, and R. M. Steinman.** 1990. Dendritic cells are the principal cells in mouse spleen bearing immunogenic fragments of foreign proteins. *J. Exp. Med. 172:*383.

35. **Inaba, K., J. Metlay, M. T. Crowley, and R. M. Steinman.** 1990. Dendritic cells pulsed with protein antigens in vitro can prime antigen-specific, MHC-restricted T cells in situ. *J. Exp. Med. 172:*631.

36. **Adorini, L., J. Moreno, F. Momburg, G. J. Hämmerling, J.-C. Guéry, A. Valli, and S. Fuchs.** 1991. Exogenous peptides compete for the presentation of endogenous antigens to major histocompatibility complex class II-restricted T cells. *J. Exp. Med. 174:*945.

37. **Sette, A., J. Sidney, F. C. A. Gaeta, E. Appella, S. Colon, M. F. del Guercio, J. C. Guéry, and L. Adorini.** 1993. MHC class II molecules bind indiscriminately self and non-self peptide homologs: effect on the immunogenicity of non-self peptides. *Int. Immunol. 5:*631.

38. **Murphy, D. B., S. Rath, E. Pizzo, A. Y. Rudensky, A. George, J. K. Larson, and C. A. Janeway.** 1992. Monoclonal antibody detection of a major self peptide MHC class II complex. *J. Immunol. 148:*3483.

39. **Guéry, J.-C., M. Neagu, G. Rodriguez-Tarduchy, and L. Adorini.** 1993. Selective immunosuppression by administration of major histocompatibility complex class II-binding peptides. II. Preventive inhibition of primary and secondary antibody responses. *J. Exp. Med. 177:*1461.

40. **Illum, L., and S. S. Davis.** 1991. Drug delivery. *Curr. Op. Biotech. 2:*254.

20

ANTIGEN PRESENTING FUNCTION OF NONCLASSICAL CLASS I MOLECULES

Mitchell Kronenberg and Hilde Cheroutre

ANTIGEN PRESENTATION BY CLASSICAL CLASS I MOLECULES

Much information has been gained recently concerning antigen presentation to T lymphocytes by both classical class I and class II molecules. Class I molecules bind peptides of approximately 9 amino acids in length, while class II molecules bind peptides which range between 14 and 22 amino acids (1,2). Both types of antigen presenting molecules bind diverse sets of peptides, but 1-3 positions in these bound peptides, called anchor residues, tend to be invariant. Anchor positions are more clearly defined for class I-binding peptides, they interact with specific residues in the pockets of the peptide binding site (1,2). The amino acids in the peptide binding pockets tend to be polymorphic, and therefore different MHC alleles will bind different sets of peptides. MHC class I and class II molecules have distinct pathways by which they acquire peptide for presentation to T cells (reviewed elsewhere in this volume). Classical class I molecules require both bound peptide antigen and noncovalent association with β_2-microglobulin (β_2m) for stable cell surface expression (3). The trimolecular complex of class I heavy chain, β_2m, and peptide associates in the endoplasmic reticulum (ER, 4). A transporter comprised of two proteins, TAP-1 and TAP-2, also encoded by genes in the MHC, is required for the transport of peptides generated in the cytoplasm into the ER (5). Class II molecules traffic through the ER as well, but rarely acquire peptide in this site. Instead, class II molecules traffic through acidic endosomes on their way to the cell surface, and it is in this compartment where peptide is acquired (6).

Nonclassical class I genes have been found in variety of species including amphibians, birds and mammals (7). The molecules encoded by these genes also are known as class Ib, or medial class I molecules. The genes for these molecules may be encoded either within or outside of the MHC. In the mouse, the majority of nonclassical class I molecules are encoded by genes located in the *T* (formerly *Tla*), *Q* (formerly *Qa*), and *M* (formerly *Hmt*) distal subregions

---

of the MHC. Mice, humans, rats and rabbits also are known to express CD1 molecules, which are non-MHC encoded nonclassical class I molecules (8).

POSSIBLE SPECIALIZED ANTIGEN PRESENTING FUNCTIONS OF NONCLASSICAL CLASS I MOLECULES

The leading hypothesis for the function of nonclassical class I molecules is that they have a specialized antigen presenting function, distinct from that of the classical class I molecules. This is consistent with the significant degree of sequence similarity between classical and nonclassical class I genes, and the numerous examples of recognition of allogeneic or self nonclassical class I molecules by T lymphocytes (9). However in most of these cases, presentation of peptides by the nonclassical class I molecule has not been demonstrated. The notion of specialized antigen presentation encompasses several possibilities. The M3 gene product provides a very clear example, as it preferentially binds peptides with a formylated methionine at their NH_2 terminus (reviewed in 9). This molecule is therefore capable of binding to peptides derived from the processed NH_2-termini of bacterial proteins as well as to a very limited set of peptides derived from self proteins encoded by the mitochondrial genome (9). A second example in which peptide presented by a nonclassical class I molecule is at least partially defined is provided by the Qa-1 molecule, which probably can present a peptide derived from a random copolymer comprised of glutamine and tyrosine to a T-cell hybridoma (10), and which also may bind to a peptide(s) derived from *Mycobacterium bovis* heat shock protein (hsp) 65 (11). In both of these cases, the exact peptide presented by Qa-1 has not yet been defined. The nonclassical class I molecules as a group probably are not adapted for presenting peptides with formylated NH_2 termini, and the notion of a specialized antigen presenting function can encompass a number of other possibilities (reviewed in 12), including presentation to lymphocytes in a restricted anatomic site such as the epithelia, presentation to a distinct subset of CD4-,CD8- T cells, presentation to T cells that express a γδ T-cell antigen receptor (TCR), and finally, presentation of peptides located in a distinct intracellular compartment, analogous to the endosomal-derived class II peptides.

EXPRESSION, STRUCTURE, AND BEHAVIOR OF THE THYMUS LEUKEMIA ANTIGEN

In this chapter, we will summarize our recent efforts to characterize the antigen presenting function of the Thymus Leukemia (TL) antigen. As noted above, there may be few generalizations that cover all the nonclassical class I molecules, as different molecules may have their own specialized niche in the world of antigen presentation. Nevertheless, there are several reasons for concentrating on the TL antigen. It was the first nonclassical class I molecule to be discovered, and it is one of the very few that can be recognized by monoclonal antibodies (mAbs). It also has a unique pattern of expression as it is expressed in all inbred mouse strains by small intestine epithelial cells (13). In addition, in some inbred mouse strains, it is expressed by a variety of cells in the hematopoietic series including cortical thymocytes, activated spleen cells, a subset of fetal liver cells and Langerhans cells in the skin (reviewed in 12, 14).

The prominent expression of the TL antigen by intestinal epithelial cells is of considerable interest. The mucosal immune system confronts a unique environment in the body because the gut is normally replete with antigens. How is it that immune responses are generated to pathogens but not to food peptides? How is oral tolerance established? Is there an immune surveillance system in the gut that can provide a rapid lymphocyte response without the need for clonal expansion of antigen reactive cells? Because of its expression in intestine epithelium, the TL antigen could play a role in one or more of these processes.

Similarities Between The TL Antigen And Classical Class I Molecules

The TL antigen is defined by reactivity with antisera and mAb (12,14). It is encoded by the $T3^b$, $T18^d$, and closely related alleles and pseudoalleles encoded in the $H-2T$ region of the MHC. The TL antigen is approximately 60% similar in amino acid sequence for the $\alpha 1$ and $\alpha 2$ domains to classical class I molecules, and more than 85% for the $\alpha 3$ domain. Furthermore, TL molecules have all of the conserved amino acids in the $\alpha 1/\alpha 2$ domain peptide antigen binding site, including Tyr^7, Tyr^{59}, Tyr^{84}, Thr^{143}, Lys^{146}, Trp^{147}, Tyr^{159} and Tyr^{171} (15,16), that are believed to contact either the peptide termini or main chain atoms of the bound peptide antigen via hydrogen bonds (1). This suggests that the peptide antigen binding groove of TL molecules might be closed at the ends, and, like the classical class I molecules, that the TL antigen might bind to nanomer peptides. The TL antigen has several other features in common with classical class I molecules. It requires $\beta_2 m$ for assembly and transport to the cell surface (17). In addition, the $\alpha 3$ domain of the TL antigen can bind to CD8 (18). This was determined using a cell-cell adhesion assay, in which transfectants that overexpress the human CD8α chain could be shown to bind specifically to C1R cell transfectants that express a high level of the TL antigen. CD8 binding by the TL antigen also could be demonstrated using a functional assay. Transfected fibroblasts were generated that express a chimeric class I gene comprised of the peptide antigen binding site ($\alpha 1$ and $\alpha 2$ domains) of the classical class I molecule L^d, and the remainder encoded by the $T18^d$ gene. CD8-dependent, L^d alloreactive CTL could lyse targets expressing this chimeric class I gene as effectively as they could lyse an L^d transfectant, in both cases the lysis could be inhibited with anti-CD8 mAbs (18). CD8 binding by the TL antigen is perhaps not surprising given the high degree of sequence conservation in the $\alpha3$ domain, most notably in those amino acids thought to constitute a minimal CD8 binding site (19).

Differences Between The TL Antigen And Classical Class I Molecules

Although there are similarities between the TL antigen and classical class I molecules in both structure and behavior, there also are striking differences. One notable difference is the lack of polymorphism. There are a total of only 21 positions in the $\alpha1$ and $\alpha2$ domains in which the six sequenced TL antigen-encoding genes differ (15,16), this is much less than one would expect from a comparable set of K or D genes. More striking is the fact that none of these differences are in amino acids likely to be important for peptide or TCR contact, assuming an overall TL antigen conformation similar to that of H-$2K^b$. In those positions likely to be important for peptide antigen contact, there are, however, six silent substitutions. The evolutionary history of the TL antigen alleles and pseudoalleles is therefore exactly the opposite of that of classical class I molecules. Classical class I molecules accumulate a high rate of nonsynonymous substitutions in those amino acids thought to interact with the peptide or the TCR. The speculative conclusion drawn from these facts is that the TL antigen is selected for the presentation of a highly conserved peptide or set of peptides. We have speculated further that a peptide derived from a heat shock protein is one possible candidate antigen, as these proteins are very highly conserved (17). It is noteworthy that a similar set of observations has recently been made concerning the Qa-1^a and Qa-1^b alleles (20), 13 of the 15 amino acid substitutions that distinguish these molecules are in positions not likely to be involved in peptide antigen binding. As mentioned above, there is indirect evidence that Qa-1 molecules can bind to peptides from heat shock proteins (11). A second striking difference between the TL antigen and classical class I molecules is the independence of surface TL antigen expression from any requirement for TAP function. We have transfected TL antigen expression constructs, and other class I gene expression constructs, into TAP-2-deficient mouse RMA-S lymphoma cells, or into *Drosophila* S2 tissue culture cells, which are presumed to lack both TAP-1 and TAP-2 proteins. Data from the RMA-S cell transfectants are summarized in Table 1. Classical class I molecules produced by RMA-S cells are unstable at 37°C, although they can be expressed on the cell surface at lower temperatures (Table 1, 21). However, there is little or no difference in TL antigen surface expression when RMA-S cells are cultured at low temperature (26-28°C) or at 37°C (see Table 1). Similar results were obtained in the *Drosophila* tissue culture cells (data not shown). An L^d/TL chimeric class I molecule behaved similar to a classical class I molecule (Table 1), suggesting that the $\alpha1$ and $\alpha2$ domains of the TL antigen might be responsible for the stable, TAP-independent expression on the cell surface. These flow cytometry experiments measure only steady-state

Table I. Surface expression of class I gene products in RMA-S and RMH cells

Cell type[a]	Class I Molecule[b]	Fluorescence Intensity[c]	
		26° C	37° C
RMH	K^b/D^b	n.d.	2,091
RMA-S	K^b/D^b	704	72
RMA-S/(TL)	K^b/D^b	n.d.	68
RMA-S/(TL)	TL antigen	799	679
RMA-S/(L/TL)	L/TL chimera	514	100

[a]RMA-S/(TL) and RMA-S/(L/TL) are stable transfectants of the RMA-S cell line that express $T18^d$ and a chimeric $L^d/T18^d$ gene, respectively.

[b]Class I molecules were detected on a FACSCAN instrument with the following mouse alloantibodies: K^b/D^b mAb 28-8-6S, TL antigen by mAb TL.m4, and L/TL by mAb 30-5-7.

[c]Fluorescence intensity calculated from mean fluorescence channel according to the manufacturer's instructions.

expression levels, and it is formally possible that the TL molecules in TAP- cells are transported to the cell surface very inefficiently, but that they were able to accumulate there because of a very long half-life. Preliminary results from pulse-chase experiments demonstrate, however, that this is not the case. TL antigen traffics similarly when mutant RMA-S cells and parental TAP-2+ RMH cells are compared (H. Holcombe, unpublished data).

There are three possible interpretations of this data. First, the TL antigen might for some reason be especially stable as a peptide-free molecule. This might occur because it binds β_2m very avidly. Second, analogous to the case of HLA-A2 expression in TAP deficient human T2 cells (22), the TL antigen might be very efficient at binding to either cleaved hydrophobic signal sequences or to some other peptides that are located in the ER even in the absence of TAP function. Third, analogous to class II molecules, the TL antigen might bind to peptides in some site other than the ER. There currently are no data that would permit discrimination between these hypotheses. It is interesting to note, however, that presentation of a mycobacterial antigen by a human nonclassical class I molecule, CD1b, is both TAP independent and chloroquine sensitive (23). In this regard, CD1b is more similar to class II than to classical class I molecules. Several other nonclassical class I molecules also may be TAP independent. Mouse CD1 surface expression at 37° C does not depend upon TAP function (M. Teitell, H. Holcombe and M. Kronenberg, unpublished data), and recognition of Qa-1 and T22 alloantigens is in at least some cases TAP independent (24, S. Hedrick, personal communication). These data are intriguing, as they suggest that many nonclassical class I molecules may be adapted to present different sets of peptides than the classical class I molecules. It is not possible to construct any general rule for nonclassical class I molecules, as in some cases T-cell recognition of these molecules was shown to require TAP-2 function (25).

DOES TL ANTIGEN PRESENT PEPTIDES TO IEL?

Intestinal Epithelial Lymphocytes

Because expression in the small intestine epithelium is an invariant feature of TL antigen expression in all inbred mouse strains, while TL antigen expression by cells in the hematopoietic series is only detected in some *H-2T* haplotypes, it is logical to suppose that any putative specialized antigen presenting function for TL antigen is carried out in the gut. Based upon this line of reasoning, it has been proposed (17) that the TL antigen presents peptide antigens that are recognized by intestinal intraepithelial lymphocytes (IEL). Mouse small intestine IEL are predominantly T cells, although they have been found to have a number of unique properties when compared to T lymphocytes found in lymphoid organs (reviewed in 26). These properties include the predominance of CD8+ cells, the relatively large fraction (20-80%) of IEL which express a TCR comprised of γ and δ chains, and the rapid and high level response these cells display in assays that measure cytotoxic activity. In addition, IEL do not proliferate strongly in response to TCR-mediated or other mitogenic signals (27). This has made studies of the functional attributes of these cells rather difficult, and the specificity and function of IEL remain incompletely characterized. Some IEL may differentiate in the intestine (26), and they are relatively solitary and sessile cells. Experiments with parabiotic mice have demonstrated that they do not recirculate (28). Based upon these features, and there relatively limited diversity of TCRs, it has been proposed that at least some populations of IEL may constitute a surveillance system for stressed epithelial cells (17). According to this view, rather than recognizing diverse antigens with a clonally distributed set of receptors, many IEL may recognize a relatively invariant stress response protein made in response to infection or cell damage. This type of response might permit a rapid elimination of infected cells prior to further penetration of the pathogen.

IEL Release Serine When Cultured With TL-Positive Epithelial Cells

Evidence has been generated recently which suggests that the TL antigen is recognized by IEL, and that it may serve as an antigen presenting molecule in the small intestine to IEL (17). IEL are highly activated, cytotoxic cells that have granules containing serine esterase and perforins (29). Granule release can be stimulated by cross linking their TCRs (27), or by culturing them with

freshly isolated intestinal epithelial cells (17). The results from four
experiments implicate the TL antigen in granule release in response to epithelial
cells. First, it could be shown that the response is site specific, in that only
small intestine epithelial cells can stimulate IEL from the small intestine. A
variety of other cells types, including epithelial cells from the colon, were
ineffective. This restricted set of stimulatory cells is consistent with the
restricted pattern of TL antigen expression, TL antigen is absent from most
adult tissues including colonic epithelium (13). Second, the response is not
MHC or strain restricted. Epithelial cells from one strain can stimulate IEL
from others, consistent with the relative lack of TL antigen polymorphism.
Third, epithelial cells from β_2m- mice were ineffective at stimulating IEL from
any strain. TL antigen is not present on intestinal epithelial cells from these
mice. These data are consistent with the involvement of a nonpolymorphic or
nonclassical class I molecule. Fourth, the response could be blocked most
effectively with anti-TL mAbs but much less so or not at all with mAbs to
other class I molecules. Results from additional experiments indicated that the
predominant responding cell type is a $\gamma\delta$ T cell with a TCR γ chain encoded by
the $V\gamma5$ gene segment In summary, the data from these experiments are
consistent with the hypothesis that the TL antigen is recognized by $\gamma\delta$ IEL.

The IEL experiments do not however clearly define if the recognition of the
TL antigen by IEL requires the presence of bound peptide antigen. We have
speculated that an inducible self antigen, which is a sign of stress or infection,
is being recognized (17). The conservation of the peptide antigen binding site
in the TL antigen could then be explained in light of the possible conservation
of this self molecule. Although there is little data to directly support these
speculations, using the serine esterase release system we have observed that

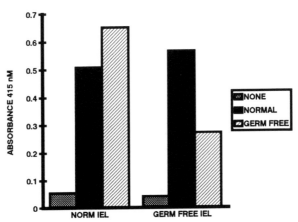

*Fig. 1. Serine esterase release, as measured in an ELISA assay, by IEL from
C3H conventional and C3H germ free mice cultured for four hours either
alone, or with epithelial cells from either germ free or conventional mice.*

epithelial cells from germ free mice are as effective as those from conventionally housed mice in the stimulation of IEL (Figure 1, 17). This can be explained if the epithelial cells in suspension are sufficiently stressed to express the hypothesized stress response protein, and therefore the presence of extensive gut bacterial flora may not be required. There are two weaknesses in these studies. First, they have all been carried out with a rather complex population of freshly isolated cells, and so it has not been possible to determine what the frequency of TL antigen reactive cells might be. Second, while TL antigen might be necessary for IEL stimulation, it clearly is not sufficient. Freshly isolated IEL do not react to TL antigen-positive cells from the thymus, nor do they kill fibroblasts transfected with the $T18^d$ gene expression construct (H.R. Holcombe, M. Teitell, unpublished data). The missing component, needed in addition to TL antigen, remains unknown although it could conceivably be a peptide from an intestine-specific stress response protein.

In Vivo Studies Of TL Reactivity By IEL

Because of the difficulty in studying IEL, and the IEL-TL antigen interaction, in vitro, we have begun to study this interaction in vivo using transgenic mice. Transgenic mice have been generated which express the TL antigen under the control of the $H-2D$ gene promoter. These have ectopic expression of the TL antigen, they express the $T18^d$ gene product on nearly all their tissues. These TL antigen transgenic mice have an approximately two-fold reduction ($p < 0.01$) in the number of $\gamma\delta$ IEL (Table 2). The $\gamma\delta$, and in particular, the $V\gamma5+$ and CD8+ IEL are selectively depleted (Table 2). The numbers of TCR $\alpha\beta+$ and CD4+ IEL are not changed, although the percentage is increased because of the $\gamma\delta$ decrease.

Table II. TCR Expression in IEL from TL antigen Transgenic Mice[a]

	TCR $\alpha\beta+$	TCR $\gamma\delta+$	$V\gamma5$ (% $\gamma\delta$)[b]
TL antigen transgenic	61	16	29
Nontransgenic	53	29	64

[a]Percent positive cells in each case, 16 transgenic and 11 control mice were analyzed.

[b]GL1 mAb positive (=$V\gamma5$ positive) as a % of total $\gamma\delta$ positive IEL.

Vγ5+, CD8+ γδ T cells are very rare outside of the intestine, there is no increase in the number of cells with this phenotype outside of the intestine in the TL antigen transgenic mice. Other γδ T-cell populations in the TL antigen transgenic mice are not grossly effected. In summary, the Vγ5 cells are not redistributed, and the data in Table 2 are consistent with negative selection of Vγ5+ IEL in TL antigen transgenic mice.

SUMMARY AND FUTURE DIRECTIONS

Several features distinguish the TL antigen from classical class I molecules, including its expression primarily in gut epithelium, a lack of dependence upon TAP molecules for stable surface expression, and conservation of the TL peptide antigen binding site. Based upon these data, we hypothesize that the TL antigen is adapted to present peptides to mucosal lymphocytes. The lack of polymorphism could result from selection for the presentation of peptides from highly conserved bacterial proteins such as heat shock proteins. Alternatively, an inducible, self stress response protein could be recognized. These alternatives are not mutually exclusive. As some intracellular pathogens can replicate in endosomes or phagolysosomes, we further speculate that TL molecules may have been selected to traffic, and acquire peptide antigens, in compartments other than the ER. A role for the TL antigen oral tolerance induction is also possible, although the role of mucosal cells in the induction of oral tolerance remains undefined.

In the absence of a molecular characterization of bound peptides, it remains formally possible that the TL antigen could be playing some role other than antigen presentation in the stimulation of IEL. For example, the TL antigen could be acting as an accessory molecule in the activation of TCR γδ+ IEL. Therefore, one aim of the current research in our laboratory is to identify the length, diversity and structural motifs of peptides bound to the TL antigen. A second aim is to determine if TL molecules expressed by RMA-S and/or Drosophila cells have bound peptides. If they do, then the structure of these peptides will be compared to those loaded into the TL antigen in RMH cells. In addition, we would investigate both the origin of these TAP-independent peptides, and their assembly with TL antigen heavy chains and β_2m. We also will try to reconstruct the stimulatory γδ IEL epitope, by adding peptide fractions, that have been acid eluted from intestinal epithelial cells, to transfectants that express high levels of the TL antigen. Future studies of the TL antigen/IEL interaction will lead to further insights into general mechanisms of antigen presentation. In addition, such studies should elucidate important features of the still poorly characterized localized immune responses in the intestinal mucosae.

ACKNOWLEDGMENTS.

Supported by NIH grants CA52511 and AG10152. We thank Mr. David Ng
for help with the preparation of the manuscript, and our colleagues Drs. Pirooz
Eghtesady, Hilda Holcombe, Chetan Panwala, Beate Sydora and Michael
Teitell for helpful discussions.

REFERENCES

1. **Fremont, D. H., M. Matsumura, E. A. Stura, P. A. Peterson, and I. A. Wilson**. 1992. Crystal structures of two viral peptides in complex with murine MHC class I H-2K^b. *Science 257*:919.

2. **Madden, D. R., J. C. Gorga, J. L. Strominger, and D. C. Wiley**. 1992. The three dimensional structure of HLA-B27 at 2.1 A resolution suggests a general mechanism for tight peptide binding to MHC. *Cell 70*:1035.

3. **Elliott, T**. 1991. How do peptides associate with MHC class I molecules? *Immunology Today 12*:386.

4. **Townsend, A., T. Elliott, V. Cerundolo, L. Foster, B. Barber, and A. Tse**. 1990. Assembly of MHC class I molecules analyzed in vitro. *Cell 62*:285.

5. **Monaco, J. J., S. Cho, and M. Attaya**. 1990. Transport protein genes in the murine MHC: possible implications for antigen processing. *Science 250*:1723.

6. **Guagliardi, L. E., B. Koppelman, J. S. Blum, M. S. Marks, P. Cresswell, and F. M. Brodsky**. 1990. Co-localization of molecules involved in antigen processing and presentation in an early endocytic compartment. *Nature 343*:133.

7. **Stroynowski, I., and K. Fischer Lindahl**. Antigen presentation by nonclassical class I molecules. *Current Opinion in Immunology*, in press..

8. **Calabi, F., and A. Bradbury**. 1991. The CD1 system. *Tissue Antigens 37*:1.

9. **Fischer Lindahl, K.** 1993. Peptide antigen presentation by nonclassical class I molecules. *Seminars in Immunology 5*:117.

10. **Vidovic, D., M. Roglic, K. McKune, S. Guerder, C. MacKay, and Z. Dembic.** 1989. Qa-1 restricted recognition of foreign antigen by a γδ T-cell hybridoma. *Nature 340*:646.

11. **Imani, F., and M. J. Soloski.** 1991. Heat shock proteins can regulate expression of the *Tla* region-encoded class Ib molecule Qa-1. *Proc. Natl. Acad. Sci. USA 88*:10475.

12. **Teitell, M., H. Cheroutre, C. Panwala, H. R. Holcombe, P. Eghtesady, and M. Kronenberg.** Structure and function of *H-2T* (*Tla*) region class I MHC molecules. *Crit. Rev. Immunol.,* in press.

13. **Hershberg, R., P. Eghtesady, B. Sydora, K. Brorson, H. Cheroutre, R. Modlin, and M. Kronenberg.** 1990. Expression of the thymus leukemia antigen in mouse intestinal epithelium. *Proc. Natl. Acad. Sci. USA 87*:9727.

14. **Chorney, M. J., H. Mashimo, K. A. Chorney, and H. Vasavada.** 1991. TL genes and antigens. In *Immunogenetics of the major histocompatibility complex*. R. Srivastava, B. P. Ram, and P. Tyle, eds. VCH, New York, NY, p.177.

15. **Chen, Y-T., Y. Obata, E. Stockert, T. Takahashi, and L. J. Old.** 1987. The *Tla* region genes and their products. *Immunol. Res. 6*:30.

16. **Mashimo, H., M. J. Chorney, P. A. Pontarotti, D. A. Fisher, L. Hood, and S. G. Nathenson.** 1992. Nucleotide sequence of the BALB/c *H-2T* region gene, *T3ᵈ*. *Immunogenetics 36*:326.

17. **Eghtesady, P., C. Panwala, and M. Kronenberg.** Recognition of the Thymus Leukemia Antigen by Intestinal γδ T Lymphocytes. *Proc. Natl. Acad. Sci. USA,* in press.

18. **Teitell, M., M. F. Mescher, C. A. Olson, D. R. Littman, and M. Kronenberg.** 1991. The thymus leukemia antigen binds human and mouse CD8. *J. Exp. Med. 174*:1131.

19. **Salter, R. D., R. J. Benjamin, P. K. Wesley, S. E. Buxton, T. P. J. Garrett, C. Clayberger, A. M. Krensky, A. M. Norment, D. R. Littman, and P. Parham, P.** 1990. A binding site for the T-cell co-receptor CD8 on the α3 domain of HLA-A2. *Nature 345*:41.

20. **Connolly, D. J., L. A. Cotterill, R. A. Hederer, C. J. Thorpe, P. J. Travers, J. H. McVey, P. J. Dyson, and P. J. Robinson.** 1993. A cDNA clone encoding the mouse Qa-1a histocompatibility antigen and proposed structure of the peptide antigen binding site. *J. Immunol. 151*:6089.

21. Ljunggren, H.-G., N. J. Stam, C. Ohlen, J. J. Neefjes, P. Hoglund, M-T Heemels, J. Bastin, T. N. Schumacher, A. M. Townsend, K. Karre, and H. L. Ploegh. 1990. Empty MHC class I molecules come out in the cold. *Nature 346*:476.

22. Wei, M. and P. Cresswell. 1992. HLA-A2 molecules in an antigen-processing mutant cell contain signal sequence-derived peptides. *Nature 356*:443.

23. Porcelli, S., C. T. Morita, and M. B. Brenner. 1992. CD1b restricts the response of human CD4-, CD8- T lymphocytes to a microbial antigen. *Nature 360*: 593.

24. Aldrich, C. J., R. Waltrip, E. Hermel, M. Attaya, K. Fischer Lindahl, J. J. Monaco, and J. Forman. 1992. T cell recognition of Qa1^b antigens on cells lacking a functional Tap-2 transporter, *J. Immunol. 149*:3773.

25. Attaya, M., S. Jameson, C. K. Martinez, E. Hermal, C. Aldrich, J. Forman, K. Fischer Lindahl, M. J. Bevan, and J. J. Monaco. 1992. *Ham-2* corrects the class I antigen-processing defect in RMA-S cells. *Nature 355*:647.

26. Lefrancois, L. 1991. Extrathymic generation of intraepithelial lymphocytes: Generation of a separate and unequal T-cell repertoire?, *Immunology Today 12*: 436.

27. Sydora, B., P. Mixter, H. R. Holcombe, P. Eghtesady, K. Williams, M. C. Amaral, A. Nel, and M. Kronenberg. 1993. Intestinal intraepithelial lymphocytes are activated and cytolytic but do not proliferate as well as other T cells in response to mitogenic signals. *J. Immunol. 150*:2179.

28. Poussier, P., P. Edouard, C. Lee, M. Binnie, and M. Julius. 1992. Thymus-independent development and negative selection of T cells expressing T cell receptor α/β in the intestinal epithelium: Evidence for distinct circulation patterns of gut- and thymus-derived T lymphocytes. *J. Exp. Med. 176*:187.

29. Guy-Grand, D. M., M. Malassis-Seris, C. Briottet, and P. Vassalli. 1991. Cytotoxic differentiation of mouse gut thymodependent and independent intraepithelial T lymphocytes is induced locally. *J. Exp. Med. 173*:1549.

21

DEFINITION OF T CELL EPITOPES OF *LISTERIA MONOCYTOGENES* AND REGULATION OF ANTIGEN PROCESSING BY THE BACTERIAL EXOTOXIN LISTERIOLYSIN-O (LLO)

H.Kirk Ziegler, Susan A.Safley,and Elizabeth Hiltbold

INTRODUCTION

The Role Of LLO In Infection With And Immunity To *Listeria Monocytogene*s.

    L. monocytogenes is a Gram-positive, facultatively intracellular bacterium capable of causing severe infections in newborns and immunocompromised individuals. Murine infection with *Listeria monocytogenes* is a model system for studying cellular immunity to intracellular parasites. The importance of T cell-mediated immunity in recovery from murine listeriosis has been established, and both CD4+ and CD8+ T cells appear to be involved in a successful anti-*L.monocytogenes* immune response (3-5). Virulent strains of *L. monocytogenes* secrete an exotoxin known as listeriolysin O (LLO), which has been identified as a major virulence factor (1,5,6,7,19)
    The structural gene encoding LLO has been cloned and sequenced. LLO is composed of 504 amino acids and contains highly conserved region near the C-terminal portion of the molecule that may be essential for the hemolytic activity of the molecule. Its hemolytic properties are activated by reducing agents and inhibited by low levels of cholesterol, by heat or by oxidizing agents. While structurally and antigenically related to other sulfhydryl-activated (oxygen-labile) toxins, such as streptolysin-o (SLO) and pneumolysin (PNY), LLO is unique in that its membrane lytic activity is maximum at low pH (about 5.5). This feature might account for its function as a major virulence factor in permitting its action within the acidic environment of macrophage phagosomes (1,8,9,26,27).
    Evidence suggests that LLO mediates bacterial virulence by favoring bacterial growth in macrophages (1) and by causing inhibition of antigen

processing via class II MHC pathways (5,6). LLO allows bacteria to escape the phagosome and replicate in the cytoplasm (19). Inhibition of class II MHC-mediated antigen presentation likely contributes significantly to the virulence of these bacteria by preventing T cell activation required for effective immune responses and clearance of bacteria. At the same time, cytoplasmic expression may permit the entrance of bacterial antigens into a class I MHC pathway which would account for the protective CD8+ T cell response. Interestingly, LLO plays a central role in immunity to *L. monocytogenes.* Immunization with Hly-positive (Hly+), but not with isogenic Hly -negative strains of *L. monocytogenes* protect mice from subsequent challenge with virulent Hly+ strains. This effect is likely due to the functions of LLO as a principal antigen, an adjuvant, and/or a factor promoting bacterial survival (5,6,16) The identification of the impact of the multiple functions of LLO on MHC functions related to macrophage and T cell responses represents an interesting and important challenge for continued studies.

   In this manuscript, we describe the major class I and class II MHC epitopes of LLO. By using T cell hybridomas specific for such epitopes together with defined transposon mutants of *Listeria* that express truncated LLO molecules (containing the relevant epitopes but not cytolytic activity), we have tested the idea that the C-terminal lytic region of LLO plays a role in directing LLO epitopes to different class I vs class II processing pathways. Finally, the role of γ/δ T cells in immunity to *Listeria* infection is discussed. (21,22)

DEFINITION OF T CELL EPITOPES OF *LISTERIA MONOCYTOGENES*

Class I MHC Epitopes Of LLO.

   The CD8+ T cell response to LLO has been recently investigated in the lab of Bevan (10,11,17,18). Definition of the antigenically dominant K^d binding peptide was achieved by first acid eluting peptides from *Listeria*-infected spleen cells from Balb/c mice, separating them by high performance liquid chromatography (HPLC) into fractions, and using each fraction to sensitize P815 target cells for lysis by a *Listeria*-specific CTL clone (17). One fraction contained the T cell activating activity. With the allele-specific peptide binding motif of the class I H-2 K^d allele recently published, it was possible to use the predictive motif to scan the amino acid sequence of LLO for peptides which fit the motif. The motif predicts that K^d binding peptides are nine amino acids long with a tyrosine residue at position two and either leucine or isoleucine at the carboxy terminus. Upon inspection of the LLO sequence, there were found to be 11 nonamer peptides which matched this motif. Each of the 11 synthetic nonamer peptides were then loaded onto the P815 cells to determine if any one would stimulate the specific lysis of the labeled targets. One peptide, LLO 91-99, was found to sensitize the P815 target cells for lysis by the *Listeria*-specific CTL clone. This peptide comigrated with the CTL targetting peptide fraction from the infected spleens on an HPLC with a shallow acetonitrile gradient.

These data suggest that this peptide (LLO 91-99) is the natural, dominant, K^d binding epitope of LLO (17).

Using this data gathered by Bevan's laboratory, we have generated T cell hybridomas specific for the LLO 91-99 epitope presented by K^d (12). We have also confirmed that the CD8+ T cell response to LLO 91-99 was a predominant reactivity and that class I and class II MHC epitopes of LLO are non-overlapping. Such class I-restricted hybridomas together with class II-restricted hybridomas have supplied convenient and well defined cellular probes for processing and presentation of LLO via the class I and class II MHC pathway (see below).

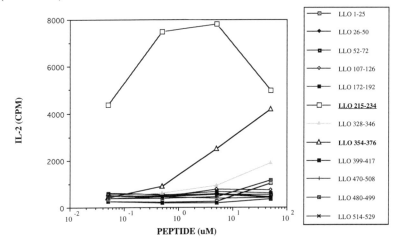

Fig.1. Polyclonal Interleukin-2 response of peritoneal exudate T cells taken from mice imunized with Listeria to a battery of LLO peptides presented by syngeneic peritoneal macrophages. Note response to LLO 215-234.

Class II MHC Epitopes Of LLO.

Concerning the CD4+ T cell response to LLO, peptide epitopes, conforming to models for the prediction of T cell epitopes, were identified from the deduced amino acid sequence of LLO (20). Of many peptides studied, covering about 50% of the LLO molecule, it was clear that a region in and around LLO 215-226 was a dominant epitope for class II MHC-restricted CD4+ T cells (Figure 1). This peptide fit algorithms used to predict T cell epitopes, and had the correct spacing of amino acids that is characteristic of I-E^k binding peptides. This peptide, LLO 215-234, could bind to I-E^k, I-E^d, and I-A^d as defined by competitive inhibition binding assays. A majority of peritoneal exudate CD4+ T cells from *Listeria*-immunized mice are specific for LLO 215-234. The minimal stimulatory T cell epitope was shown to be LLO 215-226 using a nested set of synthetic peptides (Figure 2). The immunodominance of this set of peptides is in keeping with the idea that immunodominant peptides are those that bind promiscuously and with high affinity to MHC molecules.

Sequence	Position
N S L N V	230-234
A V N N S L N V	227-234
A F K A V N N S L N V	224-234
F G T A F K A V N N S L N V	221-234
I A K F G T A F K A V N N S L N V	218-234
S Q L I A K F G T A F K A V N N S L N V	215-234
D E M A Y S E S Q L I A K F G T A F K	208-226
K I D Y D D E M A Y S E S Q L I A K F G T A F K	203-226
S E S Q L I A K F G T A F K	213-226
E S Q L I A K F G T A F K	214-226
S Q L I A K F G T A F K	215-226
Q L I A K F G T A F K	216-226
L I A K F G T A F K	217-226
I A K F G T A F K	218-226
A K F G T A F K	219-226
K F G T A F K	220-226
F G T A F K	221-226

Fig. 2. Diagram of amino acid sequences of nested synthetic peptides of LLO. The peptides which both bind to class II MHC (IE) and have significant antigenic activity are underlined. The minimal T cell epitope with full activity, LLO 215-226, is indicated.

A majority of LLO-specific T cell hybridomas analyzed (22 of 32 hybridomas), responded to LLO 215-234. Studies with a panel of monoclonal T cell hybridomas revealed that the region of 203-246 contained several dominant T cell epitopes (Safley, Reay, Jensen, and Ziegler, manuscript in preparation). The majority of T cell hybridomas tested recognized LLO 203-226, LLO 208-226, LLO 215-226, and LLO 215-234, although each exhibited a characteristic pattern of preferential reactivity. In particular, one hybridoma (IIIC5) could be stimulated by LLO 203-226 but not by LLO 208-226 nor any other synthetic LLO peptide tested. Using antigen presenting cells from B10 congenic mice and cells transfected with either IA^k or IE^k alone, we found that IIIC5 recognized an epitope in the amino-terminal half of LLO 203-226 in association with IA^k, while a different hybridoma (IB5) recognized an epitope in the carboxy-terminal half presented by IE^k. These studies help to define some of the features contributing to the molecular basis of immunodominance in a model of bacterial immunity. Such studies have particular relevance since the T cells are generated as a result of bacterial infection (as compared to contrived immunization with CFA etc). We demonstrate that the dominance of this region is related to a number of factors, including the presence of multiple T cell epitopes, sites for two restriction molecules (IE^k and IA^k), high affinity binding to IE^k, and ability to bind promiscuously to several class II MHC alleles.

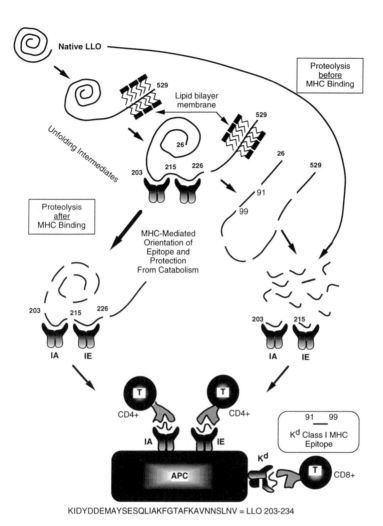

KIDYDDEMAYSESQLIAKFGTAFKAVNNSLNV = LLO 203-234

Fig. 3. Model of LLO processing directed by MHC. The carboxy terminal portion of LLO may not be available for binding to MHC due to association with membranes. Native LLO may be unfolded and then cleaved by cathepsin B at K_{226}, allowing LLO 203-226 accessibility for binding to class II MHC before other portions of the molecule become available for binding. The binding of this region of LLO to MHC may protect it from further degradation by proteases. Alternatively, the native LLO may be fully degraded to peptide fragments that compete for binding to MHC individually.

Immunodominance of the LLO 203-234 region may also be related to secondary structure of the LLO molecule. The presence of multiple T cell epitopes suggests that this region may be very accessible for processing and presentation. The sequence of LLO 203-226 contains three positively charged and five negatively charged residues (net charge -2), most of which are concentrated toward the amino-terminus. The highly charged nature of LLO 203-234 may favor a less buried or folded conformation that could make it particularly accessible to MHC binding in the native molecule or during unfolding as the pH of the endosome drops. As such, accessibility may be a key factor in MHC-directed processing and presentation as illustrated in Fig. 3. However, this issue will be answered only when three dimensional structural data is available with LLO both in solution and interacting with membranes during the execution of lytic activity.

Another secondary factor that may increase T cell reactivity with LLO 215-234 is cross-reactivity with an epitope present in LLO 354-376 (or other common epitopes that shape the T cell repertoire). It is clear that monoclonal T cell hybridomas recognize both LLO 215-234 and LLO 354-376. While the primary sequences of the two peptides are not very similar, both fit a recently developed IE^k binding motif (M. Davis and P. Reay, HHMI, Stanford, personal communication): a large hydrophobic residue at position 1, a polar residue at position 6, and a charged residue at position 9. The IE^k-binding studies demonstrated that LLO 354-376 can bind relatively weakly to IE^k (Kd > 10 μM) and this feature correlated with reactivity to this peptide since concentrations as high as 100 μM were required in vitro to elicit a response. Thus, the cross-reactivity of these two peptides may be related to formation of a conformational epitope that is recognized in association with IE^k.

As more data is accumulated it is becoming possible to predict with increasing accuracy (using primary structure) those peptide epitopes with the desired features of immunodominance and activity with a variety of MHC types. Furthermore, we have shown the value of determining the minimal epitope (e.g LLO 215-226) in that immunization with this peptide generates T cells capable of recognizing the purified LLO molecule (pLLO) in vitro. It should be noted that this is not always the case since peptide immunization often results in T cells that recognize the homologous peptide but not necessarily well the protein antigen (20). We have also experimented with the use of synthetic peptides and live bacteria delivery systems for epitopes as potential vaccine in the model of murine listeriosis. Preliminary data suggest that some protection is afforded by immunization with the class II immunodominant epitope of LLO.

DIFFERENTIAL REQUIREMENTS FOR ANTIGEN PROCESSING

Given that the dominant T cell epitopes of LLO for both CD8+ and CD4+ α/β TCR-expressing T cells have been defined, it was possible to use monoclonal T cell hybridomas specific for each epitope to study the requirements of antigen presentation involved in listerial infection. We have generated T cell hybridomas with specificity for LLO 91-99 presented by H-2 K^d, and for LLO 215-234 presented by I-E^k. These hybridomas have been tested in vitro with macrophages from (H-2^k x H-2^d)F1 mice infected with live

Listeria stains [generated using transposon mutagenesis (7)] which secrete either a full-length, lytic LLO, (Hly+) or a non-hemolytic truncated form of LLO (Hly-) which contains both epitopes (Fig. 4).

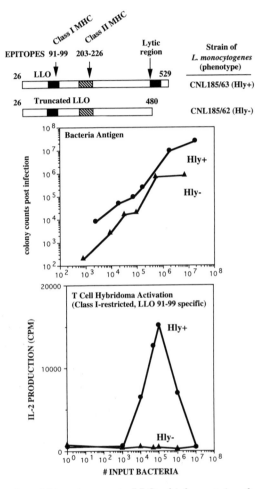

Fig. 4. Both strains of Listeria secrete LLO which contains the dominant class I and class II MHC-binding T cell epitopes. The strains differ at the carboxy termini due to the truncation of the molecule at residue 480 in strain CNL85/162. Macrophages were infected with either strain of Listeria (Hly+, or the truncated, Hly- strain) at increasing multiplicities of infection for four hours without antibiotics. The number of input bacteria is shown versus the colony counts post infection. Class I restricted T cell hybridoma responsd to macrpohages infected with Listeria expressing fully lytic LLO (Hly+) but not to truncated, non-hemolytic LLO (Hly-).

These studies have demonstrated the requirement for the lytic region of LLO for mediating efficient entry into the cytoplasmic processing pathway and presentation by class I MHC (Figure 4). This lytic region of LLO is also associated with decreased presentation by the endocytic, class II MHC pathway (data not shown, 6,12). Kinetic analyses using these hybridomas have also shown that presentation of the epitope LLO 91-99 occurs only after a two hour processing period and does not reach optimal presentation until up to eight hours after infection. This is in contrast to the rapid processing of truncated LLO via the class II MHC pathway. At later times of infection (8-24 hrs) even the Hly- bacteria that express the truncated LLO molecule can be presented via class I MHC pathway. These results indicate that while lysis of the endosome membrane by LLO favors rapid class I MHC presentation, it is not an absolute requirement. These results extend the findings of others (4,9). The model depicted in Fig 5 accounts for class I and class II MHC-mediated presentation of LLO epitopes.

OTHER EPITOPES AND PRESENTING MOLECULES

Although there is a dominant immune response directed toward peptides derived from LLO, there are several other listerial antigens that are responded to by polyclonal CD4+ T cell populations from *Listeria*-immunized mice, such as internalin, lecithinase, phospholipase, and a 60 kd protein, P60 (Ziegler and Cossart, unpublished observations).

The presentation of listerial peptide to cytolytic T cells by a non-classical MHC allele, H-2M3, has also recently been reported (18). Only peptides bearing an aminoterminal N-formyl group can compete for binding to this molecule. This feature, together with the resistance of the active antigen fraction to aminopeptidases, suggests that the relevant epitope is a formylated peptide. This suggests that bacterial peptides which are initiated with N-formylation could bind to M3. This data points to the role of nonclassical class I molecules (15,18) in the defense against bacterial pathogens such that the unique translation initiation mechanism used by procaryotic machinery can be perceived by the T cell immune system. The generality of this finding and the precise role of this recognition system in bacterial immunity remain to be well established. Ongoing studies in our lab include the use of the formylated and non-formylated peptides corresponding to the signal sequence of LLO (LLO 1-25) to test these ideas and define the relevant processing mechanism operative.

ROLE OF γ/δ T CELLS IN IMMUNITY TO *LISTERIA*

Two of most intriguing aspects of γ/δ T cell biology are their predominance at sites of infection and preferential localization in epithelial tissue at sites of first contact with invading microbes. Given the fact that a role for γ/δ T cells in primary resistance to *Listeria* infection has been firmly established (21,22,29), it is critical to address the fundamental mechanisms by which γ/δ T cells predominate in selected sites and tissues and to define the phenotype, function and activation requirements of γ/δ T cells. While both α/β T cells and γ/δ T cells have been shown to play a role in resistance to infection by *Listeria monocytogenes*. (21), it is the α/β T cells that play the predominant role in the specific secondary response to infection. While the epitopes of the α/β T cells have been defined as described above, the epitopes, presenting molecules and processing pathways operative with γ/δ T cells remain largely unknown.

The cytokine expression and responsiveness of γ/δ T cells has been studied (22,28). The γ/δ T cell has been shown to be a cellular source of IFNγ, and this cytokine is clearly a potent macrophage activating agent for antimicrobial defenses. It is apparent that γ/δ T cells respond to the macrophage-derived cytokines IL-1 and IL-12 with vigorous production of IFNγ. Both IL-1 and IL-12 are induced by exposure of macrophages to bacteria (13). The response of γ/δ T cells to HKLM and macrophages is inhibited by neutralizing antibodies to IL-12 (Skeen and Ziegler, submitted). This indictes a critical role for IL-12 in bacteria-mediated activation of γ/δ T cells. [The role and specificity of the γ/δ TCR, however, remains unclear.] Thus, an activation pathway can be envisioned in which invading bacteria cause macrophage production of IL-1 and IL-12 which in turn activates γ/δ T cells to produce IFNγ which then induces the full activation of macrophages to permit the efficient destruction of bacteria. The initial burst of IFNγ production generated in this way may also bias the response to the TH1 cell differentiation which would mediate further macrophage activation and lymphocyte recruitment via IFNγ and IL-2. In this scheme, γ/δ T cells are not only dependent upon activation of other cell types but also able to regulate the function of both macrophages and other T cells.

The proliferative capacity of γ/δ T cells is also be regulated by cytokines produced by both macrophages, α/β T cells and possibly epithelial cells. Our experiments indicate that γ/δ T cells do not generate IL-2 for an autocrine pathway. Instead, γ/δ T cells proliferate preferentially to IL-1 and IL-7 (and both α/β and γ/δ T cells proliferate in response to IL-1 and IL-2) (22,28). We are testing the hypothesis that the special cytokine environment of epithelial tissue (IL-1 and IL-7 ?) and sites of infection accounts for selective or enhanced tissue expression of γ/δ T cells.

In any rational approach to vaccine development or treatment of infections both α/β T cells and γ/δ T cells must be considered. While the molecular basis of activation of α/β T cells and immunodominace is beginning to be well defined, these fundamental questions about γ/δ T cells remain one of the major puzzles in modern immunology.

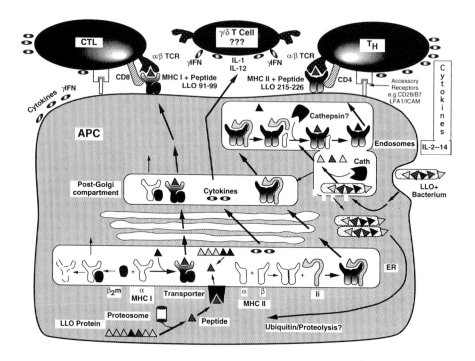

Fig. 5 Model of the antigen processing of LLO from Listeria monocytogenes by macrophages and accessory function related to α/β T cells and γ/δ T cells. Listeria are first ingested by the macrophage and taken up into an endosome. The bacteria can be degraded within this endosome and processed by the proteases there. These processed antigenic peptides can then bind to class II MHC within a compartment in the endocytic pathway and be sent to the cell surface for recognition by CD4+ T helper cells. These bacteria can also enter the cytoplasm through the secretion of LLO. LLO can disrupt the endosomal membrane and allow the bacteria to escape to the cytoplasm where they can better survive and multiply. In the cytoplasm, listerial antigens can also be processed and then transported into the endoplasmic reticulum where they can bind to newly synthesized class I MHC. This peptide-class I MHC complex can then be sent to the cell surface for recognition by CD8+ cytolytic T cells.

ACKNOWLEDGEMENTS

This chapter represents a brief review of relevant results of others and a summary of work of the laboratory of Kirk Ziegler performed by Chris Cluff, Susan Safley, Marriane Skeen, Beth Hiltbold, Kevin Pearce, Naresh Verma, and others in collaboration with Phil Reay, Mark Davis, Gary Schoolnik and others. This work was supported by grants from the NIH and the ACS.

REFERENCES

1. **Berche, P., J.L. Gaillard, and P.J. Sansonetti.** 1987. Intracellular growth of *Listeria monocytogenes* as a prerequisite for in vivo induction of T cell-mediated immunity. *J.Immunol 138*: :2266.

2. **Bishop, D.K., and D.J. Hinrichs.** 1987. Adoptive transfer of immunity to *Listeria monocytogenes*. The influence of in vitro stimulation on lymphocyte subset reqirements. *J. Immunol. 139* :2005.

3. **Bouwer, H.G., C.S. Nelson, B.L. Gibbons, D.A. Portnoy, and D.J. Hinrichs.** 1992. Listeriolysin O is a target of the immune response to *Listeria monocytogenes*. *J. Exp.Med. 175* :1467.

4. **Brunt, L.M., D.A. Portnoy, and E.R. Unanue.** 1990. Presentation of *Listeria monocytogenes* to CD8+ T cells requires secretion of hemolysin and intracellular bacterial growth." *J.Immunol 145* :3540.

5. **Cluff, C.W., M. Garcia, and H.K. Ziegler.** 1990. Intracellular hemolysin-producing *Listeria monocytogenes* strains inhibit macrophage-mediated antigen processing. *Inf. andJmm.. 58* :3601.

6. **Cluff, C.W., and H.K. Ziegler.** 1987. Inhibition of macrophage-mediated antigen-presentation by hemolysin-producing *Listeria monocytogenes*. *J. Immunol 139* :3808.

7. **Gaillard, J.L., P. Berche, and P. Sansonetti.** 1986. Transposon mutagenesis as a tool to study the role of hemolysin in the virulence of *Listeria monocytogenes*. *Inf. and Imm. 52* :50.

8. **Harding, C.V., and H.J. Geuze.** 1992. Class II MHC molecules are present in macrophage lysosomes and phagolysosomes that function in the phagocytic processing of *Listeria monocytogenes* for presentation to T cells *J. Cell Bio. 119* : 531.

9. **Harding, C.V., and E.R. Unanue.** 1991. Cellular mechanisms of

antigen processing and the function of class I and class II major histocompatibility complex molecules. *Cell Regulation 1* :499.

10. **Harty, J.T. and M.J. Bevan**. 1992. CD8+ T cells specific for a nonamer epitope of *Listeria monocytogenes* are protective in vivo. *J. Exp. Med. 175* :1531.

11. **Harty, J.T., R.D. Schrieber, and M.J. Bevan**. 1992. CD8 T cells can protect against an intracellualar bacterium in an interferon gamma-independent fashion. *Proc. Natl. Acad.Sci. USA 89* :11612.

12. **Hiltbold, E., and H.K. Ziegler**. Mechanisms of class I and class II MHC presentation of antigens of *Listeria monocytogenes* . Manuscript in preparation.

13. **Hsieh, C., S.E. Macatonia, C.S. Tripp, S.F. Wolf, A. O'Garra, and K.M. Murphy**. 1993. Developement of Th1 CD4+ T cells through Il-12 produced by Listeria-induced macrophages. *Science 260* :547.

14. **Jorgensen, J.L., P.A. Reay, E.W. Ehrich, and M.M. Davis**. 1992. Molecular components of T cell recognition. *Ann. Rev. Imm. 10* :835.

15. **Kurlander, R.J., S.M. Shawar, M.L. Brown, and R.R. Rich**. 1992. Specialized role for a murine class I-b MHC molecule in prokaryotic host defenses. *Science 257* : 678.

16. **Marshall, N.E., and H.K. Ziegler**. 1991. The role of bacterial hemolysin production in induction of macrophage Ia expression during infection with *Listeria monocytogenes. J. Immunol. 147* :2324.

17. **Pamer, E.G., J.T. Harty, and M.J. Bevan**. 1991. Precise prediction of a dominant class I MHC-restricted epitope of *Listeria monocytogenes. Nature 353* :852.

18. **Pamer, E.G., C. Wang, L. Flaherty, K. Fisher Lindahl, and M.J. Bevan**. 1992. H-2M3 presents a *Listeria Monocytogenes* peptide to cytotoxic T lymphocytes. *Cell 70* :215.

19. **Portnoy, D.A., T. Chakraborty, W. Goebel, and P. Cossart**. 1992. Molecular determinants of *Listeria monocytogenes* pathogenesis. *Inf. and Imm. 60* :1263.

20. **Safley, S.A., C.W. Cluff, N.E. Marshall, and H.K. Ziegler**. 1991. Role of listeriolysin-O in the T lymphocyte response to infection with *Listeria monocytogenes. J.Immunol. 146* :3604.

21. **Skeen, M.J., and H.K. Ziegler**. 1993. Induction of murine

peritoneal γ/δ T cells and their role in resistance to bacterial infection. *J.Exp. Med. 178* :971.

22. **Skeen, M.J., and H.K. Ziegler**. 1993. Intercellular interactions and cytokine responsiveness of peritoneal α/β T cells and γ/δ T cells from *Listeria*-infected mice:synergistic effects of IL-1 and Il-7 on γ/δ T cells. *J. Exp. Med. 178* : 985.

23. **Ziegler, H.K.** 1988. Differential requirements for the Pprocessing of bacterial antigens by macrophages: A comparison of live bacteria, particles, proteins and peptides. in *Antigen Presenting Cells: diversity, Differentiation, and Regulation,* p.83.

24. **Ziegler, H.K.** 1990. Regulation of macrophage-mediated antigen presentation by microbial products. in *Microbial Determinants of Virulence and Host Response* , p.283.

25. **Ziegler, H.K., C.A. Orlin, and C.W. Cluff.** 1987. Differential requirements for the processing andpresentation of soluble and particulate bacterial antigens by macrophages. *Eur. J. Imm. 17* :1287.

26. **Ziegler, H.K., and E.R. Unanue.** 1982. Decrease in macrophage antigen catabolism caused by ammonia and chloroquine is associated with inhibition of antigen presentation to T cells. *Proc.Natl.Acad.Sci USA 79* :175.

27. **Ziegler, H.K., and E.R. Unanue**. 1982. Identification of a macrophage antigen-processing event required for I region-restricted antigen presentation to T lymphocytes. *J.Immunol. 127* :1869.

28. **Ziegler, H.K., M.J. Skeen and K.M. Pearce.** 1994. Role of α/β and γ/δ T cells in Innate and Acquired Immunity. Proc. N.Y.Acad.Sci. In Press.

29. **Mombaerts, P., J. Arnoldi, F.Russ, S. Tonegawa, and S.H.E. Kaufmann.** 1993. Different roles of α/β and γ/δ T cells in immunity against an intracellular bacterial pathogen. Nature *365* :53.

22

SILENT CRYPTICITY AND ITS CONSEQUENCES FOR AUTOIMMUNITY AND INFECTION

Iqbal S. Grewal, Kamal D. Moudgil, and Eli E. Sercarz

INTRODUCTION

Antigen specific CD4$^+$ T-cells recognize foreign antigens in the form of a bimolecular complex comprising of major histocompatibility complex (MHC) class II molecules and a bound peptide on the surface of antigen presenting cells (APC), usually after some type of metabolic processing, involving unfolding and/or fragmentation of the native proteins (1,2). One of the processed peptide products that can be termed a "prodeterminant" then binds to a particular MHC molecule, and is subsequently trimmed before presentation to T cells (3). Many different antigenic peptides can be bound to a given MHC molecule which are related through sharing a structural motif (4).

A protein antigen usually includes numerous potential determinants, which could induce a T cell response in the form of peptides (3). But in general, only one or a very few regions within a protein molecule succeed in generating a T cell response. The prerequisites for immunogenicity of a potential determinant are the following: a) the putative immunogenic region should possess amino acid residues that are *critical for binding to a given MHC molecule* as well as interaction with the appropriate T cell receptor (TCR), b) that determinant should be *readily available*: it should be generated from the native protein following initial antigen unfolding in a form that allows efficient binding to the MHC molecule, c) the determinant should be *produced in sufficient concentration* so that the MHC/peptide-TCR interaction exceeds the threshold level for activation of T cells, and d) the determinant under consideration *should compete favorably* with other determinants generated from the same native antigen for binding to the same MHC molecule. The above set of criteria are based on the assumption that there is no complete hole in the T cell repertoire so that T cells potentially reactive to a given determinant are present in the animal.

In vivo T cell responses to many multideterminant proteins, including hen lysozyme, staphylococcal nuclease, myoglobin, bacteriophage lambda repressor, ovalbumin etc. are narrowly focused to very few determinants (5-8) . These determinants are referred to as "dominant" (3). There are three main groups of determinants within a protein antigen. The first two types of determinants, dominant and sub-dominant, are defined as those that give a strong (dominant) or moderate to weak (sub-dominant) T cell proliferative response *in vitro* when lymph node cells (LNC) primed with the native antigen are stimulated with a peptide fragment containing one of these determinants, and *vice versa* (3,9,10). The presence of a third kind of determinant, termed "cryptic" can be defined by its inability to raise a T cell proliferative response in LNC when the native antigen is used for immunization, although in its peptide form, it is either a weak or strong determinant.

Cryptic determinants can be further classified into three types: the *facultative or latent cryptic*, whose processing and presentation can be upregulated under certain special circumstances; the *absolute cryptic*, whose presence is revealed by immunization with the peptide, but never with the native protein antigen (3); and the *silent cryptic* which can only be revealed by effecting appropriate modifications in the antigen. There are several factors that can render a determinant cryptic: a) antigen processing events-- such as slow or incomplete unfolding of the native antigen molecule so that the determinant is not generated, or excessive processing resulting in destruction of the determinant; b) competitive events-- failure to compete successfully for binding to appropriate MHC molecule, either because the determinant is outcompeted by other determinants from the same native antigen for binding to a particular MHC molecule, or because a neighboring determinant is captured by an unrelated MHC molecule preempting its opportunity to bind to other MHC molecules or c) hindering events-- hinderotypy caused by residue(s) flanking the minimal determinant so that the determinant cannot bind to the appropriate MHC molecule (agretypic hinderotope) or to the TCR (epitypic hinderotope).

In this chapter, we will discuss how antigen processing plays a role in determining silent crypticity, based on our studies using the hen lysozyme (HEL) system. We will particularly focus on how a potential determinant can be rendered into a "silent cryptic" due to a C-terminal hindering residue.

HEL POSSESSES A "SILENT CRYPTIC" DETERMINANT WITHIN REGION 46-61 IN C57BL/6 MICE WHICH IS DOMINANT IN C3H.SW MICE

We have recently discovered a silent cryptic determinant within the HEL molecule in C57BL/6 mice within region 46-61 (NTDGSTDYGIL-QINSR) (3,11). The C57BL/6 mouse strain is non-responsive to HEL peptide 46-61 (p46-61), whereas another mouse strain with identical MHC (H-2^b) but different non-MHC genes, C3H.SW, gives a strong response to this peptide. In the latter strain, p46-61 is a dominant determinant. The silent cryptic determinant within this peptide in C57BL/6 can only be revealed when a test peptide used both to immunize mice and for *in vitro* recall is appropriately modified. The silent crypticity of p46-61 in C57BL/6 mice is attributable to a C-terminal arginine at position 61 (R61) which hinders the binding of this peptide to the A^b molecule. This will be discussed in detail in the following sections.

HINDEROTYPY CAN BE ATTRIBUTED TO INEFFICIENT ANTIGEN PROCESSING

As mentioned above, p46-61 was a good immunogen in C3H.SW mice but was nonimmunogenic in C57BL/6 mice of the same MHC haplotype, which only differ in their non-MHC genes. Therefore, a role for the antigen processing machinery in masking this potential determinant in C57BL/6 was postulated. For instance, C57BL/6 could differ from C3H.SW with respect to processing enzyme(s) required to reveal this determinant.

The core of the minimal determinant within p46-61 that was immunogenic in C3H.SW mice was determined by immunizing this mouse strain with p46-61 and measuring proliferative responses of LNC *in vitro* using overlapping pin peptides (shifting by one amino acid residue at a time) spanning the entire sequence of p46-61. The minimal determinant in p46-61 was found to reside within amino acid residues 50-60 (11). Thus, R61 was not critical for the immunogenicity of p46-61 in C3H.SW mice.

Taking a clue from the above results in C3H.SW mice, we predicted that HEL peptide 46-60 (NTDGSTDYGILQINS) (p46-60) might be immunogenic in C57BL/6 mice, although they were nonresponsive to p46-

61. This was indeed true: p46-60 could induce a strong T cell proliferative response in C57BL/6 mice, at the same concentration that was used for immunization of C3H.SW mice with p46-61. Thus, fine processing of p46-61 was essential to generate the immunogenic determinant, which could be p46-60 or a shorter peptide. These results clearly demonstrate that the nonimmunogenicity of p46-61 in C57BL/6 mice was due to the presence of the C-terminal arginine (R61) in that peptide. The removal of R61 rendered the same peptide highly immunogenic. Thus, R61 was acting as a "hinderotope" (12) interfering with either the binding of p46-61 to A^b (agretypic hinderotope) or the interaction between the MHC/peptide complex and the appropriate TCR (epitypic hinderotope). The phenomenon of hinderotypy was first proposed by Brett et al. (12) and we have provided the first experimental evidence to conclusively demonstrate the existence of such determinants.

HINDRANCE CAUSED BY A FLANKING RESIDUE CAN LEAD TO SILENT CRYPTICITY

To study the mechanism of hindrance caused by R61, we tested the binding of p46-61 and p46-60 to the A^b molecule using a competitive inhibition assay. In this assay, a well characterized $H-2^b$-restricted T cell hybridoma specific for HEL peptide 81-96 (p81-96) was used and the test peptide (p46-61 or p46-60) was allowed to compete with p81-96. The inhibition of interleukin-2 (IL-2) production served as the read-out in this assay. Antigen presenting cells, even from responder C3H.SW mice, treated with inhibitors of processing (leupeptin or chloroquine) or fixed with paraformaldehyde, could not utilize p46-61 to compete with p81-96 whereas p46-60 was able to inhibit IL-2 production under these conditions. However, when untreated APC from C3H.SW mice were used in the experiments, either p46-60 or p46-61 could inhibit IL-2 production, suggesting that fine processing of p46-61 to p46-60 or to a shorter peptide is needed to allow it to bind A^b and to generate a T cell response. These experiments clearly implicate R61 as responsible for hindering p46-61 from binding to the A^b molecule. The above results provide insight into the antigen processing defect mentioned earlier, that C57BL/6 mice are unable to fine process p46-61 by removing the hindering residue, R61, whereas C3H.SW mice can do so.

Furthermore, this type of hindrance leads to a change of status of a potential determinant from dominance in the C3H.SW strain to silent crypticity in C57BL/6 mice. A similar type of phenomenon in which fine processing of a determinant was critical for immunogenicity of Class I peptides has been reported earlier (13,14). Our earlier studies had indicated that residues far removed from the actual determinant could influence the processing of that determinant (15) Subsequently, we and other investigators have reported that the flanking regions around a minimal determinant can influence processing of that determinant (16-18). Similarly it has been reported that even a single amino acid change can profoundly affect antigen processing and presentation (19-21). Here, we summarize our results showing that an effect of hindrance of binding to an MHC molecule is demonstrable by a single terminal amino acid residue.

PROBLEMS IN IDENTIFICATION OF SILENT CRYPTIC DETERMINANTS

Silent cryptic determinants fail to activate T cells due to the presence of hindering residues in the peptides, and it is difficult to predict whether a non-immunogenic stretch of a dozen amino acids contains one. For example, C57BL/6 mice failed to generate T cells reactive to p46-61 derived either from the peptide itself, or from processing of the native protein, HEL, thus making it appear superficially that no determinant existed within this region. This observation emphasizes the need for careful selection of peptides when identifying T cell determinants. Accordingly, there may be some potential determinants within a native antigen whose revelation depends on the choice of the peptide that is used to elicit the response. A simple strategy to identify silent cryptics consists of testing overlapping peptides spanning a potential determinant of interest via both immunization and recall *in vitro*. A peptide lacking an amino- or carboxy-terminal hindering residue will elicit a T cell response, whereas the peptide possessing the extra hindering residue will not.

Under usual circumstances, the minimal core of a determinant can be determined by simply immunizing the mice with a large peptide and testing the T cells on a set of all the overlapping peptides. This procedure has been used extensively in our laboratory and the cores of several determinants within HEL and myelin basic protein have been identified (9,22). However this method usually will fail to determine the core of a silent cryptic determinant. First a peptide containing any hindering residue will not stimulate the T cells *in vivo*. Second, even if one uses a peptide lacking hindering residues for priming, primed T cells cannot be stimulated *in vitro* by any of the overlapping peptides containing the hindering residue, if it is an agretypic hinderotope.

SILENT CRYPTICITY CAN LEAD TO GENERAL UNRESPONSIVENESS TO A NATIVE PROTEIN

Another important aspect of silent cryptic determinants relates to unresponsiveness to the native protein antigen containing that silent cryptic determinant. In our studies discussed above, C57BL/6 mice did not respond to p46-61, and at the same time these mice were also low responders to native HEL. On the contrary, C3H.SW mice produced a vigorous response to p46-61 as well as to native HEL. From these observations, it appeared that there is a definite relationship between response to such a determinant in a particular mouse strain and its influence on response of that animal to native HEL. This relationship is easy to contemplate if the T cell response to a native protein molecule is primarily focused on one or two key determinants within that antigen. In the case where the key determinant happens to be a silent cryptic determinant, then a lack of response to that particular determinant would also lead to unresponsiveness to the whole molecule. To test this idea, we immunized C57BL/6 and C3H.SW mice with HEL intraperitoneally and the response of splenic T cells was tested against different HEL peptides *in vitro*. As expected, C3H.SW mice responded predominantly to p46-61 whereas C57BL/6 mice did not give any response to that peptide. In C3H.SW mice, there was only a minimal response to two other determinants, whereas C57BL/6 mice responded to none of the HEL determinants. From these results it can be concluded that responsiveness to a key determinant in a multideterminant protein antigen is an important factor in determining the response of the same strain of mice to the native protein antigen.

This is also an instructive case to consider from the point of view of responsiveness to HEL as a whole and the effect of regulatory mechanisms such as suppression. In draining lymph nodes following footpad injection of HEL in the C3H.SW strain, a response *does* occur to a group of determinants-- 20-35, 30-53, 46-61 and 74-88. In the early period following immunization, no CD8$^+$ T suppressor (Ts) cells appear in draining lymph nodes (23). Therefore, in the splenic response to HEL in this strain, the restriction to the 46-61 determinant appears to indicate that T cells directed against the other known determinants (20-35, 30-53 and 74-88) are suppressed (24) while 46-61-specific T helper cells manage to resist suppression. However, C57BL/6 mice which are not able to respond to this 46-61 remain totally unresponsive to the HEL molecule in the spleen where T suppressor cells are found, while a response to the other T cell determinants in the lymph node can be demonstrated. In summary, hindrance of MHC binding by a single amino acid residue, R61, coupled with a unique pattern of susceptibility to suppression, leads to complete unresponsiveness to HEL in the spleen.

SILENT CRYPTICS MAY NOT BE AVAILABLE DURING ANTIGEN PROCESSING FOR MHC CLASS II BINDING

Previously, we have reviewed evidence supporting the idea of attachment by large peptides of native antigens, approaching the size of the antigen itself, to class II binding sites (25). Viewed from this larger perspective, the important dimension about binding to MHC by the most available determinant, which underlies the dominance "choice" can be seen as a competition among several more or less available determinant regions surrounded by hindering flanking regions. During the initial competition for binding among the determinants, the principle of silent crypticity will often be employed in the context of the long unfolded protein. Later, the dangling ends protruding from the groove of the MHC molecule may render the determinant temporarily silent (epitypic hinderotypy) until they are processed away.

UNMASKING OF SILENT CRYPTIC DETERMINANTS UNDER PHYSIOLOGICAL/PATHOLOGICAL CONDITIONS

The results described above indicate that if the appropriate antigen processing machinery existed in an animal, capable of removing any hindering residue flanking a potential determinant, the presence of previously silent cryptic determinants could be revealed. In our case, the C3H.SW strain possesses the necessary non-MHC genes to fine process p46-61, whereas MHC-identical C57BL/6 mice with a different set of non-MHC genes are unable to do so. Presumably, possession of a particular non-MHC gene could favor processing certain silent cryptic determinants and not others.

These results can be extended to the diverse human population. It is likely that the susceptibility of different individuals to certain pathogens/autoantigens (actually, certain key determinants on antigens of the pathogen/autoantigen that are of importance in induction of disease) is often governed by their non-MHC genes that control the antigen processing machinery. An individual who can process a key determinant from a pathogen may survive the infection compared to another individual who lacks the ability to process and present that particular determinant. Susceptibility within a family to infection can likewise be attributed to a difference in an antigen processing allele among the offspring. A child may remain completely healthy, whereas its siblings may be susceptible to the same disease owing to differences at a single non-MHC allele contributed by one of the parents, who also is at risk. A similar argument could be made for susceptibility to an autoimmune disease.

A subtle change in MHC molecules through mutation can also be demonstrated to unmask a silent cryptic determinant. Mouse H-2^k strains (B10.A, B10.BR and C3H.HeJ) are good responders to p46-61 (3,10,25,26). The minimal determinant in these strains within p46-61 is 52-61 and R61 is essential for generating a T cell response to p46-61 in these strains (27). Thus, R61 causes no hindrance of binding of p46-61 to the A^k molecule, which indicates that a hinderotope may hinder binding to only certain MHC molecules. This view is further supported by recent studies, where naturally eluted peptides from A^k molecules show that p46-61 has the ability to bind to the A^k molecule (28). This difference in the ability of p46-61 to bind to the A^k, but not the A^b molecule, can be attributed to differences in the structure of the MHC groove between these two MHC molecules. To further explore the ability to accommodate hindering structures, we have studied the bm-12 strain, a mutant of C57BL/6 which has identical non-MHC genes but differs from C57BL/6 mice at 3 amino acid positions on the class II Aβ chain (amino acids 67, 70, 71) (29). Interestingly, a strong T cell response to p46-61 is observed in bm-12 mice. Because processing should be identical in C57BL/6 and bm-12 mice, these results indicate that the phenomenon of 46-61 binding in bm-12 relates to a change in the groove at one of the mutated positions that can accommodate R61. Thus, alteration in MHC structure can also lead to unmasking of silent cryptic determinants.

We can make an extrapolation of the above information to human subjects. It is well known that under the continuous pressure from environmental pathogens and other forces, selection of new the MHC mutations will occur. Most of these mutations are expected to be a positive asset to a given population permitting a diversification of defensive maneuvers within the species. However, if in a given section of the population the new mutation favors presentation of some silent cryptic determinants within foreign/self antigens, then that section of the population might become more (or less) susceptible to a particular autoimmune disease. Similarly, mutations within or around a silent cryptic determinant within antigens of a pathogen can unmask these determinants. However, this change may be deleterious to the host if the determinant now revealed generates an immune response that is crossreactive to a self antigen, which might lead to autoimmunity through "molecular mimicry" (22).

There is evidence to suggest that the proteasomes, a major extralysosomal proteolytic system, play an important role in antigen processing and that proteasomes are regulated by interferon-γ (IFN-γ) (20,31). Most of the evidence is from the class I system in mice, but proteasome-related genes have also been shown in the human MHC class II region (32). It can be speculated that under special circumstances (e.g., an unusual response to physiological stress, inflammation etc.), including upregulation of antigen processing by IFN-γ, there may be selective induction of a particular cellular protease(s) resulting in more efficient processing, or there may be expression of an aberrant protease which results in a qualitative change in antigen processing with the ultimate result of more efficient processing of a silent cryptic determinant. One example of a designed change in the hierarchical pattern of dominance based on new enzymatic sites comes from experiments in our laboratory. Deng et al. (unpublished data) mutated a single HEL residue adjoining the C-terminus of a subdominant determinant to create a new cathepsin B site. This altered lysozyme gave rise to a 20-fold enhancement of response to this determinant. In addition, work in our laboratory has suggested that under stressful conditions (e. g. housing density or treatment with amphetamine) there is significant alteration in the level of antigen presentation (33, 34).

CONSEQUENCES OF SILENT CRYPTICITY

The presence of a hindering residue within or flanking a potentially immunogenic determinant reflects a disadvantageous situation for several foreign determinants, particularly those within antigens of pathogens that are critical in inducing a protective T cell response for the host. At the same time, it provides the pathogen with a potential defensive strategy by not allowing a key protective determinant to be expressed at all so that the host cannot mount an immune response against the pathogen. Perhaps more important is the issue of the contribution of silent cryptic determinants to autoimmunity. The presence of a silent cryptic determinant within a self-antigen leads to a situation where T cells potentially reactive to that determinant escape tolerance induction during development of the T cell repertoire. If the presentation of the silent cryptic determinant can subsequently occur under special circumstances as described in the previous section, it might provide a focus for future autoimmune attack. However, if it happens that the silent cryptic become upregulated and presentable through malignancy, it may be that the enhanced expression actually induces a potent and protective T cell response, destroying the tumor.

ACKNOWLEDGMENTS

This work was supported by grants from the American Cancer Society (IM-626), and the NIH (AI-11183 and CA-24442). We are grateful to Dr. David B. Stevens for critically reading the manuscript.

REFERENCES

1. **Brodsky, F. M., L. E. Guagliardi**. 1991. The cell biology of antigen processing and presentation. *Ann. Rev. Immunol. 9*:707

2. **Jorgensen, J. L., P. A. Reavy, E. W. Ehrich, M. M. Davis**. 1992. Molecular components of T-cell recognition. *Ann. Rev. Immunol.*

10:835.

3. **Sercarz, E. E., P. V. Lehmann, A. Ametani, G. Benichou, A. Miller, and K. Moudgil**. 1993. Dominance and crypticity of T cell antigenic determinants. *Ann. Rev. Immunol. 11*:729.

4. **Buus, S., A. Sette, S.M. Colon, C. Miles and H.M. Grey**. 1987. The relation between major histocompatibility complex (MHC) restriction and the capacity of Ia to bind immunogenic peptides. *Science 235*:1353.

5. **Benjamin, D. C, J. A Berzofsky, I. J. East, F. R. N. Gurd, C. Hannum, S. J. Leach, E. Margoliash, J. G. Michael, A. Miller, E. M. Prager, M. Reichlin, E. E. Sercarz, S. J. Smith-Gill, P. E. Todd, and A. C. Wilson**. 1984. The antigenic structure of proteins: A reappraisal. *Ann. Rev. Immunol. 2*:67

6. **Fox, B.S., C. Chen, E. Fraga, C. A. French, B. Singh, and R. H. Schwartz**. 1987. Functionally distinct agretopic and epitopic sites. Analysis of the dominant T cell determinant of moth and pigeon cytochrome c with the use of synthetic peptide antigens. *J. Immunol. 139*:1578.

7. **Maizels, R., J. Clarke, M. Harvey, A. Miller, and E. E. Sercarz**. 1980. Ir-gene control of T cell proliferative responses: Two distinct expressions of genetically nonresponsive state. *Eur. J. Immunol. 10*:516.

8. **Wicker, L., M. Katz, U. Krych, A. Miller, and E. E. Sercarz**. 1981. Focusing in the immune response to a protein antigen. *14th Leukocyte Culture Conference*. Eds. H. Kirchner and K. Resch. Elsevier/North Holland Biomedical Press. p. 99.

9. **Gammon, G., H.M. Geysen, R.J. Apple, E. Pickett, M. Palmer, A. Ametani and E.E. Sercarz**. 1991. T cell determinant structure: cores and determinant envelopes in three mouse major histocompatibility complex haplotypes. *J. Exp. Med. 173*:609.

10. **Gammon, G., N. Shastri, J. Cogswell, S. Wilbur, S. Sadegh-Nasseri, U. Krzych, A. Miller and E. E. Sercarz**. 1987.

The choice of T cell epitopes utilized on a protein antigen depends on factors distant from, as well as at, the determinant site. *Immunol. Rev.* *98*: 53.

11. **Grewal I. S**. 1993. The molecular basis of unresponsiveness of low responder C57BL/6 mice to a model protein antigen hen lysozyme: processing, recognition and regulation. Ph.D. thesis. University of California Los Angeles.

12. **Brett, S.J., K.B. Cease and J.A. Berzofsky**. 1988. Influences of antigen processing on the expression of the T cell repertoire. Evidence for MHC-specific hindering structures on the products of processing. *J. Exp. Med. 168*:357.

13. **Sherman, L.A., T.A. Burke and J.A. Biggs**. 1992. Extracellular processing of peptide antigens that bind class I major histocompatibility molecules. *J. Exp. Med. 175*:1221.

14. **Kozlowski, S., M. Corr, T. Takeshita, L.F. Boyd, C.D. Pendleton, R.N. Germain, J.A. Berzofsky, and D. H. Margulies**. 1992. Serum angiotensin-1 converting enzyme activity processes a human immunodeficiency virus 1 gp160 peptide for presentation by major histocompatibility complex class I molecules. *J. Exp. Med. 175*:1417.

15. **Shastri, N., A. Miller, and E. E. Sercarz**. 1986. Amino acid residues distinct from the determinant region can profoundly affect activation of T cell clones by related antigens. *J. Immunol. 136*:371

16. **Vacchio, M.S., J.A. Berzofsky, U. Krzych, J. A. Smith, R.J. Hodes and A. Finnegan**. 1989. Sequences outside a minimal immunodominant site exert negative effects on recognition by staphylococcal nuclease-specific T cell clones. *J. Immunol. 143*:2814.

17. **Bhayani, H., F.R. Carbone and Y. Paterson**. 1988. The activation of pigeon cytochrome c-specific T cell hybridomas by antigenic peptides is influenced by non-native sequences at the amino terminus of the determinant. *J. Immunol. 141*:377.

18. **Bhardwaj, V., V. Kumar, H.M. Geysen and E.E. Sercarz**. 1992. Subjugation of dominant immunogenic determinants within a chimeric

peptide. *Eur. J. Immunol. 22*:2009.

19. **Liu, Z.R., K.P. Williams, Y.H. Chang and J. A. Smith**. 1991. Single amino acid substitution alters T cell determinant selection during antigen processing of *Staphylococcus aureus* nuclease. *J. Immunol. 146*:438.

20. **Finnegan, A. and C.F. Amburgey**. 1989. A single amino acid mutation in a protein antigen abrogates presentation of certain T cell determinants. *J. Exp. Med.* 170:2171.

21. **Kim, B.S. and Y.S. Jang**. 1992. Constraints in antigen processing result in unresponsiveness to a T cell epitope of hen egg lysozyme in C57BL/6 mice. Eur. *J. Immunol. 22*:775.

22. **Bhardwaj, V., V. Kumar, I. S. Grewal, T. Dao, P. Lehmann, H. M. Geysen and E. E. Sercarz**. 1994. The T cell determinant structure of myelin basic protein in B10.Pl, SJL/J and their F1s. *J. Immunol.* In Press.

23. **Araneo, B. A. R.L. Yowell, and E. E. Sercarz**. 1985. Recognition and display of the predominant idiotype among members of the regulatory circuitry controlling the anti-lysozyme immune response. *J. Immunol. 134*:1073.

24. **Sadegh-Nasseri, S., V. Dessi and E. E. Sercarz**. 1986. Selective reversal of H-2 linked genetic unresponsiveness to lysozymes: II. Alteration in the T-helper/T-suppressor balance owing to gene(s) linked to Ir leads to responsiveness in BALB.B. *Eur. J. Immunol. 16*: 486.

25. **Ametani, A., A.Sette and E. E. Sercarz**. 1993. Crypticity of T cell determinants: relationship to the molecular context of the antigen and to its processing requirements. Submitted for publication.

26. **Allen, P. M., G. R. Matsueda, R. J. Evans, J.D. Dunbar Jr., G. R. Marshall and E. R. Unanue**. 1987. Identification of the T cell and Ia contact residues of a T cell antigenic epitope. *Nature 327*:713

27. **Allen, P. M., D. J. Strydom and E. R. Unanue**. 1984. Processing of

lysozyme by macrophages: identification of the determinant recognized by two hybridomas. *Proc. Natl. Acad. Sci. 81*:2489

28. **Nelson, C. A., R. W. Roof, D. W. McCourt and E. R. Unanue**. 1992. Identification of the naturally processed form of hen egg white lysozyme bound to the murine major histocompatibility complex class II molecules I-A^k. *Proc. Natl. Acad. Sci. 89*:7380.

29. **Ronchese, F., M. A. Brown, and R. N. Germain**. 1987. Structure-function analysis of the A$_\beta^{bm-12}$ mutation using site-directed mutagenesis and DNA-mediated gene transfer. *J. Immunol. 139*:629.

30. **Monaco, J. J**. 1992. A molecular model of MHC class-I-restricted antigen processing. *Immunol. Today 13*:173.

31. **Yang, Y., J. B. Waters, K. Fruh, and P. Peterson**. 1992. Proteasomes are regulated by interferon γ: Implications for antigen processing. *Proc. Natl. Acad. Sci. 89*:4928.

32. **Kelly A., S. H. Powis, R. Glynne, E. Radley, S. Beck, and J. Trowsdale**. 1991. Second proteasome-related gene in the human MHC class II region. *Nature 353*:667.

33. **Heilig, M., M. Irwin, I. Grewal, and E. Sercarz**. 1993. Sympathetic regulation of T-helper cell function. *Brain Behv. Immu. 7*:154.

34. **Grewal, I. S., M. Heilig, E. E. Sercarz, and A. Miller**. 1993. Environmental regulation of T cell function in mice: group housing of males affects processing and/or presentation. Submitted for publication.

INDEX